AF333878

MICROWAVE RADIOMETRY OF VEGETATION CANOPIES

ADVANCES IN GLOBAL CHANGE RESEARCH

VOLUME 24

Editor-in-Chief

Martin Beniston, *Department of Geosciences, University of Fribourg, Switzerland*

Editorial Advisory Board

The titles published in this series are listed at the end of this volume.

MICROWAVE RADIOMETRY OF VEGETATION CANOPIES

by

Alexander A. Chukhlantsev

Russian Academy of Sciences, Moscow, Russia

Springer

A C.I.P. Catalogue record for this book is available from the Library of Congress.

ISBN-10 1-4020-4681-2 (HB)
ISBN-13 978-1-4020-4681-0 (HB)
ISBN-10 1-4020-4682-0 (e-book)
ISBN-13 978-1-4020-4682-7 (e-book)

Published by Springer,
P.O. Box 17, 3300 AA Dordrecht, The Netherlands.

www.springer.com

Printed on acid-free paper

Printed in the Netherlands.

TABLE OF CONTENTS

Foreword ix

Preface xi

Chapter 1. BASICS OF MICROWAVE RADIOMETRY 1

    1.1. Theory of Thermal Radiation 1

    1.2. Microwave Radiometers and Systems 12

    1.3. Radiometric Measurements 18

Chapter 2. PHYSICAL AND ELECTRICAL PROPERTIES
           OF SOILS AND VEGETATION 21

    2.1. Physical and Dielectric Properties of Soils 21

    2.2. Biometrical Features and Electrical Properties
         of Vegetation 36

Chapter 3. MICROWAVE EMISSION FROM BARE SOILS   53

    3.1. Microwave Emission Models of Bare Soils 53

    3.2. Experimental Research on Microwave Emission from
         Bare Soils.. 69

Chapter 4. THEORY OF MICROWAVE PROPAGATION
           THROUGH VEGETATION MEDIA 75

    4.1. General Approach to the Description of Electromagnetic
         wave Propagation in Vegetation 75

    4.2. The Model of Vegetation as a Continuous Medium 78

    4.3. The Model of Vegetation as a Collection of Scatterers
         (Discrete Model) 91

    4.4. Extinction and Scattering of Electromagnetic Waves by
         Plant Elements 97

4.5. Microwave Propagation Through a Vegetation Layer. Relation of Electrodynamic Parameters to Biometric Features of Vegetation 110

Chapter 5. EXPERIMENTAL STUDIES OF MICROWAVE PROPAGATION IN VEGETATION CANOPIES 119

5.1. Methods of Experimental Research 119

5.2. Experimental Results 129

Chapter 6. MODELING OF MICROWAVE EMISSION FROM VEGETATION CANOPIES 147

6.1. General Approach to the Modeling of Radiation Parameters for Vegetated Soils 147

6.2. The Emissivity of a Vegetation Canopy 154

6.3. Modeling Microwave Emission from Forests 166

6.4. Polarization Properties of Microwave Emission from Vegetation Canopies 169

6.5. Spatial Variations of Microwave Emission from the Earth's Surface 171

6.6. Global Simulation of Microwave Emission from Land 173

Chapter 7. EXPERIMENTAL RESEARCH ON MICROWAVE EMISSION FROM VEGETATION CANOPIES 177

7.1. Research on Microwave Emission from Vegetated Fields ... 177

7.2. Research on Microwave Emission from Forests 197

7.3. Statistical Properties of Microwave Emission from Vegetation Canopies 204

Chapter 8. VEGETATION EFFECT IN MICROWAVE REMOTE SENSING 207

8.1. Accounting for Vegetation Effect in Microwave Radiometry of a Surface 207

8.2. Vegetation Biomass Retrieval from Microwave Radiometric Measurements 217

8.3. Soil Moisture and Vegetation Biomass Retrieval from Multi-Configuration Microwave Radiometric Measurements 230

8.4. Vegetation Effect in Active Microwave Remote Sensing ... 236

Chapter 9. MICROWAVE RADIOMETRY OF VEGETATION
CANOPIES IN CONTEXT OF GLOBAL CHANGE
RESEARCH 241

9.1. Global Climate Problems and the Carbon Cycle 241

9.2. Assimilation of Remote Sensing Data into Global Carbon
Cycle Models 248

References 257

Index 281

FOREWORD

Research into microwave radiation from the Earth's surface in the presence of vegetation canopies, as well as the development of algorithms for retrieval of soil and vegetation parameters from microwave radiometric measurements, have been actively conducted for the last thirty years by many scientific groups and organizations all over the world. A complete bibliography of works on this problem would encompass hundreds of titles. The capability of the microwave radiometric method to determine soil moisture and vegetation biometric indices was revealed a quarter of a century ago by the author and many of his colleagues. In spite of the fact that the fundamentals and the basic physics of the microwave radiometry of soils and vegetation covers have been well developed, interest in the problem has not decreased but indeed has grown significantly in the last decade. This phenomenon has several reasons. The first one is the importance of these objects themselves in the remote ecological monitoring of land surface. In fact, soil moisture and vegetation covers play a key role in the hydrological cycle and in water and energy transfer on the border of land surface and atmosphere through evaporation and transpiration. The second reason is increased technical potentialities of microwave radiometric devices by being installed on spacecraft. In the modern design of space microwave radiometric systems, a high spatial resolution is achieved by using multi-beam antennas, synthetic aperture antennas, and big antennas with electronic and mechanical scanning that allows obtaining radio images of the Earth's surface. Accomplishment of large international projects that include global monitoring of the hydrological state of land surface (EOS Aqua, SMOS, Hydros, and others) shows that microwave radiometry of soil and vegetation more and more has become an instrument of practical application and operational use. In this respect, a systematic account of questions concerning the microwave radiometry of the Earth's surface in the presence of vegetation canopies seems to be useful and is the main objective of this book. The fundamental three-volume book by Ulaby, Moore, and Fung (1981-1986.), in which these questions were partially considered, was published a long ago. For a time after its publication, many papers on the problem appeared. A large number of papers and books were published in Russian and are not widely available for western researchers. To systemize these publications is also an objective of this book. The other objective of this monograph is to present young scientists, MS and PhD students who are involved or intending to be involved with microwave radiometric programs, with a text-book for studying the problem (and microwave radiometry in general) as a whole. Therefore, the book includes both general questions about microwave radiometry and particular practical questions of microwave radiometric measurements, design of experiments, etc.

The book is mainly based on works completed in the Institute of Radioengineering and Electronics of the Russian Academy of Sciences by the author and his colleagues during the last thirty years. Results by other researchers and groups are also reviewed, summarized, and analyzed. The author tried to review works of all researchers involved with the considered problem. But the number of works on the topic is really big, and he is sorry if someone's contribution is not mentioned in the book. The author will be glad to receive remarks and suggestions concerning the content of the book.

PREFACE

During the past twenty to thirty years, ferment in globalization processes in all sphere's of human activity has revealed numerous problems that arise from the interaction of society with nature. Among these problems, the global carbon cycle has acquired a special significance because of its now well-known greenhouse effect. To solve this and many similar problems, it is necessary to develop new concepts and approaches to our analyses of global environmental change, and in particular to select priorities in observing and assessing the existing state of natural systems. For example, at the present time, different global change models are being developed to assist in predicting climatic change. The functioning of these models and the reliability of estimates provided by them require permanent monitoring of the natural systems state. This monitoring allows us to detect real changes in environmental conditions and to compare predicted and observed trends in these changes. In addition, the observed data on the state of environmental components may be used as both model input parameters and feedback for model corrections.

Some ideas of permanent monitoring of environmental change have been realized with Earth Observing Systems that include space-borne and ground-based units to collect data on different parameters of the Earth's surface and atmosphere. Space-borne sensors play a key role in such monitoring, since they can rapidly collect a large amount of data over extensive territories. At the same time, ground-based data are important in validating and verifying the satellite information and filling possible gaps in remote observations.

The space-borne means of Earth observation that have been developed during the last decades include practically all significant functions of the nature-society system. Remote sensing data are the main source of operative information for the systems of control of global ecological, biogeochemical, hydro-physical, epidemiological, geophysical, and even demographic situations on the Earth. Remote sensing of the Earth's surface and atmosphere from space is based on receiving emitted or scattered electromagnetic radiation in different ranges of the electromagnetic wave spectrum. Operation of space-borne remote sensing means is affected by forces in the atmosphere. Particularly, optical sensors are not efficient in the presence of clouds. They also can not be used at night, since they can only detect solar radiation scattered by the Earth's surface. Infrared sensors are also influenced by the atmosphere. On the other hand, microwave remote sensors operate at decimeter, centimeter, and millimeter wavelengths. The atmosphere is practically transparent at decimeter and centimeter wavelengths, and microwave observations can be performed even in the presence of heavy cloudiness.

Microwave remote sensing methods are divided into active (radar) and passive (radiometric) ones. Microwave radiometric sensors measure the parameters of thermal electromagnetic radiation from the Earth's surface that provides their twenty-four-hour performance. Different remote sensing means are sensitive to various atmosphere and surface parameters. It appeared that microwave radiometry as applied to observations of land surface is an appropriate instrument for monitoring surface soil moisture and temperature. Soil moisture plays a crucial role in hydrology, agronomy, and meteorology. It governs the redistribution of precipitation between infiltration and runoff, it affects the development of crops through its dominance on regulating water-uptake by the plants, and it manages the partitioning of energy and water through evaporation and transpiration at the lower boundary of the atmosphere. Soil moisture is thus a key variable in the hydrological cycle. Monitoring soil moisture status on regional and global scales is of primary importance for understanding and protecting the environment, as well as for natural resources management that has been emphasized by the World Climate Research Program (1995). That is why two space missions have been planned, that are directly intended for soil moisture observations at a global scale by means of microwave radiometry.

The European Space Agency has selected the Soil Moisture and Ocean Salinity (SMOS) mission for implementation as the second mission in the line of Earth Explorer Opportunity mission (Silvestrin *et al.*, 2001; Kerr *et al.*, 2001). The goal of the SMOS mission is to observe two key variables, namely soil moisture over land and ocean surface salinity, by means of L-band (a wavelength of 21 cm) microwave imaginary radiometry. SMOS will also provide information on root zone soil moisture, vegetation, and biomass and contribute to research on the cryosphere. It is expected that the knowledge of global distribution of soil moisture and ocean salinity at adequate spatial and temporal sampling will significantly enhance weather forecasting, climate and extreme event predictions.

In the USA, the NASA Hydrosphere State (Hydros) mission was selected by NASA in 2002 as an Earth System Science Pathfinder mission for further development and is currently scheduled for launch in 2010

(Entekhabi *et al.*, 2004). Hydros is designed to provide global maps of the Earth's soil moisture and freeze/thaw state every 2-3 days for weather and climate prediction, water, energy, and carbon cycle studies, and natural hazards monitoring. Hydros utilizes a unique active and passive *L*-band microwave concept to simultaneously measure microwave emission and backscatter from the surface across a wide spatial swath. The key derived products are soil moisture at 40-km resolution for hydro-climatology obtained from the radiometer measurements, soil moisture at 10-km resolution for hydrometeorology obtained by combining the radar and radiometer measurements in a joint retrieval algorithm, and freeze/thaw state at 3-km resolution for terrestrial carbon flux dynamics studies obtained from radar measurements.

The presence of a vegetation cover on the Earth's surface affects the microwave emission from the surface in two ways. First of all, the vegetation cover screens the microwave emission from the soil surface. Secondly, the microwave emission from the vegetation layer itself is added to the emission from the soil. Vegetation effect depends on the vegetation type and vegetation biometric features as well as on the measuring configuration (frequency, polarization, observation angle). Assessment of the vegetation impact on the microwave radiometry of surface soil moisture, and examination of feasibilities of biometric parameters retrieval from microwave radiometric measurements are the main objectives of the present work.

The book contains nine chapters. In Chapter 1 introductory knowledge on the basics of microwave radiometry is given. Numerous books are available now, where the principles and fundamentals of microwave radiometry are developed and stated (e.g., Rytov, 1953; Levin and Rytov, 1967; Basharinov *et al.*, 1968; Basharinov *et al.*, 1974; Rytov *et al.*, 1978; Bogorodskii *et al.*, 1977; Bogorodskii *et al.*, 1981; Ulaby *et al.*, 1981, 1982, 1986; Tsang *et al.*, 1985; Sharkov, 2003; Armand and Polyakov, 2005), and, therefore, the microwave radiometry basics are given in brief. The material presented in Chapter 1 is taken from publications mentioned above.

Physical and microwave dielectric properties of vegetation and soil are discussed in Chapter 2. Several dielectric models of vegetation matter and soil are examined. It is shown that in the microwave band the dielectric properties of soils and vegetation material are mainly determined by their water content.

In Chapter 3, theoretical models and experimental data on the microwave emission from bare soils are presented. It is shown that the main factors that determine the intensity of microwave emission from bare soils are the moisture content of soil top layer and soil surface roughness. Other factors, such as soil type (texture), soil bulk density, etc., have a minor effect on the soil microwave emission.

The theory of microwave propagation in vegetation canopies is developed in Chapter 4. The relation between microwave propagation characteristics and vegetation biometric parameters are established. It is shown, particularly, the extinction rate of the coherent electromagnetic wave in a vegetation medium is proportional to the vegetation water content in unit volume, whereas the optical depth of a vegetation layer is mainly determined by the vegetation water content per unit area.

Chapter 5 presents a review of experimental research on microwave propagation in vegetation canopies. Different experimental techniques are discussed. Available experimental data on microwave attenuation by vegetation canopies are compared with model simulations.

Theory of microwave emission from vegetation canopies is developed in Chapter 6. A simple emission three-component radiation model (the $\tau - \omega$ model) is propounded. It is shown that the microwave emission of a vegetated soil is determined mainly by soil moisture and roughness, temperature of soil and vegetation, and vegetation water content.

An overview of experimental research on microwave emission from vegetation canopies is presented in Chapter 7.

Vegetation effect in microwave remote sensing of terrains is considered in Chapter 8. Vegetation screening in microwave remote sensing of soil moisture is discussed. Retrieval of soil moisture and vegetation water content from microwave radiometric measurements is examined.

In Chapter 9, a possibility of assimilation of microwave radiometric remote sensing data into global carbon cycle models is taken up.

Chapter 1

BASICS OF MICROWAVE RADIOMETRY

1.1. THEORY OF THERMAL RADIATION

1.1.1. Essential Definitions and Equations of an Electromagnetic Field

A full-blown theory of a unified electromagnetic field was developed by Maxwell. This theory is based on a generalization of known laws, established experimentally, and is a phenomenological macroscopic theory. It does not consider the microscopic processes in a medium in the presence of an electromagnetic field. Electric and magnetic properties of a specific medium are described by three macroscopic quantities, i.e., its relative dielectric permittivity ε, its relative magnetic permeability μ, and its conductivity σ. It is assumed that these parameters are known from experiments. Maxwell's theory solves the major task of electrodynamics, i.e., to determine the electromagnetic fields for a given system of electric currents and charges. The full system of *Maxwell's equations* is given by

$$rot\vec{E} = -\frac{\partial \vec{B}}{\partial t}, \tag{1.1}$$

$$div\vec{D} = \rho_q, \tag{1.2}$$

$$rot\vec{H} = \vec{j} + \frac{\partial \vec{D}}{\partial t}, \tag{1.3}$$

$$div\vec{B} = 0 \tag{1.4}$$

where $\vec{E}$ is the electric field strength, $\vec{B}$ is the magnetic flux density, ρ_q is the charge density, and $\vec{j}_e$ is the current density. The system is supplemented with the *material equations* that, for isotropic, non-ferroelectric, and non-ferromagnetic media, are written as

$$\vec{D} = \varepsilon\varepsilon_0\vec{E}, \tag{1.5}$$

$$\vec{B} = \mu\mu_0\vec{H}, \tag{1.6}$$

$$\vec{j}_e = \sigma\vec{E} \tag{1.7}$$

where ε_0 and μ_0 are the electric and magnetic constants, respectively.

Maxwell's theory permits the existence of electromagnetic fields in a space even if currents and charges are absent there. These fields are called electromagnetic waves that propagate through the space. In this free space, the propagation of an electromagnetic field is described by the following wave equations, which can be obtained from the above equations of Maxwell's theory:

$$\nabla^2\vec{E} - \varepsilon\varepsilon_0\mu\mu_0\frac{\partial^2\vec{E}}{\partial t^2} = 0, \tag{1.8}$$

$$\nabla^2\vec{H} - \varepsilon\varepsilon_0\mu\mu_0\frac{\partial^2\vec{H}}{\partial t^2} = 0. \tag{1.9}$$

Electromagnetic waves are transverse waves and propagate with the phase velocity $v = c/\sqrt{\varepsilon\mu}$ where $c = 1/\sqrt{\varepsilon_0\mu_0}$ is the velocity of electromagnetic waves in a vacuum. The vectors $\vec{E}$ and $\vec{H}$ are mutually perpendicular and both vectors are perpendicular to the velocity vector $\vec{v} = v[\vec{E}\times\vec{H}]/EH$. The magnitudes of vectors $\vec{E}$ and $\vec{H}$ in a wave are linked as

$$H = \sqrt{\varepsilon\varepsilon_0/(\mu\mu_0)}\,E. \tag{1.10}$$

A sinusoidal electromagnetic wave is called a monochromatic wave. If the time dependence of a field is chosen in the form $E \sim \exp\{-j\omega t\}$ where $\omega = 2\pi f$, and f is the frequency, the wave equation takes the form

$$\nabla^2 \vec{E} - \varepsilon\varepsilon_0 \mu\mu_0 \omega^2 \vec{E} = 0. \tag{1.11}$$

If the wave field depends on only one coordinate, the wave is called a plane wave. In other words, the dependence of a plane wave field on the co-ordinates is described by the multiplier $\exp\{j\vec{k}\vec{r}\}$ where $\vec{k}$ is the wave vector ($k = \omega/v$). For example, the electric field of the plane monochromatic wave propagating in z direction is found from (1.11) and is given by

$$E_x = E_{x0} \exp\{-j\omega t + k_z z), \tag{1.12}$$

$$E_y = E_{y0} \exp\{-j\omega t + k_z z + \varphi) \tag{1.13}$$

where φ is the phase difference of E_x and E_y oscillations. With an arbitrary value of φ, the plane wave is *elliptically* polarized. This means that in every point of a wave field the end of vector $\vec{E}$ describes an ellipse lying in the x-y plane. If $E_{x0} = E_{y0}$ and $\varphi = \pm(2m+1)\pi/2, (m = 0,1,2,...)$, the ellipse transforms into a circle, and the wave is *circular* polarized. If $\varphi = \pm m\pi$, $(m = 0,1,2,...)$, the ellipse turns into a line, and the wave is *linear* polarized. The plane passing through vector $\vec{E}$ and wave vector $\vec{k}$ is called the *polarization plane* of a linear polarized wave. If a linear polarized wave is incident upon a half-space at an incidence angle (the angle between $\vec{k}$ and the perpendicular to the surface of half-space) and the polarization plane is perpendicular to the surface, the wave is considered as *vertically* (*v*) polarized. If vector $\vec{E}$ of the incident wave is parallel to the surface, the wave is called *horizontally* (*h*) polarized.

The *volume density of electromagnetic field energy* is defined as the sum of volume densities of electric field energy and magnetic field energy. The volume density of electromagnetic field energy is given by

$$w_e = \frac{\varepsilon\varepsilon_0 E^2}{2} + \frac{\mu\mu_0 H^2}{2}. \tag{1.14}$$

The volume density of electromagnetic wave energy is written as (taking into account equation (1.10))

$$w_e = \varepsilon\varepsilon_0 E^2 = EH/v. \tag{1.15}$$

A moving electromagnetic wave transfers electromagnetic energy. The velocity of energy transfer is described by the Poynting's vector that is given by

$$\vec{\Pi} = w_e\vec{v} = [\vec{E}\times\vec{H}]. \tag{1.16}$$

This vector determines the density of electromagnetic energy flux, i.e., it shows what electromagnetic energy passes through a unit area in unit time.

The *intensity* of a moving monochromatic wave is defined as the mean over the period value of Poynting's vector:

$$I = \left|\langle\vec{\Pi}\rangle\right| = \langle w_e\rangle v. \tag{1.17}$$

The intensity shows what mean electromagnetic energy passes through a unit area in unit time. For a linear polarized wave, the intensity is proportional to the square of electric field amplitude E_0:

$$I = \frac{1}{2}\sqrt{\frac{\varepsilon\varepsilon_0}{\mu\mu_0}}\,E_0^2. \tag{1.18}$$

For an elliptically polarized wave, the intensity is given by

$$I = \frac{1}{2}\sqrt{\frac{\varepsilon\varepsilon_0}{\mu\mu_0}}\,(E_{x0}^2 + E_{y0}^2). \tag{1.19}$$

If a linear polarized plane wave is incident upon a body, the scattered field E_s at a large distance R from the body tends to its asymptotic form of a spherical wave ($E_s \sim 1/R$). The time average intensity of field scattered into solid angle $d\omega$ is then given by

$$dI_s = I_s(\hat{o},\hat{i})R^2 d\omega \tag{1.20}$$

where $\hat{o}$ and $\hat{i}$ are the unit vectors in the scattering and incidence directions, respectively. The ratio of dI_s to the intensity of incident wave I is called the *differential scattering cross section* $\sigma_d = dI_s / I$. In radar applications and microwave propagation modeling, the *bistatic scattering cross section* $_{bi}(\sigma\hat{o},\hat{i}) = 4\pi\sigma_d(\hat{o},\hat{i})$ and *backscattering cross section* $\sigma_{bs} = 4\pi\sigma_d (-\hat{i},\hat{i})$ are often used. The power scattered by the body in all directions is given by

$$P_s = \sigma_s I, \tag{1.21}$$

$$\sigma_s = \int_{4\pi} \sigma_d d\omega \tag{1.22}$$

where σ_s is the scattering cross section. The power absorbed by the body is given by

$$P_a = \sigma_a I \tag{1.23}$$

where σ_a is the absorption cross section. The sum of the scattering and absorption cross sections is called the *extinction cross section* of the body:

$$\sigma_e = \sigma_s + \sigma_a. \tag{1.24}$$

The scattering and absorption cross sections can be found from a solution to the diffraction problem or can be estimated from *integral representations*. The equivalent current density in Maxwell's equations can be written in the form (Ishimaru, 1978):

$$\vec{j}_e = \begin{cases} - j\omega\varepsilon_0 (\varepsilon - 1)\vec{E}, \text{ inside the body} \\ 0, \text{ outside the body} \end{cases} \tag{1.25}$$

where ε is the complex dielectric permittivity of the body. The solution of Maxwell's equations is given by

$$\vec{E}(\vec{r}) = \vec{E}_i(\vec{r}) + \vec{E}_s(\vec{r}) \tag{1.26}$$

where $\vec{E}_i(\vec{r})$ is the incident field and $\vec{E}_s(\vec{r})$ is the scattered field. The scattered field can be found with the use of Hertz's vector $\bar{A}_s$:

$$\vec{E}_s(\vec{r}) = \vec{\nabla} \times \vec{\nabla} \times \vec{A}_s(\vec{r}), \tag{1.27}$$

$$A_s(r) = -\frac{1}{j\omega\varepsilon_0} \int_V G_0(\vec{r},\vec{r}')\vec{j}_e(\vec{r}')dV' = \int_V [\varepsilon(\vec{r}')-1]\vec{E}(\vec{r}')G_0(\vec{r}',\vec{r}')dV' \tag{1.28}$$

where $G_0(\vec{r},\vec{r}') = \exp(j\vec{k}_0|\vec{r}-\vec{r}'|)/(4\pi|\vec{r}-\vec{r}'|)$ is the free space Green function, k_0 is the wave vector in the free space, and $\vec{r} \neq \vec{r}'$. From (1.26)-(1.28), we obtain

$$\vec{E}(\vec{r}) = \vec{E}_i(\vec{r}) + \vec{E}_s(\vec{r})$$

$$= \vec{E}_i(\vec{r}) + \frac{k_0^2}{4\pi} \int_V \{\vec{E}(\vec{r}')\left(1 - \frac{j}{k_0|\vec{r}-\vec{r}'|} - \frac{1}{\left(k_0|\vec{r}-\vec{r}'|\right)^2}\right) - \left(\vec{E}(\vec{r}')\cdot\frac{\vec{r}-\vec{r}'}{|\vec{r}-\vec{r}'|}\right) \times \tag{1.29}$$

$$\times (\vec{r}-\vec{r}')\left(1 + \frac{3j}{k_0|\vec{r}-\vec{r}'|} - \frac{3}{\left(k_0|\vec{r}-\vec{r}'|\right)^2}\right)\}[\varepsilon(\vec{r}')-1]\frac{e^{jk_0|\vec{r}-\vec{r}'|}}{|\vec{r}-\vec{r}'|}d\vec{r}'$$

Equation (1.29) represents the integral form of Maxwell's equations. It allows one, in principle, to find the electric field in any point if the spatial distribution of dielectric permittivity is given. To find the scattered field in the far-field zone ($k_0|\vec{r}-\vec{r}'| \gg 1$), at a distance R from the body, one can make the following substitutions:

$$\frac{1}{|\vec{r}-\vec{r}'|} \approx \frac{1}{R}, \quad |\vec{r}-\vec{r}'| \approx R - \vec{r}'\cdot\hat{o}. \tag{1.30}$$

The scattered field in the far-field zone is then presented as

$$\vec{E}_s(\vec{r}) = \vec{f}(\hat{o},\hat{i})\frac{\exp(jk_0 R)}{R}, \tag{1.31}$$

$$\vec{f}(\vec{o},\hat{i}) = \frac{k_0^2}{4\pi} \int_V [\vec{E} - \hat{o}\cdot(\hat{o}\cdot\vec{E})][\varepsilon - 1]\exp(-jk_0\vec{r}'\cdot\hat{o})dV' \tag{1.32}$$

where $f(o,\hat{i})$ is the *scattering amplitude*. Equation (1.32) gives an explicit form for the scattering amplitude through the field inside a particle. Generally speaking, this field is unknown. However, in modeling, this field can be approximated by a known function that allows one to obtain useful approximations for the scattering amplitude.

The *optical theorem* asserts that the extinction cross section of a particle is linked with the scattering amplitude as

$$\sigma_e = \frac{4\pi}{k_0} \mathrm{Im}[\vec{f}(\hat{i},\hat{i})\cdot\hat{e}_i] \tag{1.33}$$

where Im denotes the imaginary part and $\hat{e}_i$ is the unit vector characterizing the polarization of an incident wave.

If a linear polarized wave is incident from free space upon a half-space with the flat boundary, and the wave vector of the incident wave makes an angle ϑ with the normal to the interface, the *law of refraction* is written as

$$\sin\vartheta = \sqrt{\varepsilon\mu}\,\sin r \tag{1.34}$$

where r is the refraction angle, ε and μ are the permittivity and permeability of the half-space. The amplitudes of reflected and refracted fields are linked with the amplitude of the incident field by *Fresnel formulas*. For a vertically polarized incident wave, these amplitudes are given by

$$E_r = -E_0 \frac{tg(\vartheta-r)}{tg(\vartheta+r)}, \tag{1.35}$$

$$E_t = E_0 \frac{2\cos\vartheta\sin r}{\sin(\vartheta+r)\cos(\vartheta-r)} \tag{1.36}$$

where E_0, E_r, and E_t are the amplitudes of incident, reflected, and refracted waves, respectively. For a horizontally polarized incident wave, these amplitudes are given by

$$E_r = -E_0 \frac{\sin(\vartheta-r)}{\sin(\vartheta+r)}, \tag{1.37}$$

$$E_t = E_0 \frac{2\cos\vartheta\sin r}{\sin(\vartheta+r)}.$$ (1.38)

The *reflectivity* of the half-space is defined as the ratio of the reflected wave intensity to the incident wave intensity. The *transmissivity* of the half-space is defined as the ratio of the refracted wave intensity to the incident wave intensity.

If a linear polarized wave is incident upon a flat, homogeneous layer with a relative dielectric permittivity ε and thickness d, the reflectivity and transmissivity of the layer, r and t, are given by

$$r_{v,h} = \left| \frac{R_{v,h}(1-\exp\{-j2k_{z1}d\}}{(1-R_{v,h}^2\exp\{-j2k_{z1}d\})} \right|^2,$$ (1.39)

$$t_h = \left| \frac{4k_{z0}k_{z1}\exp\{j(k_{z0}-k_{z1})d\}}{(k_{z0}+k_{z1})^2(1-R_h^2\exp\{-j2k_{z1}d\})} \right|^2,$$ (1.40)

$$t_v = \left| \frac{4\varepsilon k_{z0}k_{z1}\exp\{j(k_{z0}-k_{z1})d\}}{(\varepsilon k_{z0}+k_{z1})^2(1-R_v^2\exp\{-j2k_{z1}d\})} \right|^2$$ (1.41)

where

$$k_0 = \omega/c,$$

$$k_{z0} = k_0\cos\vartheta,$$

$$k_{z1} = k_0\sqrt{\varepsilon-\sin^2\vartheta},$$

$$R_h = (k_{z0}-k_{z1})/(k_{z0}+k_{z1}),$$

$$R_v = (\varepsilon k_{z0}-k_{z1})/(\varepsilon k_{z0}+k_{z1})$$

where ϑ is the incidence angle relative to the layer normal.

There is another approach to describe wave propagation in a scattering medium that is not based on Maxwell's equations but rather on energy transfer. This is called the *transfer theory*. This theory does not originate from the wave equations (1.8), (1.9) but operates with the energy transfer in a medium containing scattering particles. This theory is built heuristically and is not rigorous in a mathematical respect. Even if the diffraction and interference effects are accounted for in the consideration of scattering by a single scatterer, the theory itself does not include diffraction effects. It is assumed in the theory that, in the summation of fields, the correlation between the fields is absent, thus, the intensities of the fields are summed but not the fields. The transfer theory operates with *ray intensity* (or *brightness*) that is the measure of energy flux. This quantity is introduced for a given direction $\hat{s}$ as the mean energy flux density in unit spectral interval and unit solid angle. The ray intensity $J(\vec{r},\hat{s})$ is measured in W m^{-2} Hz^{-1} sr^{-1}. The transfer theory takes into account the polarization effects by introducing the Stokes matrix. The vector transfer equation is of the following form: For $0 \leq \theta \leq \pi, 0 \leq \phi \leq 2\pi,$

$$\cos\theta \frac{d}{dz}\bar{I}(\theta,\phi,z) = -\hat{\kappa}_e(\theta,\phi) \cdot \bar{I}(\theta,\phi,z) + \int_0^{2\pi} d\phi' \times$$
$$\times \int_0^\pi d\theta' \sin\theta' \, \hat{P}(\theta,\phi;\theta',\phi') \cdot \bar{I}(\theta',\phi',z') \qquad (1.42)$$

where $\bar{I}(\theta,\phi,z)$ is a 4×1 column vector denoting the modified Stokes parameters in direction (θ,ϕ), $\hat{P}(\theta,\phi;\theta',\phi')$ is the 4×4 phase matrix denoting scattering from direction (θ',ϕ') into direction (θ,ϕ), and $\hat{\kappa}_e(\theta,\phi)$ is the extinction matrix which can be expressed in terms of the forward scattering amplitudes. For the case of thermal emission, to the right side of the equation is added the term $\hat{\kappa}_a \dfrac{2k_0 T(z)}{\lambda^2}$ where $\hat{\kappa}_a$ is the absorption matrix and T is the physical temperature of the medium. The vector transfer equation is usually solved numerically with the use of different approaches.

1.1.2. Blackbody Thermal Radiation

All substances radiate a continuum of electromagnetic energy owing to atomic and molecular vibrations. The nature and intensity of these vibrations depend on their temperature (thermal level of energy). The energy distribution of heated bodies' thermal emission over frequency spectrum was found with the use of the "blackbody" concept and of quantum theory assumptions.

A blackbody is an idealized body capable of absorbing all the radiation falling upon it at all frequencies, reflecting none. Furthermore, in addition to being a perfect absorber, it has to be a perfect emitter since energy absorbed by a material would increase its temperature if no energy were emitted. The blackbody emission problem was solved by Planck. His law of radiation states that the brightness B of a blackbody radiator at a temperature T and frequency f is given by

$$B = \frac{2hf^3}{c^2} \frac{1}{e^{hf/k_B T} - 1} \tag{1.43}$$

where h is Planck's constant, k_B is Boltzmann's constant, and c is the velocity of light. The blackbody brightness is a function of only frequency and temperature, and is independent of direction and position. At low frequencies, where $hf/k_B T \ll 1$, equation (1.43) reduces to

$$B = \frac{2f^2 k_B T}{c^2} = \frac{2k_B T}{\lambda^2} \tag{1.44}$$

where $\lambda = c/f$ is the wavelength. Equation (1.44) is known as the Rayleigh-Jeans radiation formula. At a room temperature of 300 K, the fractional deviation of this formula from Planck's is less than one percent for wavelengths satisfying the condition: $\lambda > 2.57$ mm ($f \approx 117$ GHz). This covers the entire radio region and the usable part of the microwave spectrum.

In keeping with the nomenclature used to describe an ideal absorber-emitter, material objects are sometimes referred to as grey bodies. Since a blackbody radiates more energy at a given temperature than any other material, the power radiated by such a material (grey body) may be thought of as the power radiated by an equivalent blackbody having a cooler temperature than the physical temperature of the material T. Such a temperature is termed as the apparent or brightness temperature T_b of the material. The emissivity of grey body e is defined as the ratio of the body's brightness to the brightness of a blackbody of the same physical temperature. If the Rayleigh-Jeans approximation is valid, the brightness temperature of grey body relates to its temperature as

$$T_b = eT. \tag{1.45}$$

From the above definitions it follows that a blackbody has a brightness temperature identical to its physical temperature and hence an emissivity of unity. Good approximations to ideal blackbodies at microwave frequencies are the highly absorbing materials; emissivities as high as 0.99 can be achieved over a limited range of frequencies and incidence angles (relative to normal). On the other extreme, a highly conductive metal plate is a perfect reflector with an emissivity close to zero.

Kirchhoff's law establishes the relation between the abilities of emitting and absorbing the electromagnetic energy by any physical body. In its most general formulation, this law states that the emissivity of a grey body is defined by its absorptive properties, i.e., the emissivity is equal to the absorption factor of the body.

1.1.3. Generalized Kirchhoff's Law

The thermal radiation appears as the result of random motion of charged particles inside a body. The fluctuation thermal electromagnetic field can be described as the field exited by random currents with the density $\vec{j}_e(\vec{r})$. The mean value of this density is equal to zero, and the spatial correlation function of its frequency spectrum is defined on the basis of the fluctuation-dissipation theorem. In the radio frequency band, the spatial correlation function is written as (Rytov, 1953)

$$\left\langle j_\alpha(\omega,\vec{r}')j_\beta(\omega,\vec{r}'')\right\rangle = \frac{\omega}{4\pi^2}\varepsilon''(\omega)k_B T\delta(\vec{r}'-\vec{r}'')\delta_{\alpha\beta} \qquad (1.46)$$

where ε'' is the imaginary part of the body's dielectric permittivity, $\delta(\vec{r}'-\vec{r}'')$ is the delta function, and $\delta_{\alpha\beta}=1\,(\alpha=\beta)$, $\delta_{\alpha\beta}=0\,(\alpha\neq\beta)$. In order to calculate the fields generated by fluctuation currents, one needs to know the Green function of the considered body, i.e., the diffraction field exited inside the body by a single current source:

$$\vec{j}_G(\vec{r}')=\hat{e}\delta(\vec{r}-\vec{r}') \qquad (1.47)$$

where $\hat{e}$ is a unit vector. The field $\vec{E}_d$, exited by the current, is the diffraction field. To determine the fluctuation field, one can use the reciprocity theorem. Levin and Rytov (1967) have found that the fluctuation field is expressed as

$$\hat{e}\cdot\vec{E}(\vec{r}')=\int_V \vec{j}_e(\vec{r})\vec{E}_d(\vec{r},\hat{e},\vec{r}')d^3\vec{r} \qquad (1.48)$$

where integration is performed over the body's volume. The mean intensity of the fluctuation field component is found from equations (1.46), (1.48):

$$\left\langle \left| \hat{e} \cdot \vec{E}(\vec{r}') \right|^2 \right\rangle = \frac{\omega k_B T}{4\pi^2} \int\limits_V \varepsilon'' \left| \vec{E}_d(\vec{r}, \hat{e}, \vec{r}') \right|^2 d^3\vec{r} . \qquad (1.49)$$

The expression under the integral is connected to the density of electromagnetic energy losses in the body. So, equation (1.49) can be written in the form

$$\left\langle \left| \hat{e} \cdot \vec{E}(\vec{r}') \right|^2 \right\rangle = \frac{2 k_B T}{\pi} Q(\hat{e}, \vec{r}') \qquad (1.50)$$

where $Q(\hat{e}, \vec{r}')$ represents the active thermal losses of the diffraction field in the body. Levin-Rytov's theory shows that to find the intensity of the fluctuation field at a given point, that is due to body's thermal emission, it is necessary to determine the thermal losses of the diffraction field exited inside the body by a unit current placed at this point and directed according to the direction of the sought fluctuation field. This approach does not apply any restrictions to the size of the body as compared to the wavelength and therefore represents a generalized form of Kirchhoff's law.

1.2. MICROWAVE RADIOMETERS AND SYSTEMS

Any microwave radiometer includes a receiving antenna and a radiometer receiver. The microwave radiometric measuring process is illustrated in Fig. 1.1.

A radiometer receiver is connected to a pencil-beam antenna pointed at a scene of interest. The objective is to measure the thermal radiation emitted by the area covered by the footprint of the antenna effective main-beam; i.e., it is desired to relate the radiometer output (represented by V_0) to the target emission that is represented by its brightness temperature T_b. To perform its function, the desired association antenna and radiometer receiver properties should be quantified.

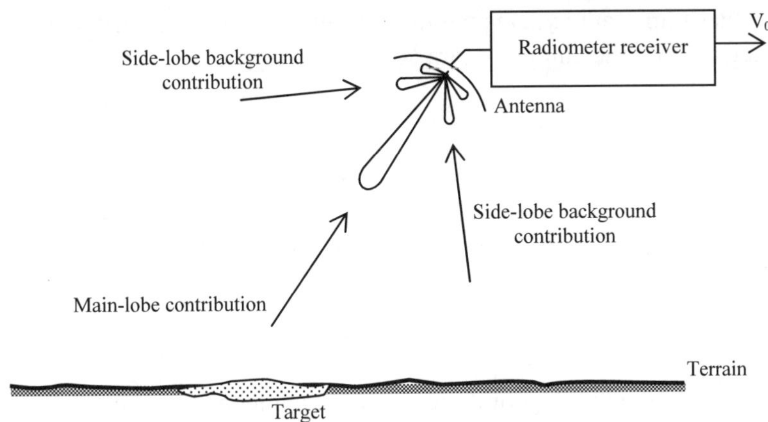

Fig. 1.1. Schematic representation of radiometric measurements.

The normalized antenna directional pattern $G(\vartheta,\varphi)$ is the angular distribution of the radiation energy flux density of the antenna in the far-field zone. For antennas of complicated configuration the antenna pattern has a multi-lobe structure. The antenna main-beam efficiency α_m and stray factor α_s are determined as follows:

$$\alpha_m = \frac{\displaystyle\int_{\Omega_m} G(\vartheta,\varphi)d\Omega}{\displaystyle\int_{4\pi} G(\vartheta,\varphi)d\Omega}, \qquad (1.51)$$

$$\alpha_s = \frac{\displaystyle\int_{\Omega_s} G(\vartheta,\varphi)d\Omega}{\displaystyle\int_{4\pi} G(\vartheta,\varphi)d\Omega} = 1-\alpha_m \qquad (1.52)$$

where Ω_m is the solid angle subtended by the main lobe of the directional pattern and $\Omega_s = 4\pi - \Omega_m$. The quantity $\int_{4\pi} G(\vartheta,\varphi)d\Omega = \Omega_A$ is known as the antenna beam solid angle. The antenna directivity is defined as

$$Directivity = 4\pi / \Omega_A. \tag{1.53}$$

The other important antenna parameter is the antenna's effective area A_e. It is connected with the antenna beam solid angle as

$$A_e = \frac{\lambda^2}{\Omega_A}. \tag{1.54}$$

The linear angular resolution of an antenna is estimated by the formula

$$\theta_{3dB} \approx 1.22 \frac{\lambda}{D_A} \tag{1.55}$$

where D_A is the linear geometric size of the antenna.

The power received by a lossless microwave antenna placed inside a blackbody in a narrow bandwidth Δf is given by

$$P_A = k_B T \Delta f \frac{A_e}{\lambda^2} \int_{4\pi} G(\vartheta, \varphi) d\Omega = k_B T \Delta f \tag{1.56}$$

where T is the blackbody and antenna radiation resistance temperature. At the same time, the noise power from a resistor R at temperature T_R is given by the Nyquist formula

$$P = k_B T_R \Delta f. \tag{1.57}$$

That is identical to (1.55) if the temperature of antenna radiation resistance is T_R. Due to this identity we can express the power received by an antenna on a temperature scale. Thus the power received by the antenna is expressed in terms of antenna temperature that is defined as

$$T_A = \frac{P_A}{k_B \Delta f}. \tag{1.58}$$

The antenna temperature for an antenna with losses is expressed as

$$T_A = \frac{\eta}{4\pi} \int_{4\pi} T_b(\vartheta, \varphi) G(\vartheta, \varphi) d\Omega + (1 - \eta) T_0 \tag{1.59}$$

where $T_b(\vartheta,\varphi)$ is the brightness temperature in the (ϑ,φ) direction, T_0 is the temperature of the medium surrounding the antenna, and η is the antenna radiation efficiency. Separating the power received in the main lobe and side lobes, one can obtain the other expression for the antenna temperature:

$$T_A = T_{bm}\alpha_m\eta + T_{bs}(1-\alpha_m)\eta + (1-\eta)T_0 \qquad (1.60)$$

where T_{bm} and T_{bs} are the weighted average brightness temperatures for the antenna main and side lobes. The radiometer output voltage V_0 can be calibrated to read the antenna temperature. Equation (1.60) provides a basis for determination of the main-lobe brightness temperature from measured values of antenna temperatures:

$$T_{bm} = \frac{T_A}{\alpha_m\eta} - \frac{T_{bs}(1-\alpha_m)}{\alpha_m} - \frac{(1-\eta)T_0}{\alpha_m\eta}. \qquad (1.61)$$

The basic function of a radiometer receiver is to measure the radiant power delivered by the antenna. For a receiver with bandwidth Δf, the power available at its input is expressed in terms of the antenna temperature as

$$P_s = k_B\Delta f[\frac{T_A}{L} + (1-\frac{1}{L})T_0] \qquad (1.62)$$

where L is the loss factor accounting for ohmic losses by the antenna and transmission line. The term in square brackets in (1.62) is considered as the receiver signal temperature T_s. Recovering the antenna temperature from the measured power requires precise knowledge of L and T_0. To avoid the influence of occasional change of the loss factor and the ambient temperature, modern radiometers are designed as a single thermostatic block containing both antenna and radiometer receiver.

With no input signal present, the noise power appearing at the receiver output (prior to detection) is assigned to an effective noise temperature of the receiver T_n. The total input power is the algebraic sum of the signal and receiver noise power:

$$P_i = k_B\Delta f(T_s + T_n). \qquad (1.63)$$

The simplest radiometer system is the total-power receiver that consists of a pre-detection high frequency section of bandwidth Δf and gain G; a square-law detector with output voltage proportional to the input power; a low-pass filter-integrator section; and indicator (Fig. 1.2).

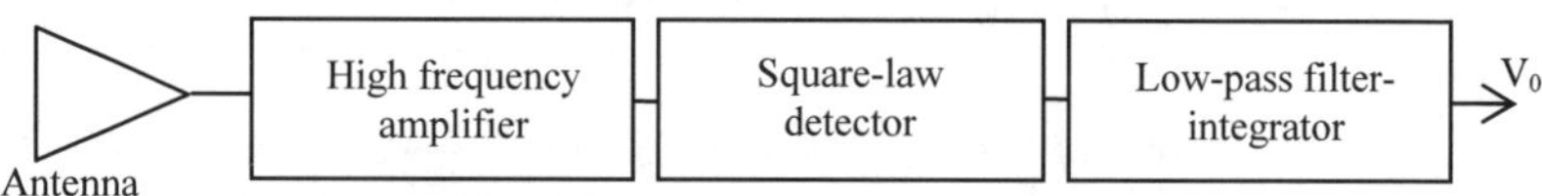

Fig. 1.2. The simplest total-power radiometer receiver.

The output voltage is related to the input power averaged over an integration time τ determined by the post-detector integrator. The detected voltage is composed of a direct current component and a fluctuating component caused by noise and gain variations. The function of the low-pass filter-integrator section is to reduce the noise variations by integrating the detector output voltage over an integration time τ. The root mean square measurement uncertainty in T_s due to noise fluctuation is estimated by

$$\Delta T_n = \frac{T_s + T_n}{\sqrt{\Delta f \tau}}. \tag{1.64}$$

Receiver gain variations contribute an additional uncertainty:

$$\Delta T_G = (T_s + T_n)\frac{\Delta G}{G} \tag{1.65}$$

where ΔG is the effective value (rms) of the detected receiver power-gain variations. These two types of variations are considered statistically independent, and therefore can be combined to define the *radiometer sensitivity*, $\Delta T_{\min}$, as follows:

$$\Delta T_{\min} = \sqrt{(\Delta T_n)^2 + (\Delta T_G)^2} = (T_s + T_n)\sqrt{\frac{1}{\Delta f \tau} + \left(\frac{\Delta G}{G}\right)^2}. \tag{1.66}$$

Radiometer sensitivity (or radiometric resolution) is the minimum input signal variation ΔT_s required to produce a direct current change in the receiver output level equal to the root mean square value of the fluctuating component.

Dicke proposed the comparison-radiometer technique to avoid the influence of gain and receiver effective noise temperature variations. He suggested: if the receiver input is periodically switched between the antenna and a comparison, or reference, noise source at a switching rate higher than the highest significant spectral components in the gain variation spectrum, then the gain variation contribution can be reduced significantly. In other words, over a period of one switching cycle, the low-frequency (slow) gain variation component will be hardly noticeable. Since the Dicke radiometer actually measures a temperature contrast between the signal temperature and the reference temperature, its output voltage is independent of the receiver effective noise temperature. However, the desired signal temperature is now observed for only half the time (as compared to the total-power receiver). Hence, the sensitivity of this radiometer (the Dicke radiometer is called also the modulation radiometer) is

$$\Delta T_n \approx 2\frac{T_s + T_n}{\sqrt{\Delta f \tau}}. \tag{1.67}$$

There are several other types of radiometers. Their description and parameters can be found in the literature.

To obtain a radiometric image of a scene of interest, it is necessary to scan with the antenna maim-beam. If the radiometer is stationary relative to the scene, line-by-line scanning is usually used. With moving airborne platforms, scanning in only the cross-track dimension can suffice. Antenna scanning, or beam steering, can be accomplished either mechanically or electronically. Mechanical beam steering involves changing the pointing direction of the antenna axis by moving (in angle) the entire antenna or its feed. Electronic beam steering is achieved by using a planar-phased array consisting of many radiating elements. The main beam can be steered in both dimensions by electronically controlling the relative phase of each element. Radiometric systems with electronic scanning have some drawbacks: complexity, losses in the phase shifters, and high cost and weight. A simpler solution is a system that includes several antennas (push-broom systems). The antennas could be conically directed to the scene providing the same observation angles for all antennas. Also a one phase array antenna can be used to obtain several main beams by connecting antenna elements with certain constant phase shift. A separate radiometer receiver can be used for every antenna (or every main beam) to increase the sensitivity. Antenna angular resolution (1.55) increases with the increase of antenna size. On the other hand, the use of very large antennas is difficult, especially in satellite radiometric systems, because of the antennas' heavy weight and complexity of assembling them on a satellite. To avoid this problem, a synthetic aperture technique has been developed. This technique restores the directional pattern of a large antenna by measurements with some separate fragments of the antenna. It is similar to the use of a lens fragment which operates as the entire lens but with a certain loss of intensity. The interested reader is referred to papers by Ruf *et al.* (1988), Le Vine (1990), Laursen and Skou (1998), and Camps *et al.* (2000) for more details.

1.3. RADIOMETRIC MEASUREMENTS

As a rule radiometric measurements start from a test of the radiometric system. This test includes measurements of basic system parameters: radiometric sensitivity, bandwidth, antenna and transmission line loss, antenna main-beam efficiency, etc.

The sensitivity of a radiometer receiver is determined by connecting to its input a matched load at different controlled temperatures T_s (cold and hot references). It allows one to establish the relation between the output voltage of the radiometer receiver and the signal temperature. The sensitivity of the radiometer receiver is estimated from these measurements as

$$\Delta T \approx \frac{1}{6} \frac{\delta T}{\Delta T_s} \qquad (1.68)$$

where ΔT_s is the known temperature contrast between the hot and cold references, δT is the peak-to-peak swing of output signal fluctuations when the hot (or cold) reference is connected to the input. (It follows from (1.67) that the sensitivity depends on T_s, and therefore different estimates of the sensitivity are obtained with δT observed for hot and cold references.)

When the temperature scale of the radiometer receiver is formed by the above procedure, the transmission line loss factor can be determined by connecting the hot and cold references to the transmission line instead of the antenna. The loss factor is determined by

$$L = \frac{\Delta T_s}{\Delta T_{mes}} \qquad (1.69)$$

where ΔT_s is the known temperature contrast between the hot and cold references and ΔT_{mes} is this contrast measured by the radiometer receiver.

Antenna radiation and main-beam efficiency, η and α_m, are determined by procedures applied in the antenna technique. When all parameters of the radiometric system are known, equations (1.61) and (1.62) are used to determine the main-lobe brightness temperature.

Some experimentalists prefer to calibrate the entire radiometric system by the use of the external references of known brightness (that is to relate the output voltage of the radiometer to the main-lobe brightness temperature). The clean sky is often used as a cold reference (the brightness temperature about 3-10 K). The absorbing plates of a large size and known temperature are usually used as hot references.

A typical configuration of radiometric measurements is presented in Fig. 1.1. In measurements, one has to take into account the effect of galactic noise and atmospheric medium on the measured values of brightness temperatures. This effect is described in detail in the literature (e.g., Ulaby *et al.*, 1981-1986).

Radiometric measurements can be divided into measurements of the absolute magnitudes of brightness temperature, relative measurements (measurements of the brightness temperature contrasts), and measurements of emissivities. Absolute measurements of brightness temperatures require careful calibration of the radiometric system by external sources, i.e., references with well-known brightness temperatures. But even if this calibration is carefully performed, variations of the received signal due to the change of background emission coming to the system through the side lobes can be significant. The total error of brightness temperature absolute measurements is usually not less than 1 K. That is much greater than the radiometric resolution (sensitivity) of radiometric systems, which can achieve 0.01-0.1 K. So, a radiometric system can detect rather small variations of radiometric signal, providing a high accuracy of relative radiometric measurements, i.e., measurements of change in the brightness temperature. Measurements of target emissivity can be performed neglecting the constant background and atmospheric emission by the use of external calibration with flat references of known emissivity. These references (usually, they are a metal plate with zero emissivity and an absorbing plate with emissivity close to unity) are put on the target providing reference signals corresponding to the first and the second reference emissivity. Then, the signal from the investigated target is measured and referred to the previously formed emissivity scale.

Spatial resolution of radiometric measurements is determined by the size of antenna footprint at the Earth's surface. Using equation (1.55), one obtains that this size is

$$\Delta \approx 1.22 \frac{\lambda}{D_A} r_t \qquad\qquad (1.70)$$

where r_t is the distance between the antenna and target. The distance r_t should be greater than $2D_A^2/\lambda$, to provide measurements in the far-field zone of the antenna. This condition imposes a restriction on the minimum potential spatial resolution of a radiometric system:

$$\Delta_{\min} \approx 2.44 D_A . \qquad\qquad (1.71)$$

The restriction is not important for measurements from aircrafts of satellites since $r_t \gg 2D_A^2/\lambda$ in this case.

Chapter 2

PHYSICAL AND ELECTRICAL PROPERTIES OF SOILS AND VEGETATION

2.1. PHYSICAL AND DIELECTRIC PROPERTIES OF SOILS

2.1.1. Physical Properties of Soils

Soils consist of solid, liquid, and gas phases. Mineral matters such as quartzes, field spars, micas, etc prevail in the solid phase. The last can include also salts of different kinds ($NaCl$, $MgCl_2$, $CaCl_2$, $MgSO_4$, Na_2SO_4, etc). The amount of salts in weakly salt soils does not exceed 0.5-1% of dry soil weight. The solid phase also includes different organic matters (1-10% of dry soil weight) and soil colloids. The liquid phase (the soil solution) is water containing dissolved salts, organic matters, and gases. The gas phase is air filling empty voids of the soil.

The density of basic soil minerals (the density of the solid phase ρ_s) varies in the limits of 2500-3000 kg m^{-3}. The density of the dry soil itself (the bulk density ρ_b) depends on its type. Sands have a density of 1300-2100 kg m^{-3}, clays have a density of 1200-2600 kg m^{-3}, and fertile soils have a density of 800-1600 kg m^{-3}. One of the most important structure features of a soil is its grain-size (granulometric) composition or the relative content of particles with different sizes in the unit volume or mass of the soil. There are several classifications of soil grading that are used in soil science, geology, mining, and agriculture. According to the classification of the US Department of Agriculture, the following soil fractions are separated: 1) sand (the size of soil particles is more than 0.05 mm); silt (the size of soil particles is 0.002-0.05 mm); clay (the size of soil particles is less than 0.002mm). The content of these fractions in a soil is expressed in percents by weight. The

form of soil particles depends on the type of composing minerals: clay particles have plate-like form; sand (quartz) particles have near to spherical form.
The specific surface area in a soil is the area of soil particles per unit volume.
Another important feature of a soil is its texture that reflects the cohesion of
particles, porosity, etc.

There is no unique classification of moisture in soils now. Generally,
specialists divide it into the bound and free water. The bound water, in turn,
is subdivided into several forms: firmly bound water, film water, and transient type water. The bound water is the water absorbed by the surface of
soil particles and kept at this surface by chemical and physical-chemical
forces, which can change its physical properties in comparison with those of
the free water. The bound water forms a film around a particle with a thickness of several molecular layers. Physical properties (the density, melting
point) of the bound water have not been thoroughly studied. Some data on
these properties are available in Boyarskii *et al.* (2002) and Tikhonov
(1996). The free water is considered as liquid water in a soil subjected to the
action of the gravity force exclusively and located in macro voids and
cracks. The quantity of water in a soil (the soil moisture) is defined by several parameters. The relative gravimetric moisture m is determined as

$$m = \frac{w_w - w_d}{w_d} \qquad (2.1)$$

where w_w and w_d are the wet and dry weights of a soil sample, respectively.
The volumetric moisture m_V is the relative volume of water in the soil
sample:

$$m_V = \frac{V_w}{V_s} = m_g \frac{\rho_{ws}}{\rho_w} = m \frac{\rho_b}{\rho_w} \qquad (2.2)$$

where V_w is the volume of water, V_s is the volume of the sample, ρ_{ws} and
ρ_w is the density of wet soil and water, respectively, ρ_b is the soil bulk density. Some hydrological constants or capacities are also used for description
of soil properties. Among them are the field capacity (FC), the wilting point
(WP), and others. The field capacity is the maximum quantity of water in a
soil held by capillary forces (FC is taken as the soil moisture at a tension of
1/3 bar). The wilting point designates a level of soil moisture (it is taken to
be the soil moisture at a tension of 15 bars (Wang and Schmugge, 1980))
below which plants begin to wilt.

The quantity of the bound water in a soil depends on its texture and amounts to 2-3% for sands and can achieve 30-40% of dry weight for heavy clays and loess soils. There are some approaches to estimate the quantity of the bound water in a soil. In works by Wang and Schmugge (1980) and Schmugge (1983), this quantity is expressed as some transition level$_t$ W, which is determined from experimental data. In work by Tikhonov (1996), it is suggested that the maximum thickness of the bound water film on the clay particles of soil does not exceed nine molecular layers, that is of the order of $2.52 \cdot 10^{-7}$ cm. The quantity of the bound water is calculated taking into account the size of clay particles and their fractional weight in the soil.

The field capacity FC in terms of volumetric moisture is estimated by (Wang and Schmugge, 1980)

$$FC = 0.3 - 0.23 \times S + 0.5 \times C \tag{2.3}$$

where S and C are the weight fractions of sand and clay in the soil.

2.1.2. Dielectric Properties of Soils in the Microwave Band

2.1.2.1. Experimental Studies

Dielectric properties of soils basically determine the intensity of microwave emission from soil surface. Soil dielectric models relating the dielectric permittivity of a soil to its physical parameters are the essential part of algorithms for retrieving soil moisture from remote sensing data. Due to this reason, numerous experimental programs have been conducted in order to investigate the dielectric behavior of soil-water mixtures in the microwave region (e.g., Abdulla *et al.*, 1988; Bobrov *et al.*, 1989; Boyarskii and Tikhonov, 1995; Boyarskii *et al.*, 2002; Calvet *at al.*, 1995a; Curtis, 2001; Hallikainen *et al.*, 1985; Hipp, 1974; Hoekstra and Delaney, 1974; Ilyin and Sosnovskiy, 1994; Jackson, 1990; Kleshchenko, 2002; Krotikov, 1962; Leschanskiy *et al.*, 1971; Mätzler, 1998; Mironov *et al.*, 1995, 1997, 2004; Peplinski *et al.*, 1995; Polyakov *et al.*, 1994; Sabburg *et al.*, 1997; Wang, 1980).

Soils are random media consisting of soil particles, air voids, and liquid water. Since in the microwave band the dimensions of soil constituents are small in comparison with the wavelength, the concept of an effective dielectric constant (or simply dielectric constant) can be used for describing the propagation of electromagnetic waves in such a medium. Dielectric properties of a soil are described then by its complex relative dielectric permittivity

$$\varepsilon = \varepsilon' + j\varepsilon''. \tag{2.4}$$

The complex refractive index

$$n = \sqrt{\varepsilon} = n' + jn''$$
$$\varepsilon' = n'^2 + n''^2 \quad \varepsilon'' = 2n'n'' \tag{2.5}$$

determines the attenuation and the phase velocity of plane waves in the medium. For plain, exponentially-damped, harmonic, transverse wave propagation through the medium, power attenuation, expressed in decibels per meter, is written

$$power\ loss = 8.6859\frac{2\pi}{\lambda}n'' = 8.6859\frac{2\pi}{\lambda}\left[\frac{\varepsilon'}{2}\left(\sqrt{1+\left(\frac{\varepsilon''}{\varepsilon'}\right)^2}\right)-1\right]^2 \tag{2.6}$$

where λ is the wavelength. Plane waves of constant phase will propagate through the medium with the phase velocity:

$$phase\ velocity = c/n', \tag{2.7}$$

$$n' = \left[\frac{\varepsilon'}{2}\left(\sqrt{1+\left(\frac{\varepsilon''}{\varepsilon'}\right)^2}\right)+1\right]^2 \tag{2.8}$$

where c is the speed of light in a vacuum. The imaginary part of the refractive index determines also the power penetration depth, which is given by

$$l_p = \frac{\lambda}{4\pi n''}. \tag{2.9}$$

Electromagnetically, a soil medium is, in general, a four-component dielectric mixture consisting of air, bulk soil, bound water, and free water. Due to the high intensity of the forces acting upon it, a bound water molecule interacts with an incident wave in a manner dissimilar to that of a free water molecule, thereby exhibiting a dielectric dispersion spectrum that is very different from that of free water. The complex dielectric constants of bound and free water are each functions of the electromagnetic frequency f, the physical temperature T, and the salinity S. Hence (Dobson *et al.*, 1985; Ulaby *et al.*, 1981-1986), the dielectric constant of the soil mixture is, in general, a function of: 1) f, T, and S, 2) the total volumetric water content m_V, 3) the relative fractions of bound and free water, which are related to the soil surface area per unit volume, 4) the bulk soil density ρ_b, 5) the shape of the soil particles, and 6) the shape of water inclusions.

Probably, the first extensive measurements of the dielectric constant of soils in the microwave band (0.133-10GHz) were performed by Leschanskiy *et al.* (1971) for sand and clay soil samples with the bulk density of about 1.75 g/cm^3. Coaxial line and bridge circuit techniques were used in the measurements. In the L-band (1 GHz), the real part of the dielectric permittivity increased from 5.1 to 18 when the relative moisture changed from 4 to 16% both for sand and clay samples.

Hoekstra and Delaney (1974) used the time domain reflection technique (sample in a coaxial line) and the slotted line determinations (sample in a rectangular waveguide) to measure the complex dielectric constant of four soils (a sand, a silt, and two clays) over the frequency range from 0.1 to 26 GHz. The relative moisture of the soils varied from 0 to 15%, and the temperature from 24°C to –20°C. The results showed that the relation between volumetric water content and the complex dielectric constant is relatively independent of soil type.

In contrast, the data reported by Wang (1980) showed significant differences in the magnitude of ε for different soil types (at the same volumetric water content). Schmugge (1983) explains the effect of soil texture by considering the behavior of water as it is added to a dry soil. The large dielectric constant of liquid water is due to the molecule's ability to align its dipole moment along an applied field; thus anything that would hinder the molecular rotation, e.g., freezing, very high frequencies, or tight binding to a soil particle will reduce the dielectric constant of the water. Since the first water molecules which are added to the soil are tightly bound to the particle's surface, they will contribute only a small increase to the soil's dielectric constant. As more water is added, above some transition level W_t, the addition molecules are farther away from the particle surface and are freer to rotate and thus make a larger contribution to the soil's dielectric constant. Since the surface area in a soil depends on its particle-size distribution or texture, clay soils, with a larger surface area, will be able to hold more of this tightly bound water than sandy soils; thus this transition point occurs at higher moisture levels in clay than in sandy soils. According to Wang and

To evaluate the microwave dielectric behavior of soil-water mixtures as a function of water content, temperature, and soil textural composition Hallikainen *et al.* (1985) conducted comprehensive dielectric constant measurements with a high degree of accuracy and precision over the 1 to 18 GHz region for several soil types. Soil texture was shown to have an effect on dielectric behavior over the entire frequency range and is most pronounced at frequencies below 5 GHz.

Several measurements of the dielectric constant of soils were made using a coaxial probe technique. Jackson (1990) found that for a variety of soils field-portable probe measurements produced results which were not as variable as those observed by Hallikainen *et al.* (1985). Peplinski *et al.* (1995) expanded data on the dielectric constant to the 0.3-1.3 GHz range.

A number of specific measurements of the complex refractive index regarding the refined sand and bentonite soil were made (Kleshchenko, 2002; Mironov *et al.*, 2004) to clarify the dielectric behavior of the bound and free water when their amount in a soil is increasing. The authors have made a conclusion that these dielectric properties remain unchangeable during the process.

The main factor influencing the dielectric constant of a soil in the microwave band is the water content of the soil. An exemplary dependence of the dielectric constant on the volumetric moisture for soils of different texture is presented in Fig. 2.1. Hoekstra and Delaney (1974) introduced the ratio $\Delta \varepsilon / \Delta m_V$ as the slope of the dielectric constant dependences versus volumetric soil moisture. It is interesting to note that this slope for the real part of ε is approximately same for different types of soil (Fig. 2.1) in the range of m_V change corresponding to the change of free water content in the soil. It gave Shutko (1986) an idea to take into account only the free water content in calculations of the soil dielectric constant and the emissivity of soil surface. He found that independently of soil type

$$\Delta \varepsilon' / \Delta m_{Vfw} \approx (0.7 \pm 0.1)\varepsilon'_w \qquad (2.10)$$

where m_{Vfw} is the volumetric water content of the free water and ε_w is the dielectric permittivity of liquid water. This approach actually neglects the change of soil permittivity (and emissivity) due to the presence of the bound water but, nevertheless, provides a good estimate of the free water content contribution.

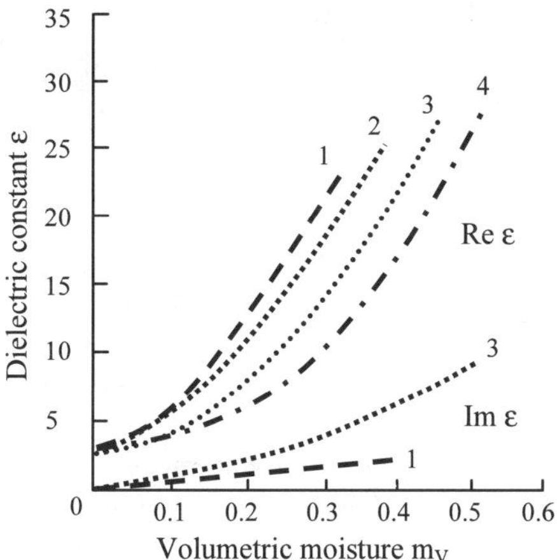

Fig. 2.1. Dielectric constant of soils versus their volumetric moisture at 1.4 GHz. 1 – Yuma sand, 2 – sandy loam, 3 – silty clay, 4 – Miller clay. T = 23°C.

Soil texture determines the content of bound water in a soil and, hence, the transition level W_t, above which the slope of the dielectric constant – volumetric moisture dependence is determined by the free water content. At any given moisture content and at all frequencies, ε' was found to be roughly proportional to sand content (and inversely proportional to clay content) (Hallikainen *et al.*, 1985).

Dependence of the dielectric constant of soils on their physical temperature was studied by many researchers. The most detailed investigations were conducted by Kleshchenko (2002). An exemplary dependence of the refractive index on the temperature for sand and bentonite (clayey soil) is presented in Fig. 2.2 and Fig. 2.3. It was observed (Hallikainen *et al.*, 1985; Polyakov *et al.*, 1994; Kleshchenko, 2002) that above 0°C, ε is very weakly dependent upon temperature. At temperatures below 0°C, temperature dependence is determined by the presence (Fig. 2.3) or absence (Fig. 2.2) of bound water in a soil, since the freezing point of the bound water in soils is much lower than 0°C. It is seen from the data of Fig. 2.3, where curves 2 and 3 relate to the case when all water in the soil is bound (soil moisture is below the transition level) and curves 1 relate to the case when the moisture is above this level.

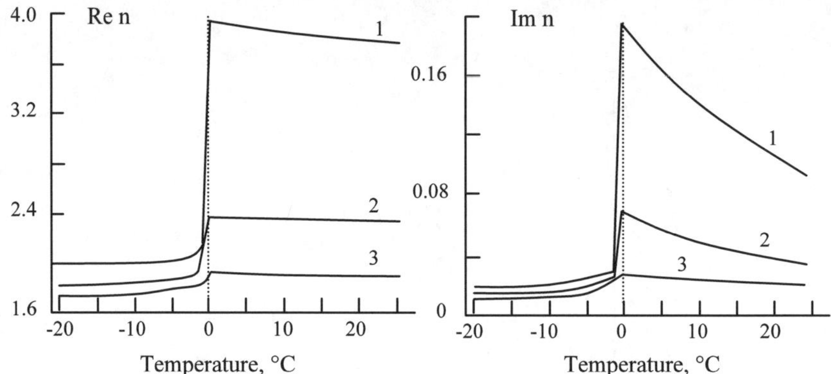

Fig. 2.2. Complex refractive index of sand versus its temperature at 1.4 GHz (Kleshchenko, 2002). Soil volumetric moisture is 0.25 (1), 0.075 (2), and 0.017 (3).

Frequency dependence of ε was studied, particularly, in Hallikainen *et al.* (1985), Mironov *et al.* (2004). It was noted (Hallikainen *et al.*, 1985) that except for the behavior of ε'' below 1.4 GHz, which is attributed to salinity effects, the spectral curves of the soil's dielectric constant are essentially compressed versions (in absolute level and slope) of the dielectric curves for water. An exemplary dependence of the dielectric permittivity on the frequency is depicted in Fig. 2.4.

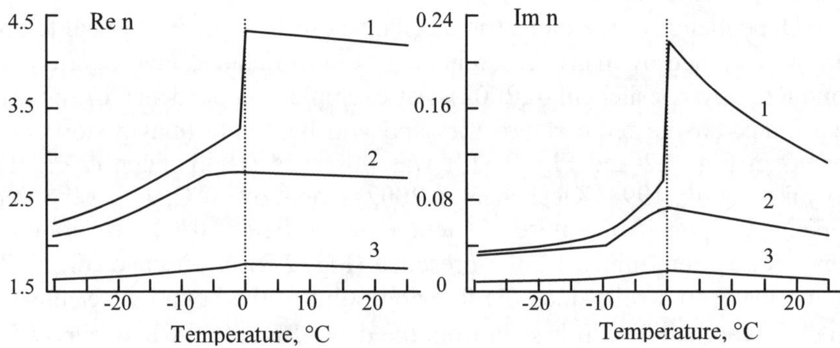

Fig. 2.3. Complex refractive index of bentonite (clayey soil) versus its temperature at 1.4 GHz (Kleshchenko, 2002). Soil volumetric moisture is 0.45 (1), 0.27 (2), and 0.035 (3).

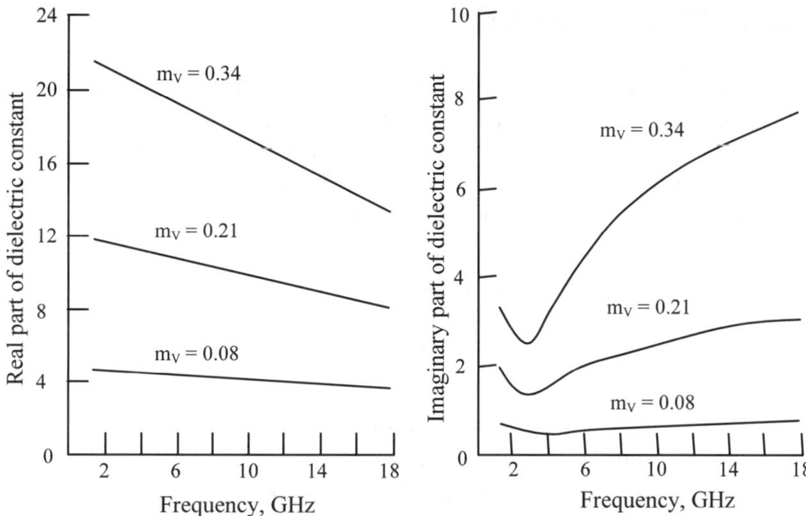

Fig. 2.4. Dielectric permittivity versus frequency for loam by data from Hallikainen *et al.* (1985).

The influence of soil bulk density on the dielectric constant of the soil diminishes significantly if the soil water content is described in terms of the volumetric moisture instead of the gravimetric one (Dobson *et al.*, 1985; Shutko, 1986). Electromagnetically, the volumetric measure is preferred because the dielectric constant of the soil-water mixture is a function of the water volume fraction in the mixture. The bulk density effect can be estimated by measuring the dielectric constant of dried soils. For some clay soils (the vertisols) the bulk density depends on their moisture (Sabburg *et al.*, 1997). The permittivity of these soils is significantly lower at any given moisture content than that of a typical sandy loam or a texturally similar, nonswelling clay soil.

Below 10 GHz, the ionic conductivity of saline water may have a marked effect on the loss factor ε''. Consequently, high soil salinity may significantly influence the dielectric properties of wet soils (Peplinski *et al.*, 1995).

Several soil parameters, like the shape of the soil particles and the shape of water inclusions, are rather difficult to measure directly and quantitatively characterized. Nevertheless, they can cause a noticeable scattering of measured data of ε at given moisture for a given soil type.

2.1.2.2. Soil Dielectric Models

Many researchers have attempted to model the microwave dielectric behavior of soil-water mixtures, ranging in complexity from simple two-component formulas to elaborate physical soil models. In most of these attempts, the authors obtained semi-empirical formulas starting from Birchak's mixing model (Birchak *et al.*, 1974). The following variables are introduced:
V_s is the volume fraction of solid phase in a soil,
V_a is the volume fraction of air in the soil,
$V_{fw} = m_{Vfw}$ is the volume fraction of free water,
V_{bw} is the volume fraction of bound water,
$V_{fw} + V_{bw} = m_V$,
ε_s is the dielectric permittivity of the solid phase,
$\varepsilon_a = 1$ is the dielectric permittivity of air,
ε_{fw} is the complex dielectric permittivity of free water,
ε_{bw} is the complex dielectric permittivity of bound water,
ρ_s is the density of the solid phase,
ρ_b is the density of the dry soil (the bulk density).
The Birchak mixing model is given as (Dobson *et al.*, 1985)

$$\varepsilon^{\alpha} = V_s \varepsilon_s^{\alpha} + V_a \varepsilon_a^{\alpha} + V_{fw} \varepsilon_{fw}^{\alpha} + V_{bw} \varepsilon_{bw}^{\alpha}. \tag{2.11}$$

Taking into account that

$$V_s = \frac{\rho_b}{\rho_s}, \; \varepsilon_a = 1, \tag{2.12}$$

and

$$V_a = 1 - V_s - m_v, \tag{2.13}$$

someone has then:

$$\varepsilon^{\alpha} = \frac{\rho_b}{\rho_s} \varepsilon_s^{\alpha} + 1 - \frac{\rho_b}{\rho_s} - m_v + V_{fw} \varepsilon_{fw}^{\alpha} + V_{bw} \varepsilon_{bw}^{\alpha}$$

$$= 1 + \frac{\rho_b}{\rho_s} \left(\varepsilon_s^{\alpha} - 1 \right) + V_{fw} \varepsilon_{fw}^{\alpha} + V_{bw} \varepsilon_{bw}^{\alpha} - m_v \tag{2.14}$$

With $\alpha = 0.5$, the model is known as the refractive model.

Shutko and Reutov (1982) examined several two-component formulas against available experimental data and concluded that the refractive formula may be used as a working model. For a two-component mixture of dry soil and free water Shutko (1986) obtained

$$\sqrt{\varepsilon} = \sqrt{\varepsilon_d}\left(1 - m_{Vfw}\right) + \sqrt{\varepsilon_{fw}}\, m_{Vfw} \tag{2.15}$$

where ε_d and ε_{fw} are the dielectric constants of dry soil and free water, respectively. In this approach, the dielectric permittivity of bound water is accepted to be approximately equal to that of ice or dry soil. The dielectric constant of the dry soil is evaluated by Krotikov's formula (Krotikov, 1962):

$$\sqrt{\varepsilon_d} \approx 1 + 0.5\rho_d \tag{2.16}$$

where $\rho_d = \rho_b$ is the density of dry soil. The input parameters of this simplest model are the dielectric constant of free water, the volumetric content of free water, and the density of dry soil. The model exhibits the correct trend for ε versus m_{fw}, but cannot account for the dependence on soil type.

Wang and Schmugge's (1980) model was proposed for 1.4- and 5-GHz frequencies. It starts from the equation (2.14)

$$\varepsilon^{\alpha} = 1 + \frac{\rho_b}{\rho_s}\left(\varepsilon_s^{\alpha} - 1\right) + V_{fw}\varepsilon_{fw}^{\alpha} + V_{bw}\varepsilon_{bw}^{\alpha} - m_v \tag{2.17}$$

with $\alpha = 1$. It is proposed that, for soil volumetric moisture below the transition moisture W_t, the water is bound to the soil solid elements. The reduction in ε for the bound water (in relation to free water) is described by another parameter: the γ coefficient. A zero value of γ corresponds to extremely bound water, with dielectric properties of ice. Higher values of γ bring the bound water dielectric constant closer to ε_{fw}. A value of 1 for γ means that there is no bound water effect for $m_V > W_t$. The soil dielectric constant is written as

$$\varepsilon = 1 + \frac{\rho_b}{\rho_s}(\varepsilon_s - 1) + m_V\varepsilon_x - m_V, \qquad \text{for } m_V < W_t \tag{2.18}$$

with

$$\varepsilon_x = \varepsilon_i + \gamma \frac{m_V}{W_t}(\varepsilon_{fw} - \varepsilon_i) \qquad (2.19)$$

and

$$\varepsilon = 1 + \frac{\rho_b}{\rho_s}(\varepsilon_s - 1) + W_t \varepsilon_x - m_V + (m_V - W_t)\varepsilon_{fw}, \text{ for } m_V > W_t \qquad (2.20)$$

with

$$\varepsilon_x = \varepsilon_i + \gamma(\varepsilon_{fw} - \varepsilon_i) \qquad (2.21)$$

where ρ_s is the density of soil solids, ε_i is the dielectric constant of ice, and ε_s is the dielectric constant of soil solids. The fitted values of the γ coefficient and of the transition moisture W_t at low frequencies are between 0.3 and 0.5 and between 0.15 and 0.35, respectively. The soil texture dependence of γ and W_t is accounted for by relating them empirically to the wilting point. For example, W_t relates to the wilting point as

$$W_t = 0.49 WP + 0.165 . \qquad (2.22)$$

In turn, Wang and Schmugge (1980) proposed to find WP in terms of volumetric moisture by the approximation formula:

$$WP = 0.06774 - 0.064 \times S + 0.478 \times C \qquad (2.23)$$

where S and C are the weight fractions of sand and clay in the soil.

Dobson *et al.* (1985) also started from the equation (2.14). An approximation was made,

$$V_{fw}\varepsilon_{fw}^\alpha + V_{bw}\varepsilon_{bw}^\alpha = m_v^\beta \varepsilon_{fw}^\alpha, \qquad (2.24)$$

and the following final formula was obtained:

$$\varepsilon^\alpha = 1 + \frac{\rho_b}{\rho_s}\left(\varepsilon_s^\alpha - 1\right) + m_v^\beta \varepsilon_{fw}^\alpha - m_v \qquad (2.25)$$

that for the real and imaginary parts of ε reads (Dobson *et al.*, 1985; Peplinski *et al.*, 1995)

$$\varepsilon'^{\alpha} = 1 + \frac{\rho_b}{\rho_s}\left(\varepsilon_s^{\alpha} - 1\right) + m_v^{\beta'}\varepsilon_{fw}'^{\alpha} - m_v, \tag{2.26}$$

$$\varepsilon''^{\alpha} = m_v^{\beta''}\varepsilon_{fw}''^{\alpha} \tag{2.27}$$

where α takes the constant value 0.65 and the parameter β is empirically related to sand and clay fractions from laboratory measurements

$$\beta' = 1.275 - 0.519 \times S - 0.152 \times C, \tag{2.28}$$

$$\beta'' = 1.338 - 0.603 \times S - 0.166 \times C \tag{2.29}$$

where S and C are the weight fractions of sand and clay in the soil.
Mironov *et al.* (2004) used the equation (2.14) with $\alpha = 0.5$ and obtained

$$\sqrt{\varepsilon} = 1 + \frac{\rho_b}{\rho_s}\left(\sqrt{\varepsilon_s} - 1\right) + m_v\sqrt{\varepsilon_{bw}} - m_v, \tag{2.30}$$

for $m_V < W_t$,

and

$$\sqrt{\varepsilon} = 1 + \frac{\rho_b}{\rho_s}\left(\sqrt{\varepsilon_s} - 1\right) + W_t(\sqrt{\varepsilon_{bw}} - 1) + (m_v - W_t)(\sqrt{\varepsilon_{fw}} - 1), \tag{2.31}$$

for $m_V < W_t$. This so-called generalized refractive mixing dielectric model is similar to the Wang and Schmugge (1980) approach except for the choice of α value.

There are several physical models for the soil dielectric constant. Dobson *et al.* (1985) considered the soil solids as a host medium containing randomly distributed and oriented inclusions. The general form of the model for calculating the dielectric constant of the mixture ε is given by De Loor (1960) as

$$\varepsilon = \varepsilon_s + \sum_{i=1}^{3} \frac{V_i}{3}(\varepsilon_i - \varepsilon_s)\sum_{j=1}^{3} \frac{1}{(1 + A_j \dfrac{\varepsilon_i}{\varepsilon^*} - 1)} \tag{2.32}$$

where ε_s and ε_i are the relative permittivity of the host (soil solids) and the inclusions (air, bound water and free water), respectively, ε^* is the effective relative permittivity near boundaries, A_j represents the depolarization ellipsoid factors, and V_i refers to the volume fractions of the inclusions. For plate-like clay inclusions and disc-shaped water inclusions, it was obtained (Dobson *et al.*, 1985) that

$$\varepsilon = \frac{3\varepsilon_s + 2V_{fw}(\varepsilon_{fw} - \varepsilon_s) + 2V_{bw}(\varepsilon_{bw} - \varepsilon_s) + 2V_a(\varepsilon_a - \varepsilon_s)}{3 + V_{fw}\left(\dfrac{\varepsilon_s}{\varepsilon_{fw}} - 1\right) + V_{bw}\left(\dfrac{\varepsilon_s}{\varepsilon_{bw}} - 1\right) + V_a\left(\dfrac{\varepsilon_s}{\varepsilon_a} - 1\right)} \tag{2.33}$$

where designations for the subscripts were given above. The models include the following input parameters.

a). The dielectric constant of soil solids. This value is estimated by (Dobson *et al.*, 1985)

$$\varepsilon_s = (1.01 + 0.44\rho_s)^2 - 0.062 . \tag{2.34}$$

Krotikov's formula (Krotikov, 1962) is also acceptable for ε_s evaluation. The known experimental data for ε_s (Boyarskii *et al.*, 2002): 4.5 + j0.05 (sand particles), 4.5 + j0.1 (silt particles), and 4.5 + j0.25 (clay particles).

b). The dielectric constant of the free water. $\varepsilon_{fw} = \varepsilon'_{fw} + \varepsilon''_{fw}$ is derived at a given frequency assuming a Debye-type relaxation as modified by Lane and Saxton (1952) to account for ionic conductivity losses

$$\varepsilon'_{fw} = \varepsilon_{w\infty} + \frac{\varepsilon_{w0} - \varepsilon_{w\infty}}{1 + (2\pi f \tau_w)^2} \tag{2.35}$$

and

$$\varepsilon''_{fw} = \frac{2\pi f \tau_w(\varepsilon_{w0} - \varepsilon_{w\infty})}{1 + (2\pi f \tau_w)^2} + \frac{\sigma}{2\pi\varepsilon_0 f} \tag{2.36}$$

where $\varepsilon_{w\infty} \approx 4.9$ is the high-frequency limit of ε_{fw}, ε_{w0} is the static dielectric constant of water that is dependent on the effective ionic conductivity σ and the relaxation time of water τ_w, f is the frequency in hertz, and ε_0 is the permittivity of free space equal to 8.854×10^{-12} F m^{-1}.

c). The dielectric constant of the bound water. This dielectric characteristic has still not been well studied. By fitting experimental data, ε_{bw} was estimated as equal to $35 - j15$ (Dobson *et al.*, 1985). Mironov *et al.* (2004) assert that ε_{bw} does not depend on the soil moisture. They report that the bound water dielectric spectrum can be found from Debye's formula with $\tau_{bw} \approx 11.33 \cdot 10^{-12}$ s and $\varepsilon_{bw0} \approx 39.35$. In contrast, Boyarskii *et al.* (2002) propose that the dielectric constant of bound water changes with the number of monomolecular layers of water around the soil particles. The relaxation time of bound water τ_{bw} changes from $5 \cdot 10^{-10}$ s for one layer to $7.7 \cdot 10^{-12}$ s for ten layers (free water). Wang and Schmugge's (1980) model also assumes the change of the bound water dielectric properties with the change of soil moisture. A paper by Hilhorst *et al.* (2001) reports more data on the dielectric relaxation of bound water in soils and its connection to the soil matric pressure.

From the theoretical point of view, study of the dielectric permittivity of a soil is an independent scientific problem. One can see from the previous text that this problem is not solved yet. All models considered can provide an acceptable fit to experimental data with an appropriate choice of their parameters. Moreover, the difference between the models is sometimes less than the scattering of experimental data due to the statistical nature of the investigated object. Experimentalists use one or another model usually reasoning from its convenience and ease for practical application. On the other hand, a rather large amount of experimental data on the dielectric properties of different soils has been collected and is available now. It allows researchers to use these data in solving a specific problem of microwave remote sensing.

2.2. BIOMETRICAL FEATURES AND ELECTRICAL PROPERTIES OF VEGETATION

2.2.1. Biometrical Features of Vegetation

Vegetation canopies represent a dynamic and complicated object of investigation. The morphology of plants changes during the vegetation period, which leads to corresponding changes in the characteristic electrodynamic parameters of the vegetation canopy. From the standpoint of remote sensing the following canopy features are most often used for modeling of electromagnetic propagation through a vegetation media (Chukhlantsev, 1992; Chukhlantsev *et al.*, 2003c).

a) Type of vegetation and stage (phase) of growth;

b) Shape and size (size distribution) of plant elements, i.e., of leaves, stalks, branches, and trunks (these parameters are determined by the type of vegetation and stage of growth);

c) Distributions of plant element orientations, i.e., of leaf, stalk, and branch angles;

d) Volumetric moisture content of plant elements m_v (part of the element volume occupied by water);

e) Vegetation volume density p (the relative volume occupied by vegetation in canopy); $p = nV_s$, where V_s is the volume of plant element, n is the number density of elements;

f) Vegetation water content per unit volume $w = pm_v\rho_w$, where ρ_w is the density of water;

g) Number of plants (trunks) per unit area N;

h) Average height of vegetation h;

i) Wet biomass of vegetation Q (the weight of vegetation per unit area);

j) Vegetation water content per unit area W; and

k) Gravimetric moisture (in wet weight basis) of vegetation $m = W/Q$.

Some biometric features of agricultural vegetation are collected from different sources and are presented in Table 2.1. These data can be useful for modeling of attenuation by main crops during the vegetation period. Data on the changes in the biometric parameters of crops during the growth process can be found, particularly, in papers by Allen and Ulaby (1989), Attema and Ulaby (1978), Basharinov *et al.* (1979), Brunfeld and Ulaby (1984, 1986), Castel *et al.* (2001), Ferrazzoli and Guerriero (1996), Ferrazzoli *et al.*, 2000; Leckie and Ranson (1998), Le Toan *et al.*, 1992; Mätzler (1990), Melon *et al.*, 2001; McDonald *et al.* (1991), O'Neill *et al.*, 2003; Saatchi and McDonald (1997), Wigneron *et al.* (1999b).

Table 2.1. Biometric features of agricultural vegetation.

Crop [reference]	The number of plants per unit area N, $1/m^2$	Vegetation average height h, m	Vegetation wet biomass Q, kg/m^2	Vegetation gravimetric moisture m	Vegetation water content per unit area W, kg/m^2
Winter wheat	400 – 1000	0.6 – 1.12	0.2 – 1.3	0.5 – 0.85	0.25 – 1.1
Winter rye	330 – 700	1.2 – 1.5	0.2 – 0.45	0.5 – 0.84	
Corn	5 – 30	1.2 – 3.2		0.81 – 0.84	
Sugar beat	4 – 8	0.3 – 0.5	1.25 – 3	0.8 – 0.9	1.2 – 2.8
Corn	6 – 7	1.8	7	0.76	5.3
Alfalfa		0.6	2.2	0.82	1.8
Soybean		0.2 – 1.04	0.25 – 2.8	0.78	
Corn		2.6	5.5		
Winter wheat	900	1.05	Volume fraction: Stalk – 0.00363 Heads – 0.01	Head – 0.45 Stalk – 0.63 Leaf – 0.16	
Soybean	26	0.82	Volume fraction: 0.0282	Leaf – 0.76 Stalk – 0.81	
Corn	Row spacing – 76 cm Average plant spacing – 19.8 cm	2.7	Volume fraction: Stalk – 0.0035 Leaf – 0.00058	Volumetric moisture: Stalk – 0.47 Leaf – 0.65	
Oat	180 – 440	1.35	6	0.7 – 0.86	
Alfalfa		0.17 – 0.73	Data on volumetric water content w are presented in Attema and Ulaby (1978)		0.23 – 1.95
Corn		0.3 – 2.7			0.2 – 10
Milo		0.3 – 1.17			0.66 – 10.9
Wheat		0.96			5.3
Tomato	7 – 10		7 – 15		

Forests occupy a third of the land surface (about fifty millions of square kilometers). They play a key role in the global circulation of carbon and nitrogen and considerably affect the energy and water balance of the biosphere. Boreal forests of Europe, Russia, Canada, and USA occupy about fifteen millions of square kilometers. Tropical forests make about 53% of the world forest store. Table 2.2 gives a general idea of the typical wet biomass values for different forests.

Table 2.2. Main features of forests.

Type of forest	Occupied square, 10^6 km^2	Annual biomass growth, kg/m^2	Biomass, kg/m^2	Dead organic matter, kg/m^2
Forest-tundra	1.55	0.65	3.8	9
Boreal taiga	5.45	0.54	10	8.1
Middle-latitude taiga	5.73	0.63	22.5	10.8
South-latitude taiga	6.6	0.65	23.5	14.5
Broadleaf coniferous forest	2.12	0.87	25	25.1
Deciduous forest	7.21	1.25	45	24.8
Subtropical forest	5.75	1.72	43	22.2
Evergreen rain forest	10.4	3.17	60	21.6
Deciduous rain forest	7.81	2.46	60	20.5
Tropical woodlands	9.18	1.42	10	15.1
Subtropical bushlands	0.9	1.96	45	21.6

A characteristic feature of forests is their stratum structure. The basic components of the forest phytocenosis are the forest stand, the undergrowth, the live soil cover, and the dead litter (Orlov and Torgashin, 1978; Smirnov, 1971).

The stand, as the main component of the forest phytocenosis, consists of one or more strata formed by trees of different heights and species. The main tree breeds of boreal forests are pine, larch, spruce, silver fir, cedar, birch, aspen, oak, beech, and ash-tree. The forest canopy has a small density of 0.03-0.3 kg/m^3, a big height of 2-25 m, a small size of elements (branches and leaves), and a relatively stable water content (gravimetric moisture). The wet biomass of the forest canopy varies from 0.2 to 1.2 kg/m^2.

The live soil cover is formed with bushes and shrubs, grass, moss, and lichen, which cover the soil under the forest canopy. Bushes have a height of 0.1-0.5 m and a wet mass of 0.1-0.5 kg/m^2. The grass cover has a height of up to 2 m, a wet mass of 0.05-0.5 kg/m^2, and a volumetric density of 0.2-1.0 kg/m^3.

In the absence of live soil cover, a layer of dead needles, leaves, small branches, and bark is formed on the soil surface. This layer has a friable structure and dries and moistens fast. The height of the layer varies from 0.01 to 0.07 m, its mass is 0.1-0.6 kg/m^2, and its density is 5-30 kg/m^3.

Compared to other strata, the stand most actively participates in the scattering and absorption of electromagnetic waves thus determining the spectral and polarization characteristics of microwave emission from a forest. To assess the spatial intensity distribution of the scattered and absorbed radiation, it is necessary to know the statistics of the upper stand stratum. On the basis of experimental data on the forest density, the statistical characteristics of the mosaic structure of the forest stand were found to obey the normal distribution

$$W(l) = \frac{1}{\sigma\sqrt{2\pi}} \exp(-\frac{1}{2\sigma^2}(l - l_0)), \tag{2.37}$$

where l is the distance between the trees, l_0 is the average distance between the trees, $\sigma = 0.01 \cdot l_0 \delta$ where δ is the distance variability coefficient (Orlov and Torgasin, 1978). The height of a tree relates to its diameter in accordance with an approximation

$$h = a_0 + a_1 d + a_2 d^2. \tag{2.38}$$

The crown diameter D, the crown length l, the height of the crown's widest part h_0, and the height of the crown beginning at h, describes the canopy structure. These values are related to each other by known regression formulas. The wet biomass of the tree components can be estimated using the stable correlation that exists between the trunk diameter (usually, at the breast height) and the mass of leaves and branches (Smirnov, 1971; Satoo, 1971; Kira and Ogava, 1971). For example, the mass of leaves of a 21-year-old fir is determined as

$$P(kg) = -0.417 + 0.128 g(cm^2) \tag{2.39}$$

where g is the trunk cross section at the breast height. Similar relations are known for the mass of branches and trunks.

Tropical forests have some special features (Richards, 1961). They have no dominant type of trees. A forest area of 1 ha may contain up to 100 different tree species. They usually form three to four strata. Single trees 50-70 m in height form the upper stratum. The main stratum consists of trees up to 35 m in height and forms a dense canopy. The undergrowth is rather weak. Tropical forests have the greatest values of biomass: up to 40-60 kg/m^2.

The modeling of electromagnetic propagation in a forest requires a big number of biometric parameters (Chauhan *et al.*, 1991, 1993; Dobson *et al.*, 1992; Ferrazzoli and Guerriero, 1996; Kasischke *et al.*, 1994; McDonald *et al.*, 1990, 1991; Sun *et al.*, 1991; Ulaby *et al.*, 1990; Wang *et al.*, 1993). The medium is subdivided into three main regions: crown, trunks, and soil. The crown is filled with scatterers representing leaves, needles, twigs, and branches. They may be positioned at various heights, according to the morphological properties of the tree species to be modeled. Usually, the scatterers of different kinds are assumed to be uniformly located within the crown. Examples of sets of biometric parameters used in the known models are shown in Tables 2.3 and 2.4. These parameters are determined from the data of *in-situ* measurements and give an idea of the biometric parameters of some tree breeds. More data on the biometric parameters of forests and on their correlations can be found in the literature cited above.

Table 2.3. Stand parameters for backscatter model simulation (Leckie and Ranson, 1998).

Species	Aspen	Hemlock
Stems/ha	1100	1100
Height (m)	15	15
Crown diameter (m)	3.0	3.0
Crown length (m)	7.0	7.0
DBH (cm)	20	20
Branches/m^3	10	105
Leaves/m^3	800	196000
Moisture content (% gravimetric)		
- trunks and branches	50	50
- leaves	60	60

Table 2.4. Stand parameters for radiometric model simulation (Ferrazzoli and Guerriero, 1996).

Species	Deciduous forest (beech, oak, maple)	Coniferous forest
Stems/ha	1500 ($h = 10$ m) and 200 ($h = 25$ m)	1200 ($h = 5$ m) and 200 ($h = 20$ m)
Height h (m)	from 0 to 25	from 0 to 20
Leaves radius (cm)	2.5	
Needle radius (mm)		0.3
Needle length (cm)		7
Leaves dry matter density (g/cm^3)	0.25	0.25
Branch volume/Total woody volume	0.5	0.2
Branch length-to radius ratio	50-100	50-100
Moisture content (% gravimetric)		
- trunks and branches	50	50
- leaves	70	70
Basal area	15 m^2/ha ($h = 10$ m) and 32 m^2/ha ($h = 25$ m)	10 m^2/ha ($h = 5$ m) and 30 m^2/ha ($h = 20$ m)

2.2.2. Electrical Properties of Vegetation Constituents

2.2.2.1. Microwave Dielectric Models of Vegetation Material

The dielectric constant of vegetation material is a significant point of the theory of electromagnetic propagation through vegetation. This quantity is determined by the water content of a given plant element and by its conductivity (salinity of water it contains). Despite their fundamental significance, the dielectric properties of plant elements in the microwave frequency range are poorly investigated. The dielectric mixture models, which are used to determine the relation of the dielectric constant ε_v of vegetation to dielectric properties of its constituents, i.e., the bulk vegetation material and the water component, vary from simple models (Fung and Ulaby, 1978) providing only approximate estimates to fairly complex models, such as the dual-dispersion model (Ulaby and El-Rayes, 1987), for example.

The simplest approximation for the dielectric constant was proposed by Peake (1959) and then used by many researchers. It reads

$$\varepsilon_v = m_v \varepsilon_w + (1 - m_v)2.5 \approx m_v \varepsilon_w \qquad (2.40)$$

where $\varepsilon_w = \varepsilon_w' + j\varepsilon_w''$ is the relative complex dielectric constant of water. The relative complex dielectric constant of saline water is given by the Debye relaxation formulas (2.35; 2.36) (Ulaby and El-Rayes, 1987), where $\varepsilon_{w\infty} \approx 4.9$, σ is the ionic conductivity of the aqueous solution in siemens per meter. Conductivity value is determined by salinity S of the solution. The salinity is defined as the total mass of solid salt in grams dissolved in 1 kg of solution and is expressed in parts per thousand ($^o/_{oo}$) on a weight basis. The relation of the model parameters ε_{w0} and τ_w to the temperature and the salinity of the solution can be found, e.g., in Stogrin (1971).

In the work by Ulaby and Jedlicka (1984), different dielectric mixing models were examined to fit experimental data on the dielectric constant of wheat and corn leaves and stalks. The propagation (refractive) two-phase model, the random-needle two-phase model, the random-disc two-phase model, the three-component random-needle model (accounting for "depressed" value of ε_w due to an action of suction forces (De Loor, 1960)), and the four-phase refractive model (where the water was subdivided into a bound water component and a free water component; bound water refers to water molecules that are tightly held to organic compounds by physical forces, and free water refers to water molecules that can move within the material with relative ease) were considered. The free-water component is assigned the dielectric properties of bulk water (the Debye equations) and the bound-water component is assumed to have dielectric properties similar to those of ice. It was noted in Ulaby and Jedlicka (1984) that until further work is performed to better determine the roles of free and bound water in heterogeneous materials, it will not be possible to uniquely specify a dielectric mixing model for a vegetation-water mixture.

To determine the dielectric dispersion properties of bound water, measurements were made by Ulaby and El-Rayes (1987) for sucrose-water solution of known volume ratios. On this basis the Debye-Cole dual-dispersion model was proposed by Ulaby and El-Rayes (1987) and was found to give excellent agreement with experimental data for corn leaves and stalks over the entire 0.2-20 GHz frequency range.

The three-component model has the form

$$\varepsilon_v = \varepsilon_{bulk} + v_{fw}\varepsilon_f + v_b\varepsilon_b \tag{2.41}$$

where ε_{bulk} is the dielectric constant of dry bulk vegetation material, ε_f and ε_b are the dielectric constants of free and bound water, respectively; v_{fw} is the volume fraction of free water, v_b is the volume fraction of the bulk vegetation – bound water mixture. At room temperature ($T = 22°C$) and salinity $S \le 10\,°/\!oo$ ε_s is given by (Ulaby and El-Rayes, 1987)

$$\varepsilon_v = \varepsilon_{bulk} + v_{fw}\left[4.9 + \frac{75.0}{1 + jf/18} + j\frac{18\sigma}{f}\right] + v_b\left[2.9 + \frac{55.0}{1 + (jf/0.18)^{0.5}}\right] \tag{2.42}$$

where f is in gigahertz and σ (the ionic conductivity of the free water solution) is in siemens per meter. One of the unknown properties of the mixture is the water distribution between free and bound fractions. The magnitudes of $\varepsilon_{bulk}, v_{fw}, v_b$, and σ and their variations with the gravimetric moisture content m_g were determined in Ulaby and El-Rayes (1987) by fitting the model to the measured data of ε_v. For green leaves ($m_g = 0.6\text{-}0.7$) volume fractions of free water and bound water ($v_{bw} = \dfrac{v_b}{3}$, $v_w + v_{bw} = m_v$) were found to be approximately equal. In general, it was proposed in McDonald *et al.* (2002) that most water is in a bound state when the water content of vegetation is low, and the fraction of free water increases with increasing water content. For example, the amount of free water for green leaves varies from 20% to 40% of total saturated weight, while the amount of bound water varies from 15% to 40%. In young xylem tissue, free water represents about 51% of the stem volume and reduces to 1% in wood older than 40 years. The amount of bound water is around 12% - 13% and remains essentially unchanged with age.

From the aforesaid, it follows that the distribution of water between the free and bound components is a critical point of the model. The relations obtained in Ulaby and El-Rayes (1987) for corn and relating the ratios of the volume fractions of free and bound water to the gravimetric moisture content may fail for other types of vegetation, e.g., for tree leaves.

Another critical point is that the model makes assumptions about the organic material which may not be appropriate for forest constituents, for instance, for forest litter (De Roo *et al.*, 1991). In particular, the dual-dispersion model, which is used to model corn leaves, assumes that the water in the organic material is composed of two parts, some of which is free and the rest is molecularly bound to sucrose sugar. While this model is more than adequate for corn leaves, other sugars may predominate in aspen and pine, for example [(De Roo *et al*, 1991). As a result, bound water can demonstrate other dispersion characteristics.

The dual-dispersion model was used in a number of radar backscatter experiments on forests to derive complex permittivity values. However, in view of the aforementioned uncertainties, some researchers prefer a more straightforward approach, i.e., direct measurements of the dielectric constants of trunks, branches, and needles (Franchois *et al.*, 1998).

A semi-empirical formula for ε_v was developed in Mätzler (1994, 1995) for broad leaves of different plants. The explicit parameters are the dry-matter fraction of the leaf

$$m_d = 1 - m = \frac{dry\ mass}{fresh\ mass} \tag{2.43}$$

and the complex permittivity of saline water ε_w. Implicit parameters are temperature, salinity, and frequency, which determines ε_w. For most plants the salinity was assumed close to $1\%_{oo}$ (that is several times less than in Ulaby and El-Rayes (1987) and Ulaby and Jedlicka (1984)). The formula is applicable in the frequency band of 1…100 GHz to fresh leaves; their m_d value is in the range 0.1…0.5 while their density is near 1 g/cm^3. The explicit form for ε_v is given by

$$\varepsilon_v = 0.522(1 - 1.32 m_d)\varepsilon_w + 0.51 + 3.84 m_d. \tag{2.44}$$

The simplicity of the equation and its validity over a large frequency range makes this formula attractive for the electromagnetic modeling of vegetation canopies.

An even simpler expression for ε_v, which was obtained by approximating experimental data for leaves and stalks of corn and wheat (Ulaby and Jedlicka, 1984)), was presented in Chukhlantsev (1992):

$$\varepsilon_v' = \left(\varepsilon_w' - \varepsilon_{bulk}\right)m_v^2 + \varepsilon_{bulk}; \quad \varepsilon_v'' = \varepsilon_w''m_v^2 \qquad (2.45)$$

where $\varepsilon_{bulk} = 3 + j0$. This approximation is very convenient for estimations.

2.2.2.2. Experimental data

Extensive studies of the dielectric properties of vegetation material began with the experiments described in Ulaby and Jedlicka (1984) and El-Rayes and Ulaby (1987). In the work by Ulaby and Jedlicka (1984), three waveguide transmission systems were used to measure the magnitude and phase of the field transmission coefficient of the sample contained in the sample holder. The sample holder was filled with chopped vegetation material. The relative complex dielectric constant of the sample was determined from measurements of the transmission coefficient and, then, was recalculated into the dielectric constant of vegetation material using the simple propagation (refractive) model. The dielectric data reported in El-Rayes and Ulaby (1987) were based on measurements of the amplitude and phase of the reflection coefficient of a coaxial probe terminated in the material under test. To avoid the difficulties with measurements of thin materials a technique was developed that included measurements of the input admittance when the probe is terminated with a thin dielectric slab placed against a background consisting of an electrically thick material with known dielectric constant.

The development of the aforementioned measuring technique resulted in the design of a Portable Dielectric Probe (PDP) by the Applied Microwave Corporation (Manual for PDP (1989)). The appearance of PDP made it possible to carry out a series of *in situ* and *in vivo* measurements of the complex permittivity of tree materials (Chauhan *et al.*, 1991, 1993; McDonald *et al.*, 1999, 2002; Dobson, 1988; Dobson *et al.*, 1991; Salas *et al.*, 1994; Way *et al.*, 1991). Some data on ε_v can also be found in Tan (1981) and of 0.5 GHz. It was shown that spatial and temporal variability in the xylem tissue dielectric constant is influenced not only by water content, but by variations in xylem sap chemistry as well. For tree A, the real part of ε_v varied along the trunk height within 6-11, while the imaginary part of ε_v varied along the trunk height within 1-2. The temporal variations of ε_v' and ε_v'' did not exceed 2 and 0.5, respectively. For tree B, the real part of ε_v varied along the trunk height within 7-22; while the imaginary part of ε_v varied along the trunk height within 0-4. The temporal variations of ε_v' and ε_v'' did not exceed 6-14 and 2, respectively. The correlation between dielectric properties and water regime and sap chemistry has important implications for microwave remote sensing of forested landscapes.

Close values of the spatial variations of the complex dielectric constant of wood were reported by Franchois *et al.* (1998). The measurements were conducted in the frequency range 1–10 GHz. The measurement method was based on an open ended coaxial probe reflection technique with a rational function approximation model for the probe tip aperture admittance. With this model, no calibration on reference liquids is required and sufficiently accurate results for the dielectric constant (ε_v') and loss factor (ε_v'') can be obtained. Results were presented for branches, parts of trunks, and needles from different tree heights. In the measurements with needles, the necessary size of the sample was obtained by making a bundle of several dozens of needles and cutting it so as to form a flat end. Values of ε_v' and ε_v'' obtained with the probe oriented along different stem directions of the trunk confirmed the anisotropic nature of wood. The longitudinal complex permittivity was roughly 1.5–3 times higher than the transverse component. Representative average values for trunks, branches, and needles were given in tables.

In the work by Shadrina (2001), the dielectric constant of xylem materials was found by measuring the complex transition coefficient of coaxial waveguide filled with a sample. Measurements were performed in the frequency range 0.4-2.0 GHz for segments of birch, pine, and poplar branches of different wetness.

Data on the complex dielectric permittivity of xylem materials were obtained in the 30-300 MHz frequency range by the use of high quality *LC*-circuit (Tsargorodtsev *et al.*, 2004). A sample was placed into the capacitor of the circuit and the shift of resonance frequency and the change of quality factor were measured.

Some available data on the dielectric permittivity of plant elements are presented in Table 2.5.

From the viewpoint of microwave attenuation in vegetation, it is expedient to consider the frequency dependence of the imaginary part ε_v'' of the dielectric constant. Some experimental data are presented in Fig. 2.5. Figure 2.6 shows the response of the imaginary part of the free and bound water dielectric constant as a function of frequency (Ulaby and El-Rayes, 1987). An analysis of Figs. 2.5 and 2.6 offers the following conclusions.

Table 2.5. The dielectric permittivity of plant elements.

Species	Moisture	Frequency (GHz)	Real part of ε_s	Imaginary part of ε_s
Corn leaves	$m_V = 0.6$	1-2	33-36	13
		3.5-6	28-30	8-12
		7.5-8.5	23	10-11
Corn stalks	$m_V = 0.6$	1-2	27-31	4-7
		3.5-6	30-31	8-12
		7.5-8.5	23	13
Wheat stalks	$m_V = 0.6$	8	23	9
Wheat leaves	$m_V = 0.6$	8	22	10
Aspen leaves	$m = 0.86$	1-7	37.5-33	12-9.5
Corn leaves	$m_V = 0.625$	1-10	50-40	
		1-2		25-13
		2-10		13-16
Fir new needles	Fresh cut	1.9	45	11
		5.5	40	13
		9.1	35.5	15
Fir old needles	Fresh cut	1.9	31	7.6
		5.5	27	9
		9.1	24	10.5
Spruce new and old needles	Fresh cut	1.9	31	7.8
		5.5	26	9
		9.1	35.5	10.4
Fir trunk xylem	Fresh cut	1.9-9.1	24-17	6-8
Fir branch xylem	Fresh cut	1.9-9.1	10-7.7	2.4-2.8
Pine branch xylem	$m_V = 0.42$	0.4-2	9.6-8.5	2
Fir, spruce, birch branch xylem	$m_V = 0.42$	1.11	8	2
Pine, fir branch xylem	Fresh cut	0.03-0.3	5-8	0.4-1

In the frequency range 2...10 GHz (the S-, C-, and X-band) the loss factor ε_v'' of tree needles and leaves amounts to 8-15. These values are higher than the highest values for corn leaves. It can be explained by the other, than for corn leaves, apportionment between free and bound water states in tree leaves tissue. The values of ε_v'' shown in Fig. 2.5 are in agreement with data of the other researchers. Trunk xylem ε_v'' varies within 7-10, which is less than ε_v'' of needles but higher than that of corn leaves. It can be noted that other researchers reported somewhat smaller values of ε_v'' for trunks. For example (Sellers *et al.*, 1995), at the frequency of 5 GHz, measured data of ε_v'' vary within 1.96-5.97 for an old aspen, within 1.2-6.34 for a spruce, within 0.93-6.13 for an old pine, and within 1.02-4.56 for a young pine. Low values of branches xylem ε_v'' in contrast to those of trunk xylem, nevertheless, are in agreement with data of ε_v'' for a young pine (Sellers *et al.*, 1995).

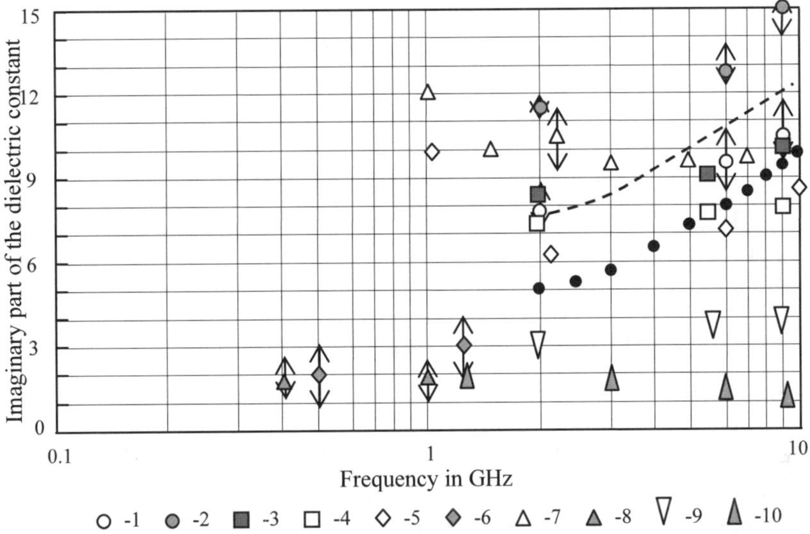

Fig. 2.5. Measured values of the imaginary part of the dielectric constant for different forest and plant constituents. Franchois *et al.* (1998): 1 – new and old needles of the spruce and old needles of the fir; 2 – new needles of the fir; 3 – trunk xylem of the fir (in the longitudinal direction); 4 – trunk xylem of the spruce (in the longitudinal direction); Ulaby and El-Rayes, 1987: 5 – corn leaves (m_g = 0.68); McDonald *et al.* (2002): 6 – Norway spruce trunk xylem; El-Rayes and Ulaby (1987): 7 – aspen leaves; Shadrina (2001): 8 – pine branches; Franchois *et al.* (1998): 9 – branches xylem of the fir (in the longitudinal direction); Sellers *et al.* (1995): 10 – trunk of balsam fir (m_v = 0.17).

The frequency dependence of ε_v'' in the frequency range 1-10 GHz are in good agreement with the dependence predicted by a dual-dispersion model with appropriate composition of volume fractions of free water v_{fw} and bulk vegetation – bound water mixture v_b. For green leaves v_{fw} is large and the contribution of a free water component is significant that leads to an increase of ε_v'' with frequency. An accurate choice of model parameters v_{fw} and v_b could be done by fitting the model to experimental data as it was done in Ulaby and El-Rayes (1987). Estimates show that the measured values of ε_v'' for leaves and the observed slope of its frequency dependence can be provided by the model with $v_{fw} = 0.2$-0.3 and $v_b \sim 0.5$. For example, the frequency dependence of ε_s'' for $v_{fw} = 0.3$ and $v_b = 0.5$ ($S = 2°/oo$) is shown in Fig. 2.5 by a dashed line. One can see that the model can well explain the spectral behavior of ε_v''. Greater values of v_{fw} produce a greater slope of the frequency dependence of ε_v'' that is in contradiction with the experiment. For low values of water content, the free water component is almost absent (see data for a balsam fir in Fig. 2.5) and the frequency dependence is close to that of bound water mixture (Fig. 2.6).

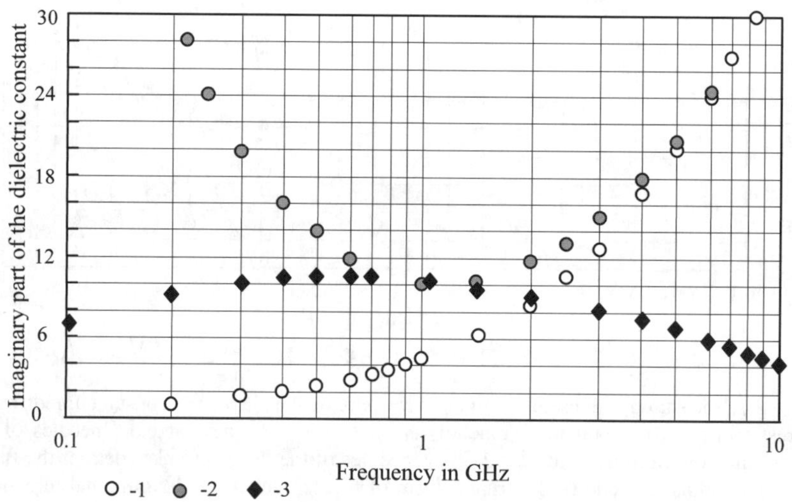

Fig. 2.6. Frequency dependence of the imaginary part of the dielectric constant for free and bound water. 1 – free water (salinity $S = 0$ $°/oo$); 2 – free water (salinity $S = 2$ $°/oo$); 3 – bound water.

Mätzler's semi-empirical formula (2.44) (Mätzler, 1994, 1995) also gives a good agreement with the experimental data in Fig. 2.5. Data of ε_v'' calculated by this formula with $m_d = 0.4$ and $S = 1°/oo$ are presented in Fig. Fig. 2.5 by black points. It should be noted that smaller values of m_d would produce a greater slope of the frequency dependence that disagrees with experiment.

Unfortunately, there are rather few data on the dielectric permittivity of leaves and branches in the frequency range 0.1–1 GHz. At the same time, these data could clarify the dielectric behavior of forest constituents in the whole microwave band and would be useful for understanding the frequency dependence of microwave attenuation by forest canopies.

2.2.2.3. Conductivity of Leaves and Branches

The aforementioned dielectric models treat vegetation material as a mixture of bulk vegetation and water (free and bound). In this case, the complex permittivity of vegetation element is a function of the dielectric constant of bulk vegetation, the dielectric constant of free and bound water, and the volume fractions of these matters. The dielectric constant of free water is calculated with the use of Debye formulas (2.35) and (2.36) taking into account the ionic conductivity of the aqueous solution. The ionic conductivity is related to the salinity of the solution or is determined from direct measurements of conductivity of liquid (sap) extracted from vegetation. In the model (2.41), the sap conductivity term and the free water permittivity is recalculated into the element conductivity and permittivity with the same coefficient v_{fw}. But the conductivity of branches is governed by spatial transfer of sap ions and represents a spatial effect, whereas the free water dielectric permittivity is governed by orientation properties of water molecules and represents a point effect. Therefore, there are no grounds to assume that the conductivity and the dielectric permittivity of branch sap should be recalculated into the conductivity and the dielectric permittivity of the branch itself with the same coefficient that is assumed by the models.

A method and results of measurements of leaves and branches conductivity is presented in Chukhlantsev *et al.* (2003a). Fresh cut coniferous branches with needles as well as branches of deciduous trees were used to measure their direct current conductivity. The scheme of measurements is presented in Fig. 2.7. Sharp blades connected to an ohmmeter were immerged into a branch tissue and a resistance between them was measured. The dependence of the resistance on a distance between the blades was examined. An example of this dependence for a pine branch with a diameter of 0.54 cm is presented in Fig. 2.8. Quite similar dependencies were obtained a branch (needle) was determined with the use of the linear part of the dependence:

$$\sigma = \frac{\Delta l}{\Delta R \cdot S} = \frac{4\Delta l}{\Delta R \cdot \pi d^2} \tag{2.46}$$

where σ is the conductivity, $\Delta R / \Delta l$ is the differential resistance, S is the branch (needle) cross section, and d is the branch (needle) diameter. To measure branch and needle diameter a vernier caliper and a micrometer was applied.

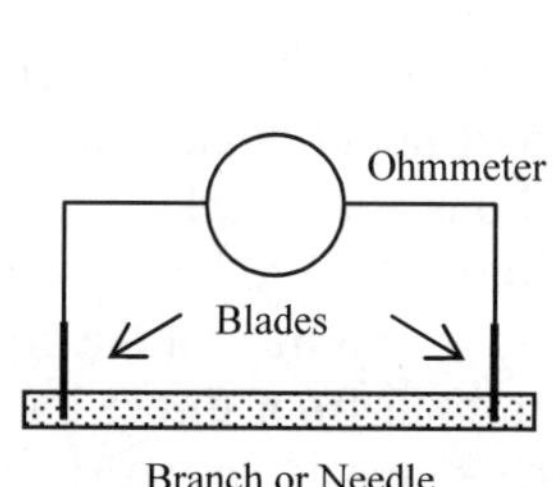

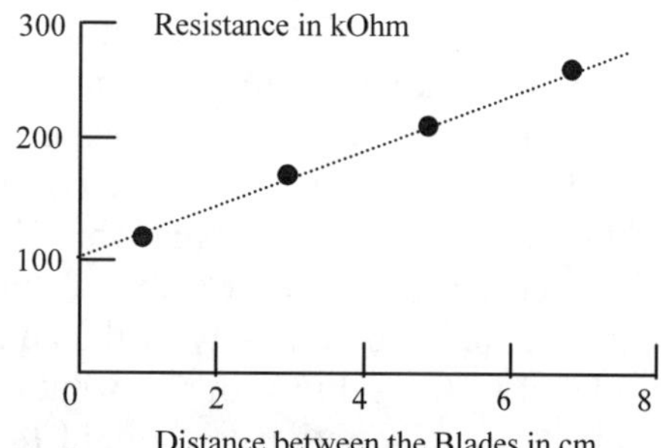

Fig. 2.7. The scheme of the conductivity measurements.

Fig. 2.8. An exemplary dependence of measured resistance (black points) on the distance between the blades.

To determine the gravimetric moisture m, branches and needles were weighted and dried at room temperature. The gravimetric moisture was found as m = (wet weight – dry weight)/(wet weight). Volumetric moisture m_v of samples was determined as $m_v = m\dfrac{\rho}{\rho_w}$, where ρ is the density of branch (or needle) and ρ_w is the density of water. The density of branch (needle) was determined by the ratio of its volume (length) immersed into water to its total volume (length) when the branch vertically floated in the water.

Some measured values of conductivity are presented in Table 2.6. The conductivity of the sap extracted from needles was measured and varied from 0.2 to 0.25 S/m. Being multiplied by the v_{fw} = 0.3 (according to equation (2.41)), this value of the sap conductivity produces the conductivity of needles and branches themselves equal to 0.06-0.075 S/m. One can see from

Table 2.6 that the measured values of branches and needles conductivity are several (3-5) times less than those predicted by the models.

Chukhlantsev *et al.* (2003a) reported that the conductivity of forest canopy elements has a great effect upon the frequency dependence of microwave attenuation by the forest canopy, especially, at frequencies below 1 GHz. In the frequency range 10-600 MHz the spectral trend of the canopy extinction rate is mainly determined by the conductivity of the canopy elements. Therefore, knowledge of exact values of the conductivity is very important.

Table 2. 6. The measured conductivity of leaves and branches.

Species	Gravimetric Moisture	Density (g/cm^3)	Volumetric Moisture	Conductivity $\times 10^2$ (S/m)
Pine needles with length of 4-8 cm	0.57-0.6	0.98-0.99	0.56-0.59	1.1-1.4
Pine branches with diameter of 0.4-1.1 cm	0.57-0.6	0.94-0.95	0.54-0.57	1.5-2.4
Thin deciduous branches (maple, aspen, etc)	0.57-0.6	0.94-0.95	0.54-0.57	1.4-2.4

where subscripts i and r stand for the incident and reflected (scattered) waves, ϑ is the incidence angle, and φ is the azimuth direction. The brightness temperature of the soil surface is expressed in the isothermal case as

$$T_{bp}(\vartheta_i,\varphi_i) = [1 - r_{sp}(\vartheta_i,\varphi_i)]T \tag{3.3}$$

where T is the physical temperature of the soil. The bistatic scattering coefficients are determined by

$$\sigma_{pq}(\vartheta_i,\varphi_i,\vartheta_r,\varphi_r) = \frac{4\pi h_r^2 I_r(\vartheta_r,\varphi_r)}{\cos\vartheta_i\, S_r I(\vartheta_i,\varphi_i)} \tag{3.4}$$

where $I_r(\vartheta_r,\varphi_r)$ is the intensity of radiation scattered by the soil area S_r at the distance h_r and $I(\vartheta_i,\varphi_i)$ is the intensity of the incident plane wave.

To compute the bistatic scattering coefficients, parameters of soil surface profile should be introduced. In general, a surface profile may consist of two height variations: a random component with certain statistical properties, and a deterministic component, e.g., related to the relief. The statistical variation of a random surface is characterized by its root mean square (rms) height σ_r and its correlation function. The surface autocorrelation function is a measure of the degree of correlation between the height at a point x and the height at a point ξ distant from x. The correlation length l_r of a surface is defined as the displacement ξ for which the autocorrelation function decreases by a factor e^{-1}. In general, the rms height σ_r is a measure of the vertical roughness of the surface and l_r is a measure of the horizontal roughness. For rough soil surface, the bistatic scattering coefficients can be computed using Kirchhoff's approximation (for slightly rough surfaces satisfying the condition $kl_r > 6$, $l^2 > 2.8\sigma_r\lambda$ where λ is the wavelength and $k = 2\pi/\lambda$ is the wave number) or using the small perturbation approach ($kl_r < 3$, $k\sigma_r < 0.3$) (Dobson and Ulaby, 1988). In the general case of surface roughness with arbitrary σ_r and l_r, a solution for the bistatic scattering coefficient is unknown.

A change of soil temperature and soil dielectric constant with the soil depth z can be accounted for by the incoherent approach, i.e., with the use of the radiative transfer equation. The brightness temperature of the soil surface is given in this approach by (Shutko, 1986; Armand and Polyakov, 2005).

$$T_{bp}(\vartheta_i,\varphi_i)=[1-r_{sp}(\vartheta_i,\varphi_i)]\int\limits_0^\infty 2kn''(z)\cos\vartheta'\,T(z)\exp\{-\int\limits_0^z 2kn''\cos\vartheta'\,dh\}dz \quad (3.5)$$

where ϑ' is the angle of refraction ($\sin\vartheta=n'\sin\vartheta'$), n' and n'' are the real and imaginary part of the soil refractive index, respectively. The penetration depth (equation (2.9)),

$$l_p=\frac{1}{2kn''}=\frac{\lambda}{4\pi\,n''},\quad (3.6)$$

is used as an estimate of the thickness of soil layer, which makes the biggest contribution to the thermal radiation from the soil surface.

In modeling of the thermal radiation from a soil surface (as in any physical modeling, though), someone meets a dilemma. On the one hand, the model should be simple enough to provide a possibility of calculations on its basis. On the other hand, the model should predict parameters of the modeled object with a prescribed accuracy. Researchers usually begin from simplest models, which may be necessarily complicated.

3.1.2. The Model of Uniform Half-Space with Flat Surface

3.1.2.1. Basic Equations

The simplest model of soil surface is the model of uniform half-space with a flat surface. In this case, only the specular reflected component of reflected waves exists and the total reflectivity equals the reflection coefficient, which is given by Fresnel formulas (see Chapter 1):

$$R_v=\left|\frac{\varepsilon\cos\vartheta_i-\sqrt{\varepsilon-\sin^2\vartheta_i}}{\varepsilon\cos\vartheta_i+\sqrt{\varepsilon-\sin^2\vartheta_i}}\right|^2,\quad (3.7)$$

$$R_h=\left|\frac{\cos\vartheta_i-\sqrt{\varepsilon-\sin^2\vartheta_i}}{\cos\vartheta_i+\sqrt{\varepsilon-\sin^2\vartheta_i}}\right|^2 \quad (3.8)$$

where ε is the complex dielectric permittivity of the soil. These formulas allow one to calculate the emissivity (the brightness temperature) of the soil surface at different frequencies, polarizations, and angles of observation. To perform the calculations, it is necessary to know the relation of ε to the soil parameters, i.e., moisture, temperature, density, etc. The soil dielectric models described in Chapter 2 can be used. Available sets of experimental data on the dielectric permittivity of soils are also used in the calculations.

3.1.2.2. Dependence of Emissivity on Soil Moisture

Typical dependencies of emissivity on the soil moisture are depicted in Fig. 3.1 and Fig. 3.2. They are calculated on the basis of experimental data on the soil dielectric constant for sandy soils (sandy loam, Field 1 in Hallikainen *et al.*, 1985) and clayey soils (silty clay, Field 5 in the same reference).

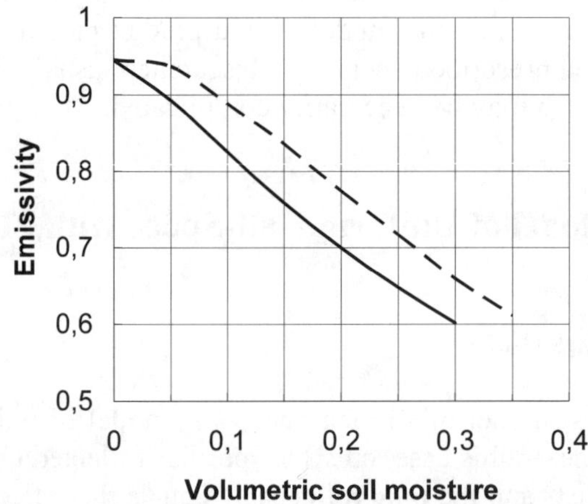

Fig. 3.1. An exemplary dependence of the emissivity of soil on its volumetric water content for sandy soils (solid line) and clayey soils (dashed line) at 1.4 GHz. T = 23°C. Angle of observation ϑ_i is 0°.

Fig. 3.1 and Fig. 3.2 demonstrate an excellent sensitivity of radiometric signal to the variations of soil moisture. With $T \approx 300$ K, a change of volumetric soil moisture by one percent leads to a change of the brightness temperature by approximately 3K. This high sensitivity of radiometric response to soil moisture makes microwave radiometry the best instrument for remote sensing of soil moisture. It should be noted that the brightness temperature contrast between very dry and very wet soil does not depend much on the soil type because the maximum value of volumetric soil moisture (field capacity) increases with the increase in clay particles fraction. This contrast achieves ~ 100 K for a smooth soil of any texture. Unfortunately, influence of other factors (soil type, surface roughness, temperature, etc.) introduces an ambiguity to the brightness temperature – soil moisture dependence. That decreases the possible number of soil moisture gradations, which can be retrieved from microwave radiometric measurements in a real situation.

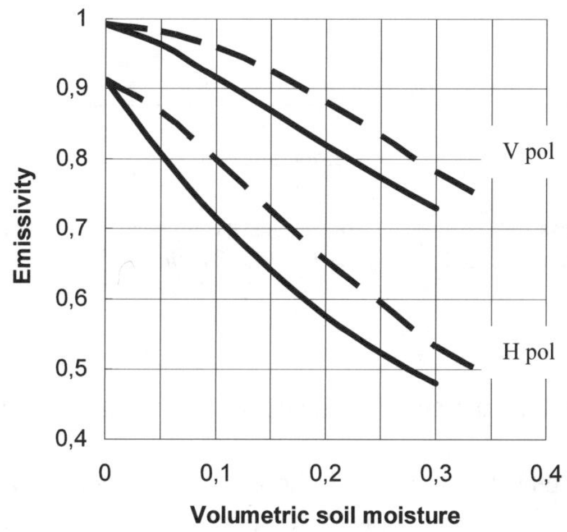

Fig. 3.2. An exemplary dependence of the emissivity of soil on its volumetric water content for sandy soils (solid line) and clayey soils (dashed line) at 1.4 GHz. T = 23°C. Angle of observation ϑ_i is 45°.

3.1.2.3. Dependence of Emissivity on Soil Temperature

This dependence is depicted in Fig. 3.3 and calculated (Polyakov *et al.*, 1994) with the use of data on the dielectric constant of soils versus its temperature (see Chapter 2). Fig. 3.3 shows that the influence of the soil temperature on its brightness temperature is rather weak in the region of positive soil temperatures. Nevertheless, it should be taken into account in retrieving the soil moisture from microwave radiometric measurements.

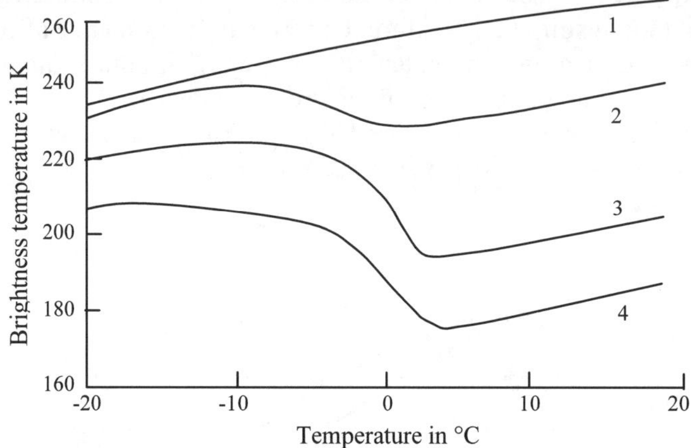

Fig. 3.3. Brightness temperature of soil versus its physical temperature at the L-band. Volumetric soil moisture is 0.05 (1), 0.1 (2), 0.2 (3), and 0.25 (4). Angle of observation ϑ_i is 0°.

3.1.2.4. Polarization Indices of Microwave Emission from Soils

At slanted observation angles, there is a significant difference between the magnitudes of soil brightness temperatures measured at vertical and horizontal polarizations. It results from the different angle dependence of Fresnel coefficients for the vertical and horizontal polarizations and is confirmed by theoretical simulations (see Fig. 3.2) and experimentally. Therefore, different polarization indices are used sometimes in microwave radiometry of land surface. The main advantage of this approach is that the temperature multiplier in equation (3.3) cancels if an index is related to the ratio of the brightness temperatures at different polarizations. The index is then determined be the ratio of soil emissivities, which are weakly dependent on the temperature. Thus, the use of polarization indices reduces the influence of soil temperature on the accuracy of the soil moisture retrieval from microwave radiometric measurements.

Different combinations of brightness temperatures measured at the same wavelength but at different polarizations (v or h) and at different observation angles ϑ_i and β_i can be proposed as the polarization indices (Pellarin et al., 2003a; Chukhlantsev et al., 2004b). Several indices are given below.

a) Normalized Polarization Difference (NPD_{ϑ_i}) is determined as

$$NPD_{\vartheta_i} = \frac{T_{bv,\vartheta_i} - T_{bh,\vartheta_i}}{T_{bv,\vartheta_i} + T_{bh,\vartheta_i}} . \tag{3.9}$$

It should be noted that, as a matter of fact, this index is usually named as the polarization index itself (e.g., Wang *et al.*, 1982a; Paloscia and Pampaloni, 1988; Owe *et al.*, 2001) and, sometimes, the half-sum of the brightness temperatures is used in the denominator of the ratio. Fig. 3.4 shows an exemplary dependence of the polarization index on the soil moisture content. It should be noted that the index changes approximately from 0.04 to 0.22 in the total range of soil moisture.

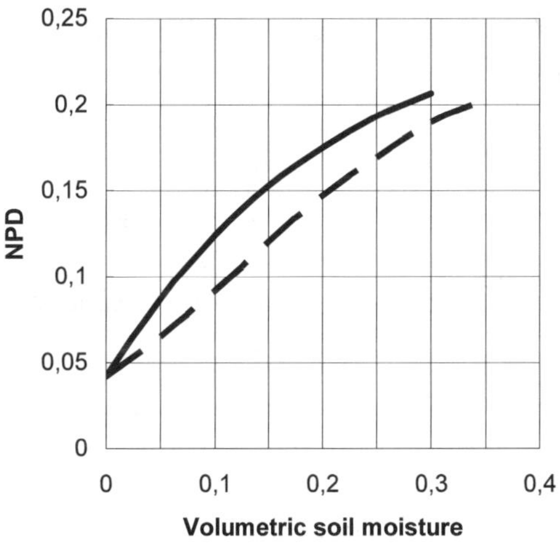

Fig. 3.4. Normalized Polarization Difference versus soil volumetric water content for sandy soils (solid line) and clayey soils (dashed line) at 1.4 GHz. T = 23°C. Angle of observation ϑ_i is 45°.

b) Polarization Ratio (PR_{ϑ_i}) is given by

$$PR_{\vartheta_i} = \frac{T_{bv,\vartheta_i}}{T_{bh,\vartheta_i}} . \tag{3.10}$$

Dependence of this index on the soil moisture is presented in Fig. 3.5.

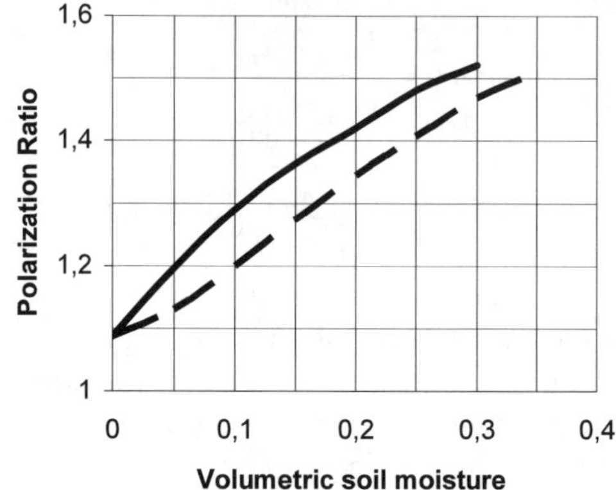

Fig. 3.5. Polarization Ratio versus soil volumetric water content for sandy soils (solid line) and clayey soils (dashed line) at 1.4 GHz. T = 23°C. Angle of observation ϑ_i is 45°.

c) Angular Ratio ($AR_{p,q,\vartheta_i\beta_i}$) is determined as the ratio of the brightness temperatures at different observation angles, for example:

$$AR_{v,\vartheta_i,\beta_i} = \frac{T_{bv,\vartheta_i}}{T_{bv,\beta_i}} . \tag{3.11}$$

Fig. 3.6 shows the dependence of this index on the soil moisture.

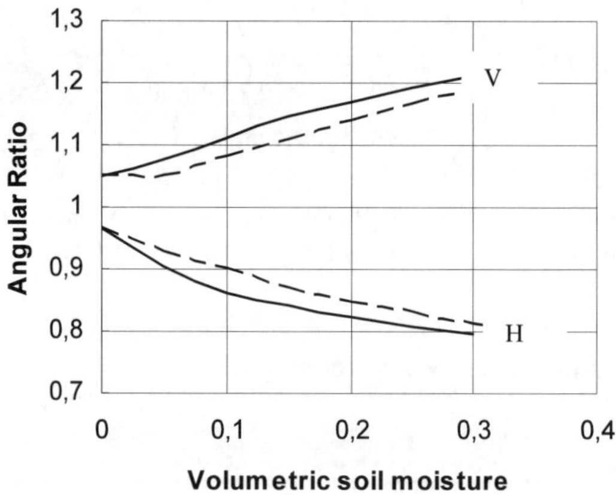

Fig. 3.6. Angular Ratio versus soil volumetric water content for sandy soils (solid line) and clayey soils (dashed line) at 1.4 GHz. T = 23°C. Angle of observation ϑ_i is 45°. V – vertical polarization, H – horizontal polarization.

It can be noted that all aforementioned indices are strongly dependent on the soil moisture and this dependence is close to the linear one. Therefore, these indices can be used in soil moisture retrieval algorithms. Several other polarization indices are presented and calculated in Chukhlantsev *et al.* (2004b).

3.1.3. Roughness Effects

Several attempts have been made to account for the effect of surface roughness on the observed brightness temperature (e.g., Choudhury *et al.*, 1979; Wang and Choudhury, 1981; Wang *et al.*, 1983; Mo and Schmugge, 1987; Wegmüller and Mätzler, 1999; Wigneron *et al.*, 2001). Physical models (Fung, 1994) are generally driven by surface characteristics derived from measurements of surface height profiles. Choudhury *et al.* (1979), using only the coherent term of the scattered field, proposed a simple model for the reflectivity

$$r_{sp} = R_p \exp\{-4k^2\sigma_r^2\cos^2\vartheta_i\} \tag{3.12}$$

where R_p is the reflectivity of a smooth surface given by equations (3.7) and (3.8), σ_r is the standard deviation of surface height. Since the model did not consider the incoherent part of the scattered field, which depends on the horizontal scale of the surface height variations, the discrepancy was found between the model and experimental data. Wang and Choudhury (1981) proposed another formulation, which is based on two semi-empirical parameters h_s and Q_s, that model the intensity of the roughness effects and the polarization-mixing effects, respectively. Then, the reflectivity is given by

$$r_{sp} = [(1-Q_s)R_p + Q_s R_q]\exp\{-h_s\cos^n\vartheta_i\} \tag{3.13}$$

where n was found (Wang *et al.*, 1983; Mo and Schmugge, 1987) to be dependent on the soil roughness and varies from 0 to 2. The model predicts properly the angular variations of the soil reflectivity with h_s and Q_s retrieved from brightness temperature measurements. However, the relation of these parameters to the soil roughness characteristics, i.e., the rms height σ_r and the correlation length l_r, is still not clear. Wigneron *et al.* (2001) attempted to clarify this relation on the base of a data set of microwave radiometric data obtained for soils with varying surface roughness conditions. They found that in the L-band (1.4 GHz) parameters Q_s and n could be set equal zero, which was generally in a good agreement with most previous studies. In this case, the reflectivity is

$$r_{sp} = R_p \exp\{-h_s\}. \tag{3.14}$$

The effective roughness parameter h_s was considered to take into account both geometric roughness effects (in relation with spatial variations in the soil surface height) and dielectric roughness effects (in relation with the variations of the dielectric constant within the soil). Mo and Schmugge (1987) proposed to express h_s in terms of the ratio σ_r / l_r that is the roughness slope parameter. This representation was also used in Wigneron *et al.* (2001). A simple formula was developed for h_s at 1.4 GHz (Wigneron *et al.*, 2001):

$$h_s = A(m_V)^B \cdot (\sigma_r / l_r)^C \tag{3.15}$$

where m_V is the volumetric soil moisture, $A = 0.5761$, $B = -0.3475$, and $C = 0.4230$. Dependence of the roughness parameter h on the soil moisture was also established in Mo and Schmugge (1987). Typical values of h_s for agricultural soils at L-band do not exceed 0.75. The presence of roughness results in a smaller slope of the emissivity-soil moisture function in comparison with this slope for a smooth soil. The slope reduction factor can be introduced from (3.14) as $\beta_r = \exp\{-h_s\}$. The reduction of the slope is demonstrated by the data presented in Fig. 3.7. The data are calculated for a sandy loam soil with the use of parameterization (3.15). Curve 1 shows the emissivity- soil moisture dependence for a smooth soil ($\sigma_r / l_r = 0$). The magnitudes $\sigma_r / l_r = 0.1$ and $\sigma_r / l_r = 0.3$ are characteristic ones for weakly and medium rough soil, respectively. One can see that the roughness effect is rather essential even for the weakly rough soil. The presence of roughness on soil surface leads also to the change of polarization properties of soil microwave emission (Fig. 3.8).

Wigneron *et al.* (2001) parameterization assumes a priory knowledge of soil roughness parameters to account for the soil roughness in retrieving soil moisture from radiometric data. However, regression coefficients A, B, and C, which were found for the L-band (Wigneron *et al.*, 2001), must depend on the frequency. For example, Coppo *et al.* (1991) proposed a simple parameterization for the equation (3.14):

$$h_s = 3\sqrt{\sigma_r} / \lambda \tag{3.16}$$

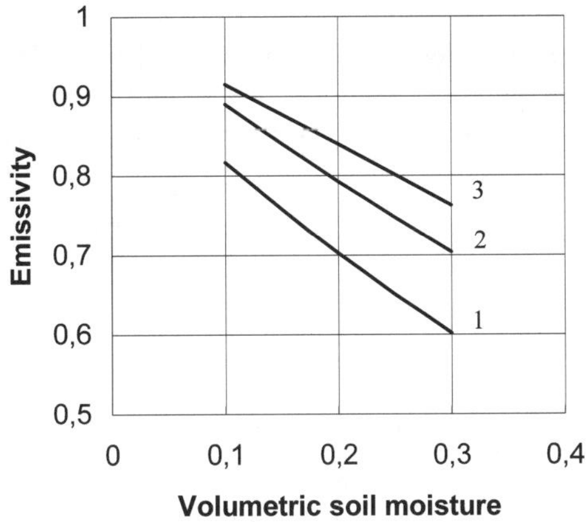

Fig. 3.7. The emissivity- soil moisture dependence for a smooth soil (1), weakly rough soil (2), and medium rough sandy soil (3) at 1.4 GHz. ϑ_i is 0°.

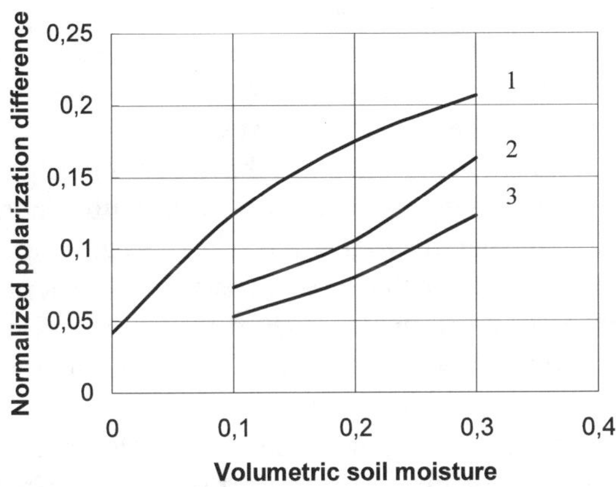

Fig. 3.8. The normalized polarization difference for a smooth soil (1), weakly rough soil (2), and medium rough sandy soil (3) at 1.4 GHz. Angle of observation ϑ_i is 45°.

where λ is the wavelength. Parameterization of h_s frequency dependence could allow one, in principle, to retrieve both the soil moisture and roughness characteristics from radiometric measurements at several frequencies. If the roughness parameters are not known and only single-frequency radiometric measurements are performed (even in the L-band), roughness effects can significantly decrease the accuracy of soil moisture retrieval from these measurements.

3.1.4. Soil Profile Effects

There are several theories describing the emissivity resulting from a soil profile with non-uniform properties (e.g., Njoku and Kong, 1977; Reutov and Shutko, 1986b; Armand and Polyakov, 2005). Numerous field studies and modeling efforts have shown that a near surface layer with the depth of 1/10 to 1/4 the wavelength dominates. It implies that microwave radiometric measurements provide moisture content of upper 0-2 cm or 0-5 cm soil layer. Observations of surface soil moisture can be utilized if there is a strong correlation to a deeper layer. This correlation is really observed for non-irrigated lands such as pastures, grasslands, steppe, etc. Statistic profiles of soil moisture exist in this case that relate surface moisture to moisture of deeper soil layers up to one meter depth. However, the use of such an approach (Reutov and Shutko, 1989) requires a priory information about precipitations, weather conditions, field moisture capacity, etc.

In a warm climate, when lands are irrigated, surface soil moisture becomes a very dynamic parameter that changes quickly. Typical averaged profiles of soil moisture for irrigated lands are presented in Fig. 3.9. These data were obtained in the Odessa region during a three year study of microwave emission from vegetable crops (Golovachev *et al.*, 1989). One can see that a transient layer is formed at the soil surface during some time period after watering. Thickness of the layer increases in the drying process. Changes of soil moisture in the transient layer have sufficiently great gradient. Some features of microwave emission from soils with a non-uniform profile of moisture are considered in the paper by Reutov and Shutko (1986b). Particularly, it was shown that the emissivity e_s of a soil with a transient layer is determined by the thickness h of the layer and by the soil moisture m_V at its bottom. Estimates of h and m_V can be obtained by measuring emissivity at several frequencies (two at least).

In radiometric measurements at a single frequency, an algorithmic dependence of emissivity on the soil moisture m_V at the bottom of the transient layer (which correlates with soil moisture of root-inhabited 0-20 cm layer) can be built on the basis of dependence of e_s on h and m_V (Golovachev *et al.*, 1989). These dependencies are presented in Fig. 3.10. The "dynamic" algorithmic dependence (it is shown by the dashed line in Fig. 3.10) is built from these dependencies, taking into account the change of h and m_V during the drying process (Fig. 3.9). This dynamic dependence well fitted the average soil moisture of a root-inhabited 0-20 cm layer during several watering-drying cycles (Golovachev *et al.*, 1989). This could be explained by a weak change of soil moisture with the depth below the transient layer and by a small contribution of over-dried transient layer moisture to the average moisture. It was shown that this dependence is stable enough for the region that enables root-layer soil moisture retrieval from radiometric measurements at one frequency. However, it should be noted that the form of the dynamic dependence will vary, obviously, with change of climatic regime, soil type, etc. The ambiguity occurring due to these reasons can be avoided by measurements of emissivity at two frequencies. A dual-frequency diagram for emissivity of bare soil with non-uniform depth moistening versus thickness of the transient layer and soil moisture at the lower border of the layer is presented in Reutov and Shutko (1986a) and Jackson and Schmugge (1989). The use of this kind of diagram allows one to make a close estimate of the thickness of the transient layer and the soil moisture at its bottom.

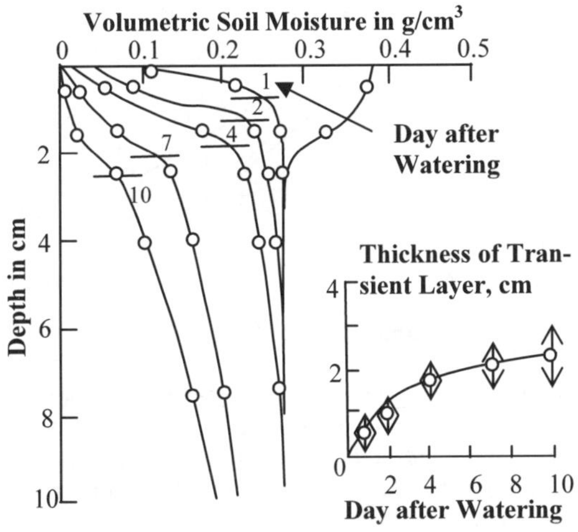

Fig. 3.9. Characteristic profiles of soil moisture and the depth of transient layer versus time after watering.

Reutov and Shutko (1986b) presented a comprehensive research report on the microwave emission of non-uniformly moistened soils. The emissivity of soils with typical vertical profiles of moisture content was analyzed. A brightness temperature contrast between a uniformly moistened soil and a soil with the same surface moisture, but with a gradient of soil moisture at its border, was studied on the basis of different theoretical models and special experiments. This contrast was found to be dependent on the surface soil moisture, the gradient of moisture at the surface, and the frequency band used. For all closely observed gradients, the contrast does not exceed 4-6 K for frequencies above 1 GHz and for values of the surface soil moisture greater than 0.1 in the volumetric basis. Nevertheless, for a dry soil surface, the contrast due to the gradient of soil moisture can reach 20-24 K in the L-band. An approach was suggested that combines microwave radiometric measurements at several frequencies and a priori data on soil hydro-physical properties, to estimate parameters of the soil moisture profile and the water content of the top one-meter soil layer.

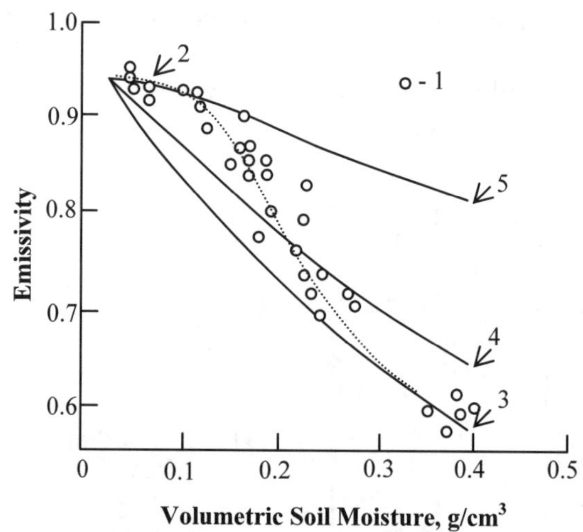

Fig. 3.10. Bare soil emissivity versus average moisture of upper 20 cm soil layer: ground truth data (1) and calculated dependence for dynamic process of drying (2). Curves 3-5 are calculated dependencies of emissivity on soil moisture at lower border of transient layer, when thickness of the layer is 0 cm (3) (uniform moistening), 1 cm (4), and 2 cm (5). Frequency is 1.9 GHz.

3.1.5. Influence of Other Soil Parameters

3.1.5.1. Influence of Soil Structure

This influence was studied, particularly, in a paper by Jackson and O'Neill (1986). Under tilled conditions the soil is actually a two phase mixture of aggregates and voids. That leads to a decrease of the effective dielectric constant of upper soil layer in comparison with that of a homogeneous three phase mixture of soil solids, air, and water peculiar to the aggregates. Tillage causes a soil loosening, an increase in soil porosity, and, as a consequence, a decrease of soil volumetric water content in the upper tilled layer with unchanged value of its gravimetric soil moisture. The influence of tilled layer in some respect is similar to the influence of a transient layer considered above. The dependence of the soil emissivity on the volumetric soil moisture in the case of tilled and tilled-and-rolled soil (Jackson and O'Neill, 1986) is quite close to that in the presence of a transient layer (the curve 2 in Fig. 3.10).

3.1.5.2. Influence of Row Structure

The effect of row structure on the microwave emission from a bare agricultural field was reported in Wang *et al.* (1980) and Promes *et al.* (1988). Those results were that both calculations and measurements showed a definite difference in the variations of the antenna temperature with angle of incidence depending on whether the antennas were scanning preferentially parallel or perpendicular to the row direction. In particular, the antenna temperature at nadir was observed to be higher in the horizontal polarization than in the vertical polarization when the antenna scanning was parallel to the row direction. As the antenna scanning was made perpendicular to the row direction, the vertically polarized antenna temperature was observed to be higher than the horizontally polarized one. These differences in the vertically and horizontally polarized antenna temperatures at nadir were enhanced with an increase in the soil moisture content. In modeling the effect of row structure, the spatial variations of the tilled rows were expressed in terms of a simple sinusoidal function. The antenna (brightness) temperature seen by the radiometers at given incident and azimuthal angles was found as the weighted sum of brightness temperatures of local soil areas over the antenna footprint.

Different functions were evaluated in Promes *et al.* (1988) for representing the spatial variations of the tilled rows. The differences between the emissivities calculated with the different functions were very small. The equations were obtained for an estimation of the difference between row and flat emissivities. This difference was expressed as a linear function of the row structure parameter (row height/row spacing). Simple row structure correction procedures were developed for three situations: 1) row structure and orientation are known, 2) row orientation is unknown, and 3) row structure and orientation are unknown. Both simulation results and the field measurements indicated that if an error of ±0.03 in estimating emissivity could be tolerated, then a reliable prediction of equivalent smooth field emissivity can be made for furrowed fields with a row-height/row-spacing ratio less than 0.2, which encompasses most dry land agricultural planting practices.

3.1.5.3. Influence of Temperature Profile

The influence of temperature profile on the emission from bare soils was considered in Schmugge and Choudhury (1981), Liou and England (1996), Liou *et al.* (1999), and Crosson *et al.* (2002). The problem is important because the soil surface temperature and moisture and their surface gradients are key parameters in that they are products of the energy balance between the land and atmosphere. Forward radiobrightness models accounting for the soil temperature profile can be separated into two classes – multilayer radiative transfer models and single-layer models (Crosson *et al.*, 2002). Single-layer models treat the soil medium as a single homogeneous slab characterized by "effective" temperature and moisture values. Radiative transfer models may be further classified as coherent or incoherent. An example of incoherent approach is the equation (3.5). A coherent model considers the phase associated with reflections between soil layers of varying dielectric constant. A comparison of coherent and incoherent models indicated that the models agreed to within about 4 K at L-band. The largest differences between models occurred when the near-surface moisture gradient was greatest. Crosson *et al.* (2002) examined the multilayer coherent radiative transfer model and Fresnel reflectance-based single-layer model by the comparison of model radiobrightness in L-band with field measurements. They found that each model agrees quite well with dielectric constant and brightness temperature measurements since there is some degree of uncertainty in model parameters and variables, as well as in the measurements used as model input and validation. Nevertheless, they conclude that a multiple layer radiative transfer model is more effective than a single-layer model in situations where the near-surface moisture or temperature gradients are strong.

Several thermal models have been developed to predict microwave emission signatures of moist bare soils over diurnal and annual periods (Liou and England, 1996; Liou *et al.*, 1999). The models are based on a solution of the heat flow equation within soil. This solution provides diurnal and annual changes of the temperature gradient at the ground surface. The brightness temperature is found as the product of the emissivity given by Fresnel formulas and the effective ground temperature. The effective ground temperature is determined as a sum of temperature at the ground surface and a term related to the temperature gradient at the ground surface. This approach seems to be prospective because it establishes a bridge between remotely sensed parameters (land surface brightness temperatures) and climatic variables.

3.2. EXPERIMENTAL RESEARCH ON MICROWAVE EMISSION FROM BARE SOILS

The earliest publications related to the microwave radiometry of bare soil moisture appeared in the beginning of the 1970s.

Basharinov and Shutko (1971) and Shutko (1982, 1986) presented a comprehensive overview of experimental research on microwave emission from bare soils conducted in the Institute of Radioengineering and Electronics of the Russian Academy of Sciences. The research projects were performed under laboratory conditions (Reutov and Shutko, 1986b), from ground moving platforms (Chukhlantsev *et al.*, 1989), and from airplanes (Kirdiashev *et al.*, 1979). In the laboratory, measurements were conducted for soil samples placed in a large photo-cuvette. The dependence of emissivity on soil moisture, soil profile, and soil roughness was studied at the wavelengths 2.25, 18, and 30 cm. Field measurements were performed with a microwave radiometer mounted on a small truck in the frequency range 1-4 GHz. Measurements from an airplane were conducted at different wavelengths of the microwave band.

Wang *et al.* (1982) reported data on ground level measurements of the microwave emission from bare fields at 1.4-GHz and 5-GHz frequencies. Both radiometers measured the brightness temperature of a target in both vertical and horizontal polarizations simultaneously. A strong correlation between the emissivity and the gravimetric soil moisture content in the top 2.5-cm layer was established. The polarization factor (equation (3.9) with the right side multiplied by 2) was an interesting parameter to study. At the observation angle of 40°, this factor for bare fields varied from 0.16 to 0.48 at 1.4 GHz over the entire soil gravimetric moisture range of 4%-24%, in agreement with calculated values for a smooth soil (Fig. 3.4).

Burke and Schmugge (1982) presented radiometric data taken at 21, 2.8, and 1.67 cm during a flight over bare and vegetated agricultural fields. Data taken over bare fields agreed well with theoretical estimates. With the surface moisture content ranging between 5 and 35 percents by weight, the emissivity ranges between 0.9 and ~0.7. It is possible to estimate the normalized polarization difference (the polarization index) from Burke and Schmugge (1982) data. It varies from 0.02 for very dry soils to 0.1 for very wet soils at 21 cm that is consistent with calculated data for rough soils (Fig. 3.8). Even lower values of the polarization index are observed at 2.8 cm and 1.67 cm. For very wet soils, *NPD* is less than 0.04 at these high frequencies. Ferrazzoli *et al.* (1992a) also reported that the observed values of *NPD* are typically in the range 0.015-0.035 for frequencies higher than 10 GHz

Schmugge (1983) presented an overview of research in the field of microwave remote sensing of soil moisture where some aircraft results on the emissivity of moist soils were given. He referred to polarization measurements of rough soils brightness temperatures conducted by Eom and Fung. For wet soils ($m_v = 0.29$), it can be found from their data that, at 50° observation angle and at 1.4 GHz, *NPD* is about 0.1 for the medium rough soil and about 0.15 for the weakly rough soil. For a smooth wet soil, their data produce *NPD* value that is about 0.22. In the subsequent overview by Jackson and Schmugge (1989), some more experimental data on the emissivity of moist soils are presented.

Theis *et al.* (1984) performed measurements of bare and vegetated fields' emissivity with the use of airborne microwave radiometers operating at 1.4 and 4.99 GHz. The passive microwave system gathered data at near nadir view angles. Scatter plots of the *C*- and *L*-band radiometer's emissivity versus soil moisture (in percent of field capacity) were plotted for bare fields only. From data presented in Theis *et al.* (1984), it is possible to estimate the slope reduction factor of the emissivity-soil moisture function due to soil roughness effect: $\beta_r = \exp\{-h_s\}$. In the *L*-band, this factor is 0.5-0.6 for the roughest field. That corresponds to $h_s = -(0.5\text{-}0.7)$ and $\sigma_r / l_r \sim 0.3$ in Wigneron *et al.* (2001) parameterization. At the same conditions, in the *C*- band, $\beta_r = 0.25\text{-}0.375$ that corresponds to $h_s = -(1\text{-}1.4)$. These data show that h_s depends on the frequency approximately as $1/\sqrt{\lambda}$ while Coppo *et al.* (1991) proposed that $h_s \sim 1/\lambda$.

Chanzy *et al.* (2000) presented a database for the emissivity of bare soils at different frequencies (1.4-90 GHz) under different soil roughness conditions. These data allow one to estimate the soil roughness factor and its spectral behavior.

Microwave emission from tilled soils was studied by Jackson and O'Neill (1986). Measurements were performed with a truck-mounted microwave radiometer operating at 1.4 GHz. A significance of agricultural row structure on the microwave emissivity of soils was studied by Promes *et al.* (1988) with truck-mounted 21-cm radiometer and Wang *et al.* (1980) with both *L*- and *X*-band radiometers mounted on a mobile truck.

The effects of soil moisture, surface roughness, and other parameters on the microwave emission from soils were also experimentally studied in Burke *et al.* (1979), Chanzy *et al.* (1997), Choudhury *et al.* (1982), Eagman and Lin (1976), England *et al.* (1992), Engman and Chauhan (1995), Ferrazzoli *et al.* (1992b), Jackson (1993, 1997), Jackson and O'Neill (1987a, 1987b), Jackson *et al.* (1984, 1986), Newton and Rouse (1980), Njoku and O'Neill (1982), Raju *et al.* (1995), Schmugge (1998), Schmugge *et al.* (1974, 1986), Tsang and Newton (1982), Wang (1983), Wang *et al.* (1987) Wegmüller *et al.* (1989) with both ground based and airplane radiometers.

In most early publications, measured values of soil brightness temperatures were compared to the moisture content by weight in the top 2-2.5 cm layer. The use of gravimetric soil moisture as a measure of soil water content caused a large spread of experimental points in the dependence of the brightness temperature on the soil moisture. It was so because of the influence of soil texture and soil density on the dielectric properties of moist soils, which becomes significant when the soil water content is described in terms of the gravimetric moisture (Dobson *et al.*, 1985). Development of adequate models for the soil dielectric constant (Wang and Schmugge, 1980; Dobson *et al.*, 1985) and experimental research of dielectric behavior of moist soils showed the advantage of transfer from gravimetric to the volumetric moisture units and stimulated researchers to use the volumetric soil moisture as the measure of soil water content in microwave radiometric measurements. When the volumetric moisture was used, the spread of experimental points in the aforementioned dependence was certainly less (Shutko, 1986). Of course, there still are the effects of soil roughness, soil profile, soil structure, etc. These effects also cause the scattering of experimental points in the dependence of the brightness temperature on the soil water content. To illustrate this scattering, some available data on variations of the soil emissivity with the volumetric soil moisture content in the top 2-3 cm layer are presented in Fig. 3.11.

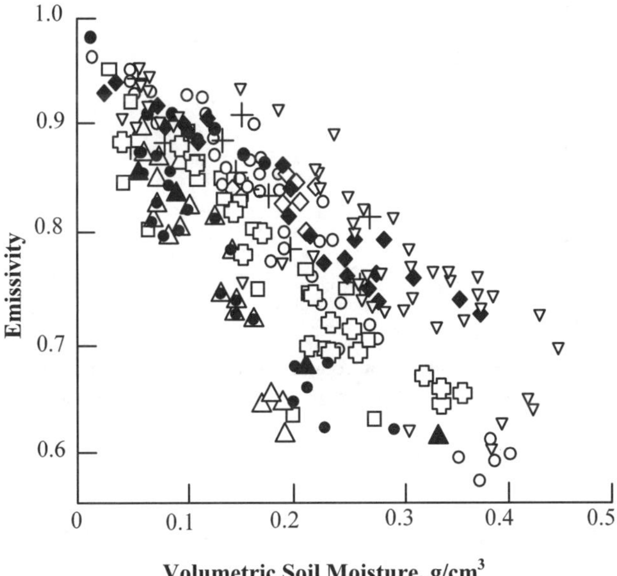

Fig. 3.11. The measured emissivity plotted against the moisture content for bare soils at 1.4 GHz.

The data are taken by the author from papers cited above and obtained by different researchers in a broad range of change in soil texture, soil structure, soil density, soil roughness, and soil profile. One can see that, if the only parameter (the volumetric soil moisture in the top soil layer) is used for the description of microwave emission from soils, the scatter of experimental points is really big and an uncertainty due to aforementioned factors is great. Nevertheless, most experimental points in Fig. 3.11 lie within an area bracketed by curves 1 and 3 in Fig. 3.7. It implies that the soil roughness effects are really significant even at *L*-band. To take the aforesaid effects into account by the use of one or another procedure, an increase in the number of independent radiometric measurements (multi-frequency or/and multi-polarization measurements) is required. For example, the frequency dependence of the soil roughness factor, depicted in Fig. 3.12 and obtained on the basis of available experimental data, distinctly shows that multi-frequency measurements can be used to discriminate soil roughness effects.

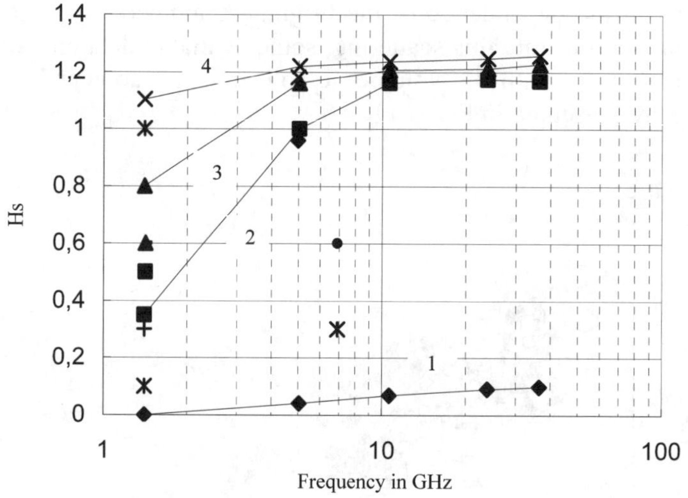

Fig. 3.12. The spectral dependence of soil roughness factor for very smooth soil (1) $\sigma_r/l_r \sim 0.01$, smooth soil (2) $\sigma_r/l_r \sim 0.1$, rough soil (3) $\sigma_r/l_r \sim 0.3$, and very rough soil (4) $\sigma_r/l_r \sim 0.8$.

The frequency dependence of the soil roughness factor depicted in Fig. 3.12 can be approximated as (Chukhlantsev *et al.*, 2005)

$$h_s = 1.2(1 - \exp[-2.5 \cdot \frac{\sigma_r}{l_r} \cdot f]) \tag{3.17}$$

where f is the frequency in GHz.

Unfortunately, multi-configuration measurements are not always possible and, in practice, researchers have to obtain an algorithmic dependence for the retrieval of soil moisture from microwave radiometric measurements at a single frequency. This dependence is validated by the use of ground truth measurements at test sites and is applicable for a given territory. Complicated algorithms based on multi channel measurements are being developed.

Chapter 4

THEORY OF MICROWAVE PROPAGATION THROUGH VEGETATION MEDIA

4.1. GENERAL APPROACH TO THE DESCRIPTION OF ELECTROMAGNETIC WAVE PROPAGATION IN VEGETATION

The modeling of electromagnetic wave propagation, attenuation, and scattering in vegetation canopies encounters some difficulties. From the theoretical point of view, vegetation canopies are randomly inhomogeneous media with inhomogeneities of different shape and size. In the microwave frequency band, the dimensions of leaves, stalks, branches, and trunks are comparable to the wavelength, which makes the modeling of electromagnetic wave propagation in this medium rather difficult because of the necessity to describe the wave scattering by a single inhomogeneity in terms of the theory of diffraction. A rigorous solution of the problem of electromagnetic wave propagation through a vegetation layer is extremely complicated (if possible at all). Due to this reason approximate models have to be used. These models either treat vegetation as a collection of randomly distributed lossy scatterers (discrete approach) or they consider the canopy as a slab with random dielectric permittivity (continuous approach). In order to understand the relation between these two conceptually different approaches and to determine the limits of their validity. the problem of electromagnetic wave propagation in vegetation should be considered from the position of random media propagation theory (Chukhlantsev, 1992). The electric field $\vec{E}(\vec{r})$ in a random medium is given by the integral equation (Ryzhov *et al.*, 1965; Ryzhov and Tamoikin, 1970):

$$\vec{E}(\vec{r}) = \vec{E}_0(\vec{r}) + k_0^2 \int_V [\varepsilon(\vec{r}_1) - 1]\vec{E}(\vec{r}_1)G(\vec{r},\vec{r}_1)d\vec{r}_1 \qquad (4.1)$$

where $\vec{E}_0(\vec{r})$ is the incident field, k_0 is the wave number in free space ($k_0 = 2\pi/\lambda$), $\varepsilon(\vec{r}_1)$ is the random dielectric permittivity, and $G(\vec{r},\vec{r}_1)$ is the free space dyadic Green's function. To obtain the moments of the random electric field, the multiple scattering series is derived from equation (4.1), and, then, this series is averaged over an ensemble of random realizations. Neglecting some of the terms of the series (the weakly connected scattering diagrams (Rytov *et al.*, 1978)), it is possible to obtain a Dyson equation for the mean field and a Bethe – Salpeter equation for the covariance. The principal point of the theory is the appropriate choice of the initial field inside the inhomogeneity in the multiple scattering series. The formal choice of $\vec{E}_0(\vec{r})$ as such a field yields the following expression for the case of a continuous medium:

$$\vec{E}(\vec{r}) = \vec{E}_0(\vec{r}) + k_0^2 \int_V [\varepsilon(\vec{r}_1) - 1]\vec{E}_0(\vec{r}_1)G(\vec{r},\vec{r}_1)d\vec{r}_1 +$$
$$+ k_0^4 \int_V [\varepsilon(\vec{r}_1) - 1][\varepsilon(\vec{r}_2) - 1]\vec{E}_0(\vec{r}_2)G(\vec{r},\vec{r}_1)G(\vec{r}_1,\vec{r}_2)d\vec{r}_1 d\vec{r}_2 + \dots \qquad (4.2)$$

It is important to emphasize that, in equation (4.2), the integration for a truly discrete medium (like vegetation, for example) is to be performed only over those volume elements where $\varepsilon(\vec{r}) \neq 0$, i.e. the elements of the scatterers. Hence, separating the volumes of scatterers, the expression (4.2) can be rewritten as

$$\vec{E}(\vec{r}) = \vec{E}_0(\vec{r}) + \sum_{i=1}^{N} k_0^2 \int_{V_i} [\varepsilon(\vec{r}_i) - 1]\vec{E}_i(\vec{r}_i)G(\vec{r},\vec{r}_i)d\vec{r}_i \qquad (4.3)$$

where the index i indicates that this value is related to the i-th scatterer; N is the total number of scatterers. For the field $E_i(\vec{r}_i)$, inside the i-th scatterer we can obtain from (4.3)

$$\vec{E}_i(\vec{r}_i) = \vec{E}_0(\vec{r}_i) + \sum_{i=1}^{N} k_0^2 \int_{V_i} [\varepsilon(\vec{r}_i') - 1]\vec{E}_i(\vec{r}_i')G(\vec{r}_i,\vec{r}_i')d\vec{r}_i' +$$

$$+ \sum_{k=1,k\neq i}^{N} k_0^2 \int_{V_k} [\varepsilon(\vec{r}_k) - 1]\vec{E}_k(\vec{r}_k)G(\vec{r}_i,\vec{r}_k)d\vec{r}_k \tag{4.4}$$

From equations (4.3) and (4.4) the multiple scattering series can be written by substituting (4.4) into (4.3) and iterating this process

$$\vec{E}(\vec{r}) = \vec{E}_0(\vec{r}_i) + \sum_{i=1}^{N} k_0^2 \int_{V_i} [\varepsilon(\vec{r}_i) - 1]\vec{E}_i^*(\vec{r}_i)G(\vec{r},\vec{r}_i)d\vec{r}_i +$$

$$+ \sum_{i=1}^{N} \sum_{k=1,k\neq i}^{N} k_0^4 \int_{V_i} [\varepsilon(\vec{r}_i) - 1]G(\vec{r},\vec{r}_i)d\vec{r}_i \int_{V_k} [\varepsilon(\vec{r}_k) - 1]\vec{E}_i^*(\vec{r}_k)G(\vec{r}_i,\vec{r}_k)d\vec{r}_k + ... \tag{4.5}$$

In equation (4.5), $\vec{E}_i^*(\vec{r}_i)$ is the electric field inside the isolated scatterer which is given by the scattering operator

$$\vec{E}_i^*(\vec{r}_i) = \vec{E}_0(\vec{r}_i) + k_0^2 \int_{V_i} [\varepsilon(\vec{r}_i') - 1]\vec{E}_i(\vec{r}_i')G(\vec{r}_i,\vec{r}_i')d\vec{r}_i'$$

$$= \vec{E}_0(\vec{r}_i) + k_0^2 \frac{(\varepsilon_v - 1)}{4\pi} \int_{V_i} \left\{ \vec{E}_i(\vec{r}_i')\left(1 - \frac{j}{k_0|\vec{r}_i - \vec{r}_i'|} - \frac{1}{(k_0|\vec{r}_i - \vec{r}_i'|)^2}\right) - \right. \tag{4.6}$$

$$\left. \left(\vec{E}_i(\vec{r}_i')\cdot\frac{\vec{r}_i - \vec{r}_i'}{|\vec{r}_i - \vec{r}_i'|}\right)(\vec{r}_i - \vec{r}_i')\left(1 + \frac{3j}{k_0|\vec{r}_i - \vec{r}_i'|} - \frac{3}{(k_0|\vec{r}_i - \vec{r}_i'|)^2}\right) \right\} \frac{e^{jk_0|\vec{r}_i-\vec{r}_i'|}}{|\vec{r}_i - \vec{r}_i'|} d\vec{r}_i'$$

where ε_v is the dielectric permittivity of the scatterer material. One can see that, in the discrete approach due to the separation of particle volumes the multiple scattering series begins with the field inside the isolated inhomogeneity. In the theory of strong fluctuations of the dielectric permittivity in a continuous medium (Ryzhov *et al.*, 1965; Ryshov and Tamoikin, 1970), the separation of the field inside an inhomogeneity is performed by separating the singularity of the Green's function in the form of a delta function and performing the principal-value integration in equation (4.1). This allows one to solve the problem of determining the effective dielectric constant of a randomly inhomogeneous medium with $\varepsilon(\vec{r})$ widely deviating from unity, but with electric dimensions of the inhomogeneity small compared to the wavelength of electromagnetic wave (the quasi-static approximation). In the discrete approximation, the field inside a scatterer can be determined from the solution to the diffraction problem.

A situation when the continuous and discrete approaches yield close results was considered in Chukhlantsev (1992). From equations (4.2), (4.5), and (4.6), one can see that this occurs when the scattering operator of an isolated particle can be expanded in a convergent Born series (the case of weak scatterers). From equation (4.6) it follows that this is possible under the following conditions: first, $|\varepsilon_v - 1| \ll 1$, which corresponds to the Rayleigh – Gans scattering in the discrete model and the theory of small perturbations of continuous media, and, second, $k_0 d |\varepsilon_v - 1| \ll 1$, where d is the characteristic size of the scatterer (e.g., the leaf thickness), which corresponds to the Rayleigh scattering for particles and to the theory of strong fluctuations in a continuous medium. If the scatterers are strong, i.e., the Born series formally obtained from equation (4.6) does not converge, the application of the continuous model to an actually discrete medium makes no sense. From the sufficient condition

$$k_0 d |\varepsilon_v - 1| \ll 1 \tag{4.7}$$

one can estimate the limits of applicability of the continuous model to vegetation. For leaves, one has $d \sim 0{,}2$ mm (the leaf thickness of most crops and trees) and $\varepsilon_v \sim 20$; setting $k_0 d (\varepsilon_v - 1) = 0{,}3$, one obtains $\lambda \sim 8$ cm. This means that both approximations can be used for modeling the wave propagation in the leaf component of vegetation up to the C-band. For higher frequencies, the use of the discrete model is more preferable.

4.2. THE MODEL OF VEGETATION AS A CONTINUOUS MEDIUM

4.2.1. Propagation of Electromagnetic Waves in a Random Continuous Medium

In the continuous model, a vegetation canopy is considered as a uniform medium whose electrodynamic characteristics are described by the effective dielectric permittivity ε_{ef} or by the effective permittivity tensor $\hat{\varepsilon}_{ef}$ (Allen and Ulaby, 1989; Chukhlantsev, 1988; Du and Peake, 1969); Fung and Fung, 1977; Fung and Ulaby, 1978; Fung, 1979; Milshin and Grankov, 2000; Redkin *et al.*, 1973; Redkin and Klochko, 1977; Tamasanis, 1992; Tsang and

Kong, 1981; Ulaby and Bush, 1976). To determine $\hat{\varepsilon}_{ef}$, Allen and Ulaby (1989), Redkin *et al.* (1973), and Redkin and Klochko (1977) used the formulas for a mixture of dielectrics obtained in the electrostatic approach. In Du and Peake (1969) and Tamasanis (1992), the continuous medium concept was applied to determine the radio wave attenuation by foliage. The wave corrections to the effective permittivity were obtained in Fung and Fung (1977), Fung and Ulaby (1978), and Fung (1979) on the basis of the small perturbation theory. The vector problem, which has to take into account the singularity of the Green's function, was considered in Tsang and Kong (1981) and Chukhlantsev (1988) using the results of strong fluctuation theory.

A general approach to a determination of the effective permittivity tensor $\hat{\varepsilon}_{ef}$ for a random continuous medium was developed in Ryzhov and Tamoikin (1970). The vector wave equation for the electric field E exited by a point source in the random medium is written as

$$\Delta\vec{E} - \vec{\nabla}\times\vec{\nabla}\vec{E} + k_0^2\varepsilon(\vec{r})\vec{E} = \vec{q}\,\delta(\vec{r}) \tag{4.8}$$

where k_0 is the wave number in a free space, $\varepsilon(\vec{r})$ is the random dielectric permittivity, $\vec{q}$ is the unit vector, and $\delta(\vec{r})$ is the delta function. A quasi-static part of $\hat{\varepsilon}_{ef}$, the deterministic uniaxial permittivity tensor $\hat{\varepsilon}_0$, is then introduced in equation (4.8):

$$\Delta\vec{E} - \vec{\nabla}\times\vec{\nabla}\vec{E} + k_0^2\hat{\varepsilon}_0\vec{E} = \vec{q}\,\delta(\vec{r}) + k_0^2[\hat{\varepsilon}_0 - \varepsilon(\vec{r})\hat{I}]\vec{E} \tag{4.9}$$

where $\hat{I}$ is the unit matrix. The integral equation corresponding to equation (4.9) is given by

$$\vec{E}(\vec{r}) = \hat{G}(\vec{r})\vec{q} - k_0^2\int\hat{G}(\vec{r}-\vec{r}_1)\hat{\varepsilon}(\vec{r}_1)\vec{E}(\vec{r}_1)d\vec{r}_1 \tag{4.10}$$

where $\hat{G}(\vec{r})$ is the dyadic Green's function of the wave equation (4.9) and

$$\hat{\varepsilon}(\vec{r}) = \hat{\varepsilon}_0 - \varepsilon(\vec{r})\hat{I} . \tag{4.11}$$

To determine the components of $\hat{\varepsilon}_0$, the following procedure is used (Ryzhov and Tamoikin, 1970). The Green's function of the vector wave equation (4.9) is represented in the form of the sum of the delta function, which accounts for the singularity of the Green's function at zero argument, and the regular part, with which the principal-value integration is performed,

$$\hat{G}(\vec{r},\vec{r}_1) = P.S.\hat{G}(\vec{r},\vec{r}_1) + \hat{R}\delta(\vec{r}-\vec{r}_1) \qquad (4.12)$$

where $P.S.$ denotes the principal value and $\hat{R}$ is a constant uniaxial matrix determined by the shape of the inhomogeneity. Then, new variables are introduced:

$$\vec{F} = \vec{E} + k_0^2 \hat{R}\hat{\varepsilon}\vec{E}$$
$$\hat{\xi} = \hat{\varepsilon}(\hat{I} + k_0^2\hat{R}\hat{\varepsilon})^{-1} \qquad (4.13)$$

The permittivity tensor $\hat{\varepsilon}_0$, which is to be determined in the problem, represents the quasi-static part of $\hat{\varepsilon}_{ef}$ and should be chosen so as to take into account the singularity of the Green's function most completely. The separation of the Green's function singularity allows one to consider the field $\vec{E}$ as the field inside the inhomogeneity of the medium, and the field $\vec{F}$, as the external field with respect to the inhomogeneity. In this representation, the tensor $\hat{\xi}$ characterizes the random polarizability of the medium, which makes it possible to determine the tensor $\hat{\varepsilon}_0$ from the condition:

$$<\hat{\xi}>=0 \qquad (4.14)$$

where brackets denote ensemble averaging. The integral equation in the new variables is used to find the mean field and $\hat{\varepsilon}_{ef}$ by conventional methods (Rytov *et al.*, 1978). In the low frequency limit, the effective permittivity is given by

$$\hat{\varepsilon}_{ef} = \hat{\varepsilon}_0 + \hat{\xi}_{ef}$$
$$\hat{\xi}_{ef} = -k_0^2 \int \Gamma(\vec{r})G(\vec{r})d\vec{r} \qquad (4.15)$$

where $\Gamma(\vec{r}) = <\hat{\xi}(\vec{r}_1)\hat{\xi}(\vec{r}_1-\vec{r})>$ is the correlation function of $\hat{\xi}$.

The results of strong fluctuation theory stated above were applied to find the effective permittivity of vegetative media in Tsang and Kong (1981) and Chukhlantsev (1988). The success in the determination of ε_{ef} depends on the appropriate choice of the matrix R that allows canceling completely the singularity of the Green's function. In Tsang and Kong (1981), it was chosen in such a way that the frequency independent terms of ξ_{ef} were equal to zero. In this case, ξ_{ef} represents only wave corrections to the static part of the effective permittivity. At the same time, it is necessary to "guess" the parameters of the field inside the inhomogeneity or to use the solution to a diffraction problem for its determination. In Chukhlantsev (1988), it was shown that an expedient way to find the matrix R and the tensor $\hat{\xi}$ is to consider the electrostatic problem for a single inhomogeneity of the medium. If in this problem the fields $\vec{F}$ and $\vec{E}$ are related as

$$\vec{F} = \hat{\Lambda}\vec{E} \tag{4.16}$$

where the components of the tensor $\hat{\Lambda}$ are determined by the shape of the inhomogeneity and by its permittivity, one can derive

$$\hat{R} = \frac{1}{k_0^2}\left(\hat{\Lambda} - \hat{I}\right)\hat{\varepsilon}^{-1}, \ \hat{\xi} = \hat{\varepsilon}\hat{\Lambda}. \tag{4.17}$$

It should be noted that the tensor $\hat{\varepsilon}_0$ represents the quasi-static part of the effective permittivity tensor (but not the static part). It just reflects the fact that the field inside the inhomogeneity is close to the uniform field (the electric size of the inhomogeneity is small compared to the wavelength). In principle (Chukhlantsev *et al.*, 2003c), the tensors $\hat{R}$ and $\hat{\varepsilon}_0$ can also contain frequency dependent terms, which are usually determined as wave corrections to $\hat{\varepsilon}_0$.

A medium with inhomogeneities in the form of spheroids like discs (leaves) and needles (small brunches, needles of conifers) was considered in Chukhlantsev (1988). The system of coordinates, the axes of which coincide with the semi-axes of an arbitrary oriented spheroid, can be formed by the rotation of the initial system of coordinates around the x axis by an angle ϑ and around the z axis by an angle φ. The coordinates of fields $\vec{E}$ and $\vec{F}$ in the coordinate system of the spheroid are determined by

$$\vec{E}' = \hat{A}\hat{B}\vec{E}, \tag{4.18}$$

$$\vec{F}' = \hat{A}\hat{B}\vec{F} \tag{4.19}$$

where $\hat{A}$ and $\hat{B}$ are the matrices, which describe the above indicated rotations:

$$\hat{A} = \begin{pmatrix} 1 & 0 & 0 \\ 0 & \cos\vartheta & -\sin\vartheta \\ 0 & \sin\vartheta & \cos\vartheta \end{pmatrix}, \tag{4.20}$$

$$\hat{B} = \begin{pmatrix} \cos\varphi & -\sin\varphi & 0 \\ \sin\varphi & \cos\varphi & 0 \\ 0 & 0 & 1 \end{pmatrix}. \tag{4.21}$$

Let the field inside the spheroid $\vec{E}'$ relate to the external field $\vec{F}'$ as

$$\vec{E}' = \hat{\alpha}\vec{F}' \tag{4.22}$$

where $\hat{\alpha}$ is a tensor, only diagonal terms of which are not equal to zero (the explicit form of this tensor will be discussed below). Then, in the initial system of coordinates,

$$\vec{F}' = \hat{C}^{-1}\hat{\alpha}^{-1}\hat{C}\vec{E}, \tag{4.23}$$

$$\hat{C} = \hat{A}\hat{B}. \tag{4.24}$$

In accordance with equations (4.13), $\hat{\xi}$ and $\hat{R}$ is then given by

$$\hat{\xi} = \hat{\varepsilon}\,\hat{C}^{-1}\hat{\alpha}\,\hat{C}, \tag{4.25}$$

$$\hat{R} = \frac{1}{k_0^2}\left(\hat{C}^{-1}\hat{\alpha}\,\hat{C} - \hat{I}\right)\hat{\varepsilon}^{-1}. \tag{4.26}$$

The explicit form for $\hat{\xi}$ is presented by the next expressions:

$$\xi_{11} = \alpha_1(\varepsilon-\varepsilon_0^1)\cos^2\varphi + \alpha_1(\varepsilon-\varepsilon_0^1)\sin^2\varphi\cos^2\vartheta + \alpha_3(\varepsilon-\varepsilon_0^1)\sin^2\varphi\sin^2\vartheta, \tag{4.27}$$

$$\xi_{12} = -\alpha_1(\varepsilon-\varepsilon_0^1)\sin\varphi\cos\varphi + \alpha_1(\varepsilon-\varepsilon_0^1)\sin\varphi\cos\varphi\cos^2\vartheta + \\ + \alpha_3(\varepsilon-\varepsilon_0^1)\sin\varphi\cos\varphi\sin\vartheta\cos\vartheta \tag{4.28}$$

$$\xi_{13} = -\alpha_1(\varepsilon - \varepsilon_0^1)\sin\varphi\sin\vartheta\cos\vartheta + \alpha_3(\varepsilon - \varepsilon_0^1)\sin\varphi\sin\vartheta\cos\vartheta, \tag{4.29}$$

$$\begin{aligned}\xi_{21} &= -\alpha_1(\varepsilon - \varepsilon_0^2)\sin\varphi\cos\varphi + \alpha_1(\varepsilon - \varepsilon_0^2)\sin\varphi\cos\varphi\cos^2\vartheta + \\ &\quad + \alpha_3(\varepsilon - \varepsilon_0^2)\sin\varphi\cos\varphi\sin\vartheta\cos\vartheta\end{aligned} \tag{4.30}$$

$$\begin{aligned}\xi_{22} &= \alpha_1(\varepsilon - \varepsilon_0^2)\sin^2\varphi + \alpha_1(\varepsilon - \varepsilon_0^2)\cos^2\varphi\cos^2\vartheta + \\ &\quad + \alpha_3(\varepsilon - \varepsilon_0^2)\cos^2\varphi\sin^2\vartheta\end{aligned} \tag{4.31}$$

$$\xi_{23} = -\alpha_1(\varepsilon - \varepsilon_0^2)\cos\varphi\sin\vartheta\cos\vartheta + \alpha_3(\varepsilon - \varepsilon_0^2)\cos\varphi\sin\vartheta\cos\vartheta, \tag{4.32}$$

$$\xi_{31} = -\alpha_1(\varepsilon - \varepsilon_0^3)\sin\varphi\sin\vartheta\cos\vartheta + \alpha_3(\varepsilon - \varepsilon_0^3)\sin\varphi\sin\vartheta\cos\vartheta, \tag{4.33}$$

$$\xi_{32} = -\alpha_1(\varepsilon - \varepsilon_0^3)\cos\varphi\sin\vartheta\cos\vartheta + \alpha_3(\varepsilon - \varepsilon_0^3)\cos\varphi\sin\vartheta\cos\vartheta, \tag{4.34}$$

$$\xi_{33} = \alpha_1(\varepsilon - \varepsilon_0^3)\sin^2\vartheta + \alpha_3(\varepsilon - \varepsilon_0^3)\cos^2\vartheta \tag{4.35}$$

where $\alpha_1 = \alpha_{11} = \alpha_{22}$, $\alpha_3 = \alpha_{33}$, $\varepsilon_0^1 = \varepsilon_0^{11} = \varepsilon_0^{22}$, $\varepsilon_0^3 = \varepsilon_0^{33}$.

The components of $\hat{\varepsilon}_0$ are determined from the condition (4.14). When the distribution of scatterers over the azimuth angle is uniform, averaging over φ gives

$$\xi_1 = \xi_{11} = \xi_{22} = \frac{1}{2}(\varepsilon - \varepsilon_0^1)[\alpha_1(1 + \cos^2\vartheta) + \alpha_3\sin^2\vartheta], \tag{4.36}$$

$$\xi_3 = \xi_{33} = (\varepsilon - \varepsilon_0^3)[\alpha_1\sin^2\vartheta + \alpha_3\cos^2\vartheta]. \tag{4.37}$$

All other components of $\hat{\xi}$ are equal to zero. The distribution of plant elements over angle ϑ depends on the type of vegetation canopy. It is expedient to consider two cases: 1) plant elements are oriented chaotically (leaves and stems of some plants); 2) plant elements have the same orientation (stalks). In the first case, averaging over ϑ produces

$$\xi_1 = \xi_{11} = \xi_{22} = \xi_{33} = \frac{1}{3}(\varepsilon - \varepsilon_0)[2\alpha_1 + \alpha_3]. \tag{4.38}$$

As it is noted above, the tensor $\hat{\xi}$ represents the random polarizability of the medium. To find the components of $\hat{\varepsilon}_0$ one has to average $\hat{\xi}$ over the ensemble of random realizations of $\varepsilon(\vec{r})$. It can be done on the assumption that

$$P(\varepsilon = \varepsilon_v) = p, \tag{4.39}$$

$$P(\varepsilon = 1) = 1 - p \tag{4.40}$$

where P denotes the probability, ε_v is the dielectric permittivity of the spheroid, p is relative volume occupied by spheroids (the relative volume density of vegetation). It can be assumed also that

$$\varepsilon = 1, \quad \alpha = 1 \tag{4.41}$$

for a space, which is not occupied by the inhomogeneity. Generally speaking, the last assumption is valid for not very big values of p that is inherent to vegetation canopies. Putting the mean value of ξ to zero, one obtains for the chaotic orientation of plant elements

$$\varepsilon_0 = \frac{1 - p + \frac{1}{3} p \varepsilon_v [2\alpha_1 + \alpha_3]}{1 - p + \frac{1}{3} p [2\alpha_1 + \alpha_3]} = 1 + \frac{\frac{1}{3} p(\varepsilon_v - 1)[2\alpha_1 + \alpha_3]}{1 - p + \frac{1}{3} p [2\alpha_1 + \alpha_3]} . \tag{4.42}$$

It is important to note the following.
1. The obtained equation is not linear with respect to the volume density of scatterers. It means that the theory accounts for their interference.
2. If solutions of the electrostatic problem for a spheroid are used for α, the obtained equation will reduce to known Maxwell Garnett, Polder-van Santen, Bruggeman, and de Loor formulas, which were obtained in the electrostatic approximation. Particularly, for small spherical inhomogeneities

$$\alpha_1 = \alpha_2 = \alpha_3 = \frac{3\varepsilon_0}{2\varepsilon_0 + \varepsilon_s} \tag{4.43}$$

and, under the condition $\varepsilon_0 \sim 1$, equation (4.42) reduces to the classical Maxwell Garnett formula

$$\varepsilon_0 = 1 + \frac{3p(\varepsilon_v - 1)}{2 + \varepsilon_s - p(\varepsilon_v - 1)} \tag{4.44}$$

Thus, known formulas for the dielectric permittivity of dielectric mixtures can be obtained as particular cases of the theory.

3. The obtained equation satisfies the condition that the dielectric constant of mixture is equal to the dielectric constant of inclusions ($\varepsilon_0 = \varepsilon_v$) when $p = 1$. However, one has to remember that for big values of p, in general, $\alpha \neq 1$, that should be taken into account in averaging.

4. In the quasi-static approach, the components of $\hat{\alpha}$ contain frequency dependent terms. These terms can be determined by accounting for more precisely the relation between the external and inner fields in the problem of diffraction by a small spheroid. Including the frequency dependent terms to equation (4.42) allows one to estimate the effects of electromagnetic wave scattering by the inhomogeneities under the condition of their interference. Earlier, these effects were accounted for by calculating ξ_{ef} in accordance with equation (4.15), which represented wave corrections to the static part of the effective permittivity $\hat{\varepsilon}_0$ (Chukhlantsev, 1988). In this case, the frequency dependent term of $\hat{\varepsilon}_{ef}$ was proportional to the volume density p, i.e., the interference of inhomogeneities was not taken into account.

In the electrostatic approximation, the components of $\hat{\alpha}$ are given by (Tsang *et al.*, 1985)

$$\alpha_1 = 1/[1 + A_1(\frac{\varepsilon_v}{\varepsilon_0} - 1)], \tag{4.45}$$

$$\alpha_3 = 1/[1 + A_3(\frac{\varepsilon_v}{\varepsilon_0} - 1)], \tag{4.46}$$

$$A_1 = \frac{a^2 c}{2} \int_0^\infty \frac{ds}{(s + a^2)^2 (s + c^2)^{1/2}}, \tag{4.47}$$

$$A_3 = \frac{a^2 c}{2} \int_0^\infty \frac{ds}{(s + a^2)(s + c^2)^{3/2}} \tag{4.48}$$

where a and c are the semi-axes of the spheroid. For a small sphere, $A_1 = A_2 = 1/3$; for a thin disk, $A_1 = 0$, $A_3 = 1$; for needles, $A_1 = 1/2$, $A_3 = 0$. For small value of the volume density, $\varepsilon_0 \sim 1$ and the following expressions for the static part of $\hat{\varepsilon}_{ef}$ were obtained (Chukhlantsev, 1988):

a) for chaotically oriented thin disks,

$$\varepsilon_0 = 1 + \frac{\dfrac{1}{3} p(\varepsilon_v - 1)(2 + 1/\varepsilon_v)}{1 - p + \dfrac{1}{3} p(2 + 1/\varepsilon_v)}, \tag{4.49}$$

b) for chaotically oriented thin needles,

$$\varepsilon_0 = 1 + \frac{\dfrac{1}{3} p(\varepsilon_v - 1)\dfrac{5 + \varepsilon_v}{1 + \varepsilon_v}}{1 - p + \dfrac{1}{3} p \dfrac{5 + \varepsilon_v}{1 + \varepsilon_v}}, \tag{4.50}$$

c) and for vertically oriented thin needles ($\hat{\varepsilon}_0 = diag\,[\varepsilon_0^1, \varepsilon_0^2, \varepsilon_0^3]$),

$$\varepsilon_0^{1,2} = 1 + \frac{2p\dfrac{\varepsilon_v - 1}{\varepsilon_v + 1}}{1 - p + \dfrac{2p}{\varepsilon_v + 1}}; \qquad \varepsilon_0^3 = 1 + p(\varepsilon_v - 1). \tag{4.51}$$

When $p \ll 1$ and $\varepsilon_v \gg 1$, these expressions are reduced to Polder-van Santen formulas often used by experimentalists:

for chaotic disks,

$$\varepsilon_0 \approx 1 + \frac{1}{3} p(\varepsilon_v - 1)(2 + 1/\varepsilon_v) \approx 1 + \frac{2}{3} p(\varepsilon_v - 1), \tag{4.52}$$

for chaotic needles,

$$\varepsilon_0 \approx 1 + \frac{1}{3} p(\varepsilon_v - 1)\frac{5 + \varepsilon_v}{1 + \varepsilon_v} \approx 1 + \frac{1}{3} p(\varepsilon_v - 1), \tag{4.53}$$

and for vertical needles (thin vertical stalks),

$$\varepsilon_0^{1,2} \approx 1 + 2p\frac{\varepsilon_v - 1}{\varepsilon_v + 1}; \qquad \varepsilon_0^3 = 1 + p(\varepsilon_v - 1). \qquad (4.54)$$

In Chukhlantsev (1992), it was shown that the wave corrections to ε_0 correspond to the scattering by Rayleigh particles and, when the scatterers are small compared to the wavelength, the continuous and the discrete models yield the same results for attenuation of electromagnetic waves by vegetation. An important advantage of the continuous model is that it provides ε_0 for dense ($p < 1$) media (e.g., snow, soil, or dense vegetation canopies). The introduction of new variables with the subsequent determination of ε_0 is equivalent to the summation of a part of the infinite series of multiple scattering, i.e., to taking into account mutual influences of the scatterers. A further development of the model may consist in the introduction of frequency dependent terms into ε_0, these terms being obtained from a refined relationship between $\vec{F}$ and $\vec{E}$. This approach should allow one to estimate the effect of scattering by inhomogeneities on the effective dielectric constant of vegetation as a function of the "packing" density of scatterers. The relation between $\vec{F}$ and $\vec{E}$ could be found from a rigorous solution of the diffraction problem. This solution is known for spherical particles. For spheroids, the aforementioned relation can be obtained by the expansion of the scattering operator (4.6) into a series by a small parameter. For example, for a small disk of radius a and thickness d, we obtain

$$\alpha_1 \approx 1 + \frac{i}{12}(k_0 d)(k_0 a)^2(\varepsilon_v - 1), \quad \alpha_3 \approx 1/\varepsilon_v \qquad (4.55)$$

and, for a medium filled by chaotic disks,

$$\varepsilon_0 = 1 + \frac{\dfrac{1}{3}p(\varepsilon_v - 1)[2 + \dfrac{i}{12}(k_0 d)(k_0 a)^2(\varepsilon_v - 1) + 1/\varepsilon_v]}{1 - p + \dfrac{1}{3}p[2 + \dfrac{i}{12}(k_0 d)(k_0 a)^2(\varepsilon_v - 1) + 1/\varepsilon_v]}. \qquad (4.56)$$

This expression allows one to analyze the dependence of the dielectric permittivity and refractive index of random media with small inhomogeneities on the density of their package.

The discrete model provides estimates of characteristics of electromagnetic wave propagation through media filled not only by Rayleigh particles but also by strong scatterers. However, this model is developed for rather sparse media. The mutual influence of densely packed scatterers has not been properly studied theoretically in the framework of the discrete model.

4.2.2. Effective Permittivity of a Vegetation Medium

Equations (4.52) – (4.54) were used by researchers to calculate the effective dielectric constant of vegetation and the related quantities. In work by Du and Peake (1969), attenuation properties of foliage media in a VHF-band were studied using the continuous model with p = 0.001-0.005 and $\varepsilon_v = m_v \varepsilon_w$, where ε_w is the dielectric constant of saline water contained in leaves at the ionic conductivity values of $(2\ldots4)\cdot10^{-3}\,\mathrm{Ohm}^{-1}\mathrm{cm}^{-1}$. The reflectivity of vegetation canopies at small grazing angles was calculated in Redkin *et al.* (1973) by Fresnel formulas with the use of expressions (4.52), (4.53) for ε_0. The tensor of dielectric permittivity of foliage media was derived in Redkin and Klochko (1977) and some data on the conductivity of sap and on the volume fraction of vegetation were presented. The effective dielectric constant of foliage was calculated in Milshin and Grankov (2000) using data from Redkin and Klochko (1977). For wheat and soybean, the effective dielectric constant was calculated in Allen and Ulaby (1989), and for corn leaves, in Ulaby et al. (1987). Tamasanis (1992) reported the effective dielectric constant of foliage calculated in the approximation of spherical inhomogeneities. In Kerr and Wigneron (1994), the effective dielectric constant of vegetation was modeled by the expression

$$\varepsilon_{ef}^{\beta} = 1 + p(\varepsilon_v^{\beta} - 1) \tag{4.57}$$

with $\beta=1$ (the linear model), $\beta=0.5$ (the refractive model), and $\beta=0.33$ (the cubic model). The refractive mixing was found to provide the best results. In Borodin *et al.* (1976) and Basharinov *et al.* (1979), formulas (4.52), (4.53) were used to calculate the attenuation in a forest canopy. Unfortunately, no data on any direct measurements of the effective dielectric constant of vegetation could be found in the literature. In the cited papers, ε_{ef} was used to calculate the attenuation (or the reflectivity). The calculated attenuation (reflection) characteristics were in good agreement with experimental data. It points to the possibility of applying the continuous model to vegetation media.

Data on the effective dielectric constant of vegetation obtained by different researchers are presented in Table 4.1. One can see that, because the vegetation medium is rather sparse, its effective dielectric constant differs only slightly from unity. This determines very small values of the vegetation reflectivity for observation angles close to nadir. The reflectivity noticeably differs from zero only for small grazing angles (Redkin and Klochko, 1977). At the same time, in the centimeter wave range, the reflectivity of a vegetation layer can reach 0.05...0.15 (Chanzy and Wigneron, 2000) due to the volume scattering of electromagnetic waves by plant elements. In the continuous approach, these scattering effects are usually neglected (or taken into account as wave corrections to ε_0), since the size of scatterers is proposed to be very small compared to the wavelength. This fact imposes a limitation on the application of the continuous approach (as it was stated above).

Table 4.1. Effective dielectric constant of vegetation.

Vegetation type	Frequency band	Effective dielectric constant
Wheat stalks	X-band (10.2 GHz)	$1.036 + j0.01644$ (ε_0^3) $1.006 + j0.00042$ ($\varepsilon_0^{1,2}$)
Soybean	X-band (10.2 GHz)	$1.038 + j0.01707$
Coniferous forest	L-band (1.4 GHz)	$(1.00095\text{-}1.0028) + j(0.00019\text{-}0.014)$
Deciduous forest	L-band (1.4 GHz)	$(1.0028\text{-}1.19) + j(0.00058\text{-}0.038)$
Foliage	50 MHz 200 MHz 500 MHz 800 MHz 1300 MHz	$1.03 + j0.00073$ $1.03 + j0.00046$ $1.03 + j0.00028$ $1.03 + j0.00021$ $1.03 + j0.00020$

The equations for ε_0 obtained in the previous section allow one not only to estimate the effective dielectric permittivity of vegetation in the microwave band but also to investigate its dependence on the vegetation volume density p. In Fig. 4.1, calculated values of the real and imaginary parts of ε_0 versus the vegetation volume density are presented. The real and imaginary parts of ε_0 are calculated by formulas (4.49) and (4.52) at $\varepsilon_v = 20 + j10$. It is seen from Fig. 4.1 that, at characteristic for vegetation values of p less than 0.01, equations (4.49) and (4.52) produce the same values of ε_0 close to those reported by other researchers (Table 4.1). A discrepancy between values of ε_0 produced by (4.49) and (4.52) is observed at values of p more than 0.3 (Fig. 4.1), which are not observed for natural vegetation.

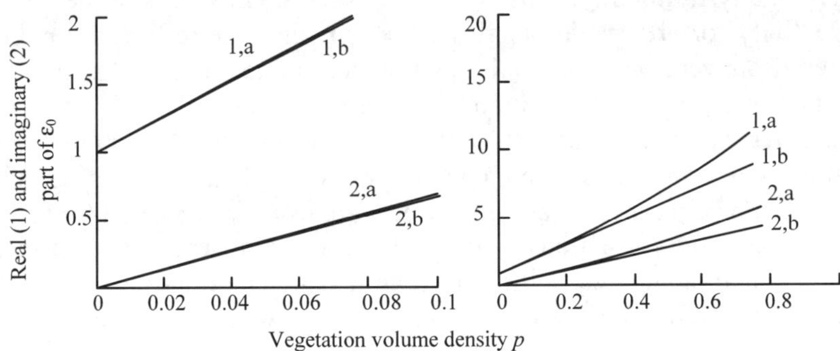

Vegetation volume density p

Fig. 4.1. Chaotic disks. Equation (4.49) and (4.52) – *a* and *b*, respectively.

Introduction of the frequency dependent terms into the quasi-static part of ε_{ef} (equation (4.56)) enables one to estimate effects of electromagnetic wave scattering by the inhomogeneities on an extinction of the mean electric field in the medium. The extinction rate γ (the attenuation per unit length) of the mean field intensity is determined by the imaginary part of vegetation refractive index $n'' = \mathrm{Im}\sqrt{\varepsilon_0}$:

$$\gamma = \frac{4\pi}{\lambda} n'' . \tag{4.58}$$

Fig. 4.2 represents the data for the imaginary part of refractive index at the wavelength 3 cm, which are calculated by equations (4.49) and (4.56) for a medium containing chaotic disks with a radius of 1 cm and a thickness of 0.03 cm (characteristic dimensions of leaves). One can see that the contribution of scattering into the total attenuation can reach 10-30% at centimeter wavelengths and should be taken into account. In the decimeter band, the scattering by leaves is small compared with the absorption.

The model (4.56) makes it possible to reveal the influence of package density of scatterers on the attenuation of the mean field due to the scattering (Fig. 4.2). The attenuation due to the scattering first increases with an increase in the density and, then, at big densities, diminishes, which can be explained by the mutual influence of the scatterers. Earlier (Chukhlantsev, 1988), the scattering by inhomogeneities of a medium were accounted for by wave corrections to ε_0 that produced a linear dependence of the frequency dependent term of ε_{ef} on the volume density p.

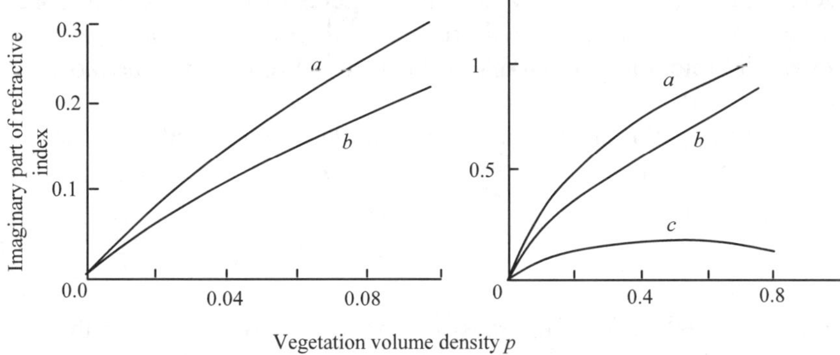

Fig. 4.2. Imaginary part of refractive index for a medium of chaotic disks at 10 GHz versus volume density as calculated by (4.56) – *a*, by (4.49) – *b*, and contribution of scattering – *c*.

It is important to note that previous formulations relate to the case of uniform distribution of scatterers in space. In forest media, scatterers fill the space with changing volume density: there are voids and places occupied by branches with leaves in a forest crown. The volume density of vegetation in these places is relatively big compared to the average volume density of the crown. Since the dependence of the refractive index on the volume density is not linear (Fig. 4.2), calculations of the extinction rate with the use of the average crown density may overestimate the attenuation. The use of "real" volume density for places occupied by scatterers (that will take into account their mutual influence) with consequent weighting of the extinction rate, accounting for the relative volume of voids, may produce better results for the attenuation at high frequencies of the microwave band.

4.3. THE MODEL OF VEGETATION AS A COLLECTION OF SCATTERERS (DISCRETE MODEL)

The discrete model was applied in Attema and Ulaby (1978), Brown and Curry, 1982; Chauhan and Lang (1989, 1994), Chauhan *et al.* (1991, 1994), Chuah *et al.* (1997), Chuang *et al.* (1982), Chukhlantsev (1981, 1989a,b), De Roo *et al.* (2001), Eom and Fung (1984, 1986), Ferrazzoli and Guerriero (1996); Karam and Fung (1983,1988); Karam *et al.* (1992), Kirdiashev *et al.* (1979), Lang (1981, 2004), Lang and Sidhu (1983), Lang *et al.* (1986), Macelloni *et al.* (2001), Sun *et al.* (1991), Tsang and Ding (1991),

Ulaby *et al.* (1987, 1990), and Wang *et al.* (1993) to determine characteristics of propagation, emission, and scattering of electromagnetic waves in vegetation canopies. To clarify specific features of the discrete approach, it is necessary to consider the main results of the theory of wave propagation through discrete random media (Finkelberg, 1968; Barabanenkov, 1975; Ishimaru, 1978).

The propagation of the mean field through such a medium is described by the dispersion equation for the wave number (Barabanenkov, 1975):

$$k^2 = k_0^2 + \widetilde{M}(k) \tag{4.59}$$

where $\widetilde{M}(k)$ is the Fourier transform of the mass operator of the Dyson equation for the mean field. In Finkelberg's correlation group approximation (Finkelberg, 1968), one obtains

$$M = \sum_{s=1}^{\infty} \int d\vec{r}_1 \ldots \int d\vec{r}_s \, g_s(\vec{r}_1,\ldots,\vec{r}_s) T_s(\vec{r}_1,\ldots,\vec{r}_s) \tag{4.60}$$

where g_s is the s-point correlation function and T_s is the scattering operator for a system of s scatterers positioned at the points $\vec{r}_1,\ldots,\vec{r}_s$. In the case of independent (uncorrelated) scatterers, $g_2 = g_3 = \ldots = 0, g_1 = n$, and

$$M = nT_1 \tag{4.61}$$

where n is the number of scatterers in a unit volume and T_1 is the scattering operator of a single particle. This case corresponds to the Foldy-Twersky approach for independent point scatterers (Barabanenkov, 1975). The replacement of $\widetilde{M}(k)$ by $\widetilde{M}(k_0)$ in the dispersion equation (4.59) implies neglect of spatial dispersion for the mean field and leads to the radiation transfer equation for the field covariance (Barabanenkov, 1975). This approximation is valid when

$$\left|\widetilde{M}(k_0)\right|/k_0^2 \ll 1. \tag{4.62}$$

Under this condition, the dispersion equation takes the form

$$k = \sqrt{k_0^2 + \widetilde{M}(k_0)} \approx k_0 + \widetilde{M}(k_0)/2k_0. \tag{4.63}$$

In this case (i.e., when the scatterers are in the far-field zones of each other), the parameters of electromagnetic wave propagation in a discrete medium are determined by the scattering and attenuation of electromagnetic waves by a single scatterer. The field scattered by a scatterer in the far-field zone is expressed as

$$\vec{E}_s = \vec{f}(\hat{o},\hat{i})\frac{e^{jkr}}{r} \tag{4.64}$$

where $\vec{f}(\hat{o},\hat{i})$ is the complex scattering amplitude. The extinction cross section σ_e and the scattering cross section σ_s are determined by

$$\sigma_e = \frac{4\pi}{k_0}\mathrm{Im}\left\{\vec{f}(\hat{i},\hat{i})\cdot\vec{e}_i\right\}, \tag{4.65}$$

$$\sigma_s = \int_{4\pi}\left|\vec{f}(\hat{o},\hat{i})\right|^2 d\omega \tag{4.66}$$

where $\vec{e}_i$ is a unit vector determining the polarization of the incident wave, $d\omega$ is the solid angle element, and Im denotes the imaginary part. The absorption cross section σ_a is determined by the Ohmic loss of the field $\vec{E}_{in}$ inside the scatterer

$$\sigma_a = \frac{k_0 \int_V \left|\vec{E}_{in}\right|^2 \varepsilon_s'' dV}{\left|\vec{E}_0\right|^2} \tag{4.67}$$

where ε_v'' is the imaginary part of scatterer dielectric permittivity. According to the optical theorem, the following relation is valid:

$$\sigma_e = \frac{4\pi}{k_0}\mathrm{Im}\left\{\vec{f}(\hat{i},\hat{i})\cdot\vec{e}_i\right\} = \sigma_a + \sigma_s. \tag{4.68}$$

In Twersky's theory (Ishimaru, 1978), propagation of the mean field in a layer of scatterers is described by a dispersion equation for the wave number:

$$k = k_0 + \frac{2\pi n f(\hat{i},\hat{i})}{k_0} . \tag{4.69}$$

Equations (4.63) and (4.69) establish a relation between Twersky's theory and the approximation (4.62) of the correlation group theory. In this approximation, one has:

$$\widetilde{M}(k_0) = 4\pi n f(\hat{i},\hat{i}) \tag{4.70}$$

and both approaches provide the same result for the wave number of the mean field in a scattering medium. More detail on the derivation of the dispersion equation for the mean field wave number can be found in a paper by Lang (2004).

Parameters of radiation transfer theory are the extinction coefficient γ, the single scattering albedo ω, and the phase function $\varsigma(\hat{o},\hat{i})$. These values are given by the expressions

$$\gamma = n\sigma_e , \tag{4.71}$$

$$\omega = \frac{\sigma_s}{\sigma_e} , \tag{4.72}$$

$$\zeta(\hat{o},\hat{i}) = \frac{4\pi \left| \vec{f}(\hat{o},\hat{i}) \right|^2}{\sigma_e} . \tag{4.73}$$

Practically all studies of electromagnetic wave propagation in vegetation that are based on the discrete approach use radiation transfer theory. The applicability of this theory is justified when individual scatterers are located in the far-field zone with respect to each other (Barabanenkov, 1975). In this case (with the correlation of the scatterer positions being neglected), the extinction coefficient is a linear function of the number of scatterers in a unit volume n (see equation (4.71)). The effect of the scatterer positioning in the near-field zone with respect to each other is proportional to the ratio of a certain scale parameter R_0 of the near-field zone to the extinction length $1/\gamma$ (Barabanenkov, 1975). The parameter R_0 is chosen so that, for particles located at distances greater than R_0, the far-zone approximation is valid. The neglect of the near-zone effect imposes limitations on the quantity R_0 from above and from below, which leads to limitations for the scattering medium.

Estimates of the effect of scatterer positioning in the near-field zone, as applied to vegetation canopies, are given in Chukhlantsev (1981, 1992). The effect was shown to be small when $1/\gamma \gg R_0 \gg \lambda,\ \sigma_e/\lambda$. These conditions are satisfied for vegetation canopies with the volume density $p \leq 3 \cdot 10^{-3}$. For dense vegetation canopies, the dependence of the extinction coefficient on the volume density is nonlinear, which is observed experimentally in the centimeter wave band where the attenuation of electromagnetic radiation by plant elements is strong (Chukhlantsev, 1986; Chukhlantsev and Golovachev, 1989). In this case, the effect of mutual screening of scatterers can be taken into account by expanding the extinction coefficient into a power series by the ratio of the wavelength to the extinction length (Chukhlantsev, 1992; Chukhlantsev and Golovachev, 1989):

$$\gamma = n\sigma_e\left(1 - cn\sigma_e / k_0 + ...\right) \approx \gamma_0\left(1 - c\gamma_0 / k_0\right) \qquad (4.74)$$

where c is the expansion coefficient, which is chosen from the regression analysis of experimental data. (In Chukhlantsev and Golovachev (1989), it was shown that, up to $p \sim 10^{-2}$, the correction to the extinction coefficient in the centimeter band does not exceed 20%; for grain crops, alfalfa, and clover, the aforementioned coefficient was found to be $c = 1.34$ dB^{-1}.) The nonlinear dependence of the extinction coefficient on the number density of scatterers that follows in the discrete model can be explained by follows. As it was noted in Chukhlantsev (1992), it is more rigorous to determine the extinction coefficient of a scattering medium from the dispersion equation (4.63) and not from equation (4.69):

$$k = \sqrt{k_0^2 + \widetilde{M}(k_0)}, \qquad (4.75)$$

$$\gamma = 2\operatorname{Im}k = 2\operatorname{Im}(k_0\sqrt{1 + \widetilde{M}(k_0)/k_0^2}) = 2\operatorname{Im}(k_0\sqrt{1 + 4\pi n f(\hat{i},\hat{i})/k_0^2}). \quad (4.76)$$

In this approach, the extinction coefficient is a non-linear function of the number density of scatterers. The expansion of square root in (4.76) into a power series by $\widetilde{M}(k_0)/k_0^2$ gives:

$$\gamma = 2k_0\operatorname{Im}\left[\frac{1}{2}\frac{4\pi n f(\hat{i},\hat{i})}{k_0^2} - \frac{1}{8}\left(\frac{4\pi n f(\hat{i},\hat{i})}{k_0^2}\right)^2 + ...\right]$$

$$= \gamma_0\left(1 - \frac{1}{2}\frac{\operatorname{Re}[f(\hat{i},\hat{i})]}{\operatorname{Im}[f(\hat{i},\hat{i})]}\frac{\gamma_0}{k_0} + ...\right) \qquad (4.77)$$

and one comes to the expression (4.74). This approach allows one to specify the extinction coefficient for "dense" media. In particular, for a medium of small spheroids,

$$\frac{\mathrm{Re}[f(\hat{i},\hat{i})]}{\mathrm{Im}[f(\hat{i},\hat{i})]} \sim \frac{\varepsilon_v' - 1}{\varepsilon_v''} \tag{4.78}$$

and the extinction coefficient calculated by (4.77) coincides with that calculated by (4.58) in the continuous approach when $p < 0.05$. This fact is evidence of the equivalence of the discrete and continuous approaches in the case considered. Indeed, the radicand in (4.76) can be formally considered as the effective dielectric constant of the scattering medium. If the scattering amplitude averaged by an ensemble of scatterers, e.g., of small chaotic disks (Chukhlantsev, 1986),

$$\left\langle f(\hat{i},\hat{i}) \right\rangle = \frac{2}{3} \frac{k_0^2 (\varepsilon_v - 1) V}{4\pi}, \tag{4.79}$$

is substituted into (4.76), it will lead to equation (4.52) obtained in the continuous approach. Thus, in the case of "soft" scatterers, the discrete and continuous approaches produce the same result for extinction coefficient. A transfer to the continuous model for a truly discrete medium is justified in this case because it is theoretically proved. Besides, the continuous approach is developed enough to account for the mutual influence of scatterers (multiple scattering). For a medium containing scatterers of arbitrary size and shape, it is more correct to use equation (4.76) obtained in the discrete approach. A formal transfer to the continuous approach in this case seems to be artificial and unjustified. However, in the last case, the determination of scattering amplitude of a single scatterer is required that is an independent problem.

The electrodynamic characteristics of a dense vegetation medium can also be found using the "dense medium radiative transfer theory" (Wen *et al.*, 1990) or the "dense medium phase and amplitude correction theory" (Chuah *et al.*, 1997).

4.4. EXTINCTION AND SCATTERING OF ELECTRO-MAGNETIC WAVES BY PLANT ELEMENTS

4.4.1. Theoretical Models

In the discrete model, electrodynamic parameters of vegetation media is characterized by the scattering operator (or the scattering amplitude, or the scattering and extinction cross sections σ_s and σ_e, respectively) of a single scatterer, and, hence, it is important to examine scattering and extinction by different individual plant elements. In the microwave frequency band, the sizes of leaves, stalks, branches, and trunks are comparable with the wavelength (resonance region), and, therefore, the scattering and extinction cross sections should be determined from diffraction models that take into account the shape and size of a given element with the highest possible accuracy. Since the plant elements are similar in shape to flat disks and strips (leaves), and to circular cylinders (stalks, branches, and trunks), the diffraction problems for bodies with the above shapes are usually discussed. The known solutions of the diffraction problem for perfectly conducting infinitely thin disks and strips are not applicable to find the emissivity of vegetation layer, because the last is determined by the active losses in a medium (however, these solutions may be useful for calculating the scattering from vegetation (Senior *et al.*, 1987)). Solutions of the diffraction problems for dielectric disks and strips in general case are not known. In this case, the cross sections σ_s and σ_e can be found only under some restrictive assumption imposed on the relation between the size of element and the wavelength. Below, several models used by researchers are described.

A. Small particles (low-frequency approximation). A general formulation for the scattered field $\vec{E}_s$ can be obtained via the Helmholtz integral equation. For a scatterer located at the origin, the Helmholtz integral equation relates the far zone scattered field to the field inside the scatterer $\vec{E}_{in}(r')$ through the relation (Schiffer and Thielheim, 1979; Karam *et al.*, 1988; Tsang *et al.*, 1985)

$$\vec{E}_s = \frac{e^{jk_0 r}}{r} \frac{k_0^2}{4\pi} (\bar{I} - \hat{s}\hat{s}) \iiint (\varepsilon_v - 1) \cdot \vec{E}_{in}(r') e^{jk_0 \hat{s}\vec{r}'} d\vec{r}' \qquad (4.80)$$

where $\bar{I}$ is the unit dyad and $\hat{s}$ is the observation direction. For small particles

$$\vec{E}_{in}(r') = \overline{a} \cdot \vec{E}_i \tag{4.81}$$

where $\overline{a}$ is the polarizability tensor and $\vec{E}_i$ is the incident wave at a given point. For small spheroids and principle directions, one obtains (Van de Hulst, 1957).

$$\sigma_{ai} = 4\pi k_0 \,\mathrm{Im}(a_i) \quad \sigma_{si} = \frac{4}{3}\pi k_0^4 |a_i|^2 , \tag{4.82}$$

$$a_i = \frac{\varepsilon_v - 1}{(\varepsilon_v - 1)A_i + 1} \cdot \frac{V_s}{4\pi} \quad i = 1, 2, 3 \tag{4.83}$$

where V_s is the volume of the spheroid and factors A_i are given by (4.47) and (4.48). For small disks,

$$\sigma_{a1} << \sigma_{a2,3} = k_0 \varepsilon_v'' V_s = k_0 \varepsilon_v'' d\sigma$$

$$\sigma_{s1} << \sigma_{s2,3} = \frac{1}{12}(k_0 d)^2 (k_0 a)^2 |\varepsilon_v - 1|^2 \sigma \tag{4.84}$$

where d is the thickness, a is the radius, and $\sigma = \pi a^2$ is the area of the disk. For small needles,

$$\sigma_{a2,3} << \sigma_{a1} = k_0 \varepsilon_v'' V_s$$

$$\sigma_{s2,3} << \sigma_{s1} = \frac{\pi}{12}(k_0 l)^2 (k_0 a)^2 |\varepsilon_v - 1|^2 a^2 \tag{4.85}$$

where l is the length and a is the radius of the needle. The absorption cross section of a small particle is proportional to its volume. The scattering cross section is much less than its absorption cross section.

   B. Very large plane particles (high-frequency approximation). The absorption and scattering cross sections of these particles can be expressed (neglecting edge effects) in terms of the reflection coefficient R and the transmission coefficient T of an infinite layer of the same thickness (Le Vine *et al.*, 1985; Chukhlantsev, 1981, 1986)

$$\sigma_a = (1 - R - T)\sigma; \quad \sigma_s = 2R\sigma \tag{4.86}$$

where σ is the geometrical cross section of the particle. When $k_0 d \sqrt{\varepsilon_v} << 1$ (for leaves this condition is satisfied up to the X-band) and the electromagnetic wave is normally incident upon the layer, the considered cross sections are given by (Chukhlantsev, 1981, 1986)

$$\sigma_a = \frac{k_0 d \varepsilon_v'' \sigma}{1 + \frac{1}{4}(k_0 d)^2 |\varepsilon_v - 1|^2 + k_0 d \varepsilon_v''} \, , \qquad (4.87a)$$

$$\sigma_s = \frac{\frac{1}{2}(k_0 d)^2 |\varepsilon_v - 1|^2 \sigma}{1 + \frac{1}{4}(k_0 d)^2 |\varepsilon_v - 1|^2 + k_0 d \varepsilon_v''} \, . \qquad (4.87b)$$

C. Plane thin particles (resonance case). The scattering and extinction cross sections are found by the use of equation (4.80) with an appropriate choice of the field inside the scatterer. There are several approaches to estimate this field. The simplest case when $\vec{E}_{in}(r') = \vec{E}_0(r')$ is valid only for very thin particles (Chukhlantsev, 1986). The next approach, see Chukhlantsev (1981, 1986), Karam and Fung (1989, and Karam *et al.* (1988), considers the inner field as the quasi-static field inside a small ellipsoid (the generalized Rayleigh-Gans approximation). This approach does not take into account the attenuation of the incident field inside the particle and is valid only for the case of very thin particles when $|k_0 d(\varepsilon_v - 1)| << 1$, where d is the thickness of the leaf. When $|k_0 d(\varepsilon_v - 1)| < 1$, integral equation (4.6) for the field inside a particle can be expanded in a perturbation series (Chukhlantsev, 1986). The series expansion parameter is the quantity $k_0 d(\varepsilon_v - 1)$, which imposes a limitation on the particle thickness and permittivity and on the wavelength of the electromagnetic wave. The last approach requires an exhausting computation of the inner field by a summation of the perturbation series. The computation was performed in Chukhlantsev (1981, 1986) and demonstrated the resonant character of the extinction cross section for particles whose size (the diameter of a disk or the width of a strip) is comparable with the wavelength. The presence of resonance effects in the extinction by leaves was also confirmed experimentally (see Chapter 5). A simpler approach for the inner field was proposed in Le Vine *et al.* (1985), where it was shown that, for a physically and electrically thin disk, the field inside a slab of the same thickness can be used as the inner field. This approach yields good agreement with experiments and is convenient for the modeling due to its relative ease.

To obtain simple expressions for the field scattered by plane thin leaves and for the scattering cross sections, the case can be considered when the wave vector of the incident electromagnetic wave is normal to the plane of the leaf. The general case of arbitrary angle of incidence and polarization leads to some computational problems, makes it difficult to obtain simple and convenient for analysis expressions, but does not change significantly the spectral behavior of the scattering. This case was considered in Karam *et al.* (1988), where the influence of leaf orientation was studied in detail.

In the case of normal incidence, when the leaf is electrically thin the changes of the phase and field can be neglected in the direction of the wave propagation. Equation (4.80) is reduced to the form

$$\vec{E}_s(\vec{r}) = \frac{e^{jk_0 r}}{r} \frac{k_0^2(\varepsilon_v - 1)d}{4\pi} (\vec{E}_{in} - (\vec{E}_{in} \cdot \hat{r})\hat{r}) \iint_S \exp\{-jk_0 \vec{r}_1 \cdot \hat{r}\} d\vec{r}_1 \quad (4.88)$$

where $\vec{E}_{in}$ is the field inside the slab of the same thickness and $\hat{r}$ is a unit vector in $\vec{r}$ direction. It is very important to note that in the case considered the cross-polarized scattered field appears. That does not take place in the case of very large plane particles (B) (no depolarization at all) and in the case of small particles (A) (the cross-polarized scattering (and the scattering generally) is very small). The co-polarized E_{co} and cross-polarized E_{cr} scattered fields are obtained from (4.88) as

$$E_{co}(\vec{r}) = \frac{e^{jk_0 r}}{r} \frac{k_0^2 d(\varepsilon_v - 1)a^2 E_{in}}{2} (1 - \cos^2\varphi\sin^2\vartheta)\frac{J_1(k_0 a\sin\vartheta)}{k_0 a\sin\vartheta}, \quad (4.89)$$

$$E_{cr}(\vec{r}) = \frac{e^{jk_0 r}}{r} \frac{k_0^2 d(\varepsilon_v - 1)a^2 E_{in}}{2} \sin^2\vartheta\sin\varphi\cos\varphi\frac{J_1(k_0 a\sin\vartheta)}{k_0 a\sin\vartheta} \quad (4.90)$$

for a disk with radius a, and

$$E_{co}(\vec{r}) = \frac{e^{jk_0 r}}{r} \frac{k_0^2 d(\varepsilon_v - 1)abE_{in}}{4\pi} (1 - \cos^2\varphi\sin^2\vartheta)\times$$
$$\times\sin c\left(\frac{k_0 a\sin\varphi\sin\vartheta}{2}\right)\sin c\left(\frac{k_0 b\sin\varphi\sin\vartheta}{2}\right) \quad (4.91)$$

$$E_{cr}(\vec{r}) = \frac{e^{jk_0 r}}{r} \frac{k_0^2 d(\varepsilon_v - 1)abE_{in}}{4\pi} \sin^2 \vartheta \sin \varphi \cos \varphi \ \times$$

$$\times \ \operatorname{sin}c\!\left(\frac{k_0 a \sin \varphi \sin \vartheta}{2}\right) \operatorname{sin}c\!\left(\frac{k_0 b \sin \varphi \sin \vartheta}{2}\right) \tag{4.92}$$

for a strip with the dimensions a and b (J_1 is the Bessel function of the first order and φ and ϑ are the azimuth and meridian angles, respectively). The scattering cross sections are found from (4.89) – (4.92) by integrating the scattering intensity over the solid angle and taking into account that (4.87)

$$\left|E_{in}\right|^2 = \frac{1}{1 + \dfrac{1}{4}(k_0 d)^2 \left|\varepsilon_v - 1\right|^2 + k_0 d \varepsilon_v''}. \tag{4.93}$$

The scattering cross sections are given by

$$\sigma_{sco} = \sigma_{sSL}(\Phi_1 + \Phi_2), \tag{4.94}$$

$$\sigma_{scr} = \sigma_{sSL}\,\Phi_3, \tag{4.95}$$

$$\Phi_1 = \frac{1}{4}(k_0 a)^2 \int_0^\pi F^2(k_0 a \sin \vartheta)\cos^2 \vartheta \sin \vartheta\, d\vartheta, \tag{4.96}$$

$$\Phi_2 = \frac{3}{32}(k_0 a)^2 \int_0^\pi F^2(k_0 a \sin \vartheta)\sin^5 \vartheta\, d\vartheta, \tag{4.97}$$

$$\Phi_3 = \frac{1}{3}\Phi_2 \quad F(x) = \frac{2}{x}J_1(x) \tag{4.98}$$

for a disk, and

$$\sigma_{sco} = \sigma_{sSL}\,\Phi_4, \tag{4.99}$$

$$\sigma_{scr} = \sigma_{sSL}\,\Phi_5, \tag{4.100}$$

$$\Phi_4 = \frac{k_0^2 ab}{8\pi^2} \int\limits_0^{2\pi} \int\limits_0^{\pi} (1 - \cos^2 \varphi \sin^2 \vartheta)^2 \times$$

$$\times \sin c^2 \left(\frac{k_0 a \sin \varphi \sin \vartheta}{2} \right) \sin c^2 \left(\frac{k_0 b \sin \varphi \sin \vartheta}{2} \right) \sin \vartheta \, d\vartheta \, d\varphi \tag{4.101}$$

$$\Phi_5 = \frac{k_0^2 ab}{8\pi^2} \int\limits_0^{2\pi} \int\limits_0^{\pi} \sin^2 \varphi \cos^2 \varphi \times$$

$$\times \sin c^2 \left(\frac{k_0 a \sin \varphi \sin \vartheta}{2} \right) \sin c^2 \left(\frac{k_0 b \sin \varphi \sin \vartheta}{2} \right) \sin^5 \vartheta \, d\vartheta \, d\varphi \tag{4.102}$$

for a strip, where σ_{sSL} is the scattering cross section for very large plane particles (4.87b). The absorption cross section in the considered case is given by equation (4.87a).

The functions Φ_1, Φ_2, Φ_3, Φ_4, and Φ_5 have been calculated and are presented in Fig. 4.3 versus the normalized frequency $k_0 a$, where a is the radius of the disk or the width of the strip.

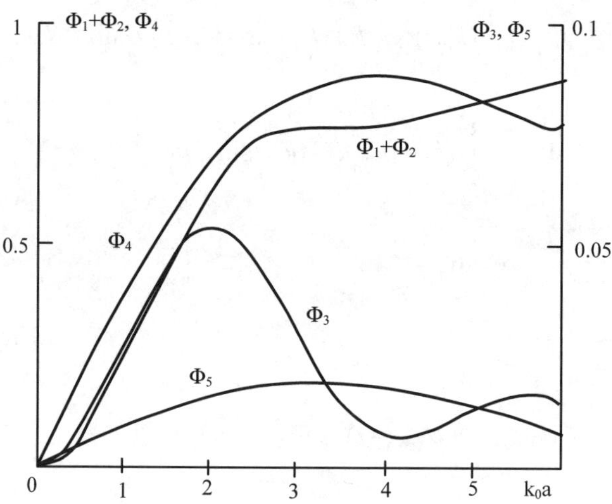

Fig. 4.3. Functions Φ_1, Φ_2, Φ_3, Φ_4, and Φ_5.

Considering Fig. 4.3, one can draw the following conclusions. The co-polarized scattering cross section tends to σ_{sSL} with increasing $k_0 a$. The cross-polarized term firstly increases but then goes to zero for large values of $k_0 a$. It is seen from Fig. 4.3 that disks are better depolarizers than strips. The obtained analytical expressions for cross sections are quite convenient for a model analysis and also allow one to reveal the relations between all approximations under study. When $k_0 a \gg 1$, they transform into expressions for the high-frequency approximation, and when $k_0 a \ll 1$, to the expressions for the low-frequency approximation.

D. Dielectric cylinders. Expressions for the extinction and scattering cross sections of circular dielectric cylinders are given in Ulaby *et al.* (1987), Van de Hulst (1957), and Chukhlantsev (1986). As a rule, these publications use the solution to the diffraction problem for an infinite cylinder (Wait, 1955, 1965). The expressions for the extinction and scattering cross sections have the form of series containing Bessel and Hankel functions. Particularly, these co- and cross-polarized cross sections for a long cylinder with a length l and a radius a are given by (Van de Hulst, 1957)

$$\sigma_e = \frac{4al}{k_0 a} \mathrm{Re}[T(0)] , \tag{4.103}$$

$$\sigma_s = \frac{2al}{\pi k_0 a} \int\limits_0^{2\pi} |T(\varphi)|^2 d\varphi \tag{4.104}$$

where $T(\varphi)$ is the scattering amplitude of an infinite length cylinder. When the vector $\vec{E}_0$ of incident field is in a plane passing through the cylinder axis and the wave vector makes an angle ϑ with the cylinder axis,

$$T_{co}(\varphi) = a_0 + 2\sum_{n=1}^{\infty} a_n \cos n\varphi , \tag{4.105}$$

$$T_{cr}(\varphi) = b_0 + 2\sum_{n=1}^{\infty} b_n \cos n\varphi , \tag{4.106}$$

$$a_n = \frac{J_n(v)}{H_n^{(2)}(v)} \times$$

$$\times \left\{ n^2 \cos^2 \vartheta \left(\frac{1}{v^2} - \frac{1}{u^2}\right)^2 - \left[\frac{H_n^{(2)'}(v)}{vH_n^{(2)}(v)} - \frac{J_n'(u)}{uJ_n(u)}\right]\left[\frac{J_n'(v)}{vJ_n(v)} - \frac{\varepsilon_v J_n'(u)}{uJ_n(u)}\right] \right\} \frac{1}{D}$$

(4.107)

$$b_n = -in \cos \vartheta \frac{J_n(v)}{H_n^{(2)}(v)} \left\{ \left(\frac{1}{v^2} - \frac{1}{u^2}\right)^2 \left[\frac{J_n'(v)}{vJ_n(v)} - \frac{H_n^{(2)'}(v)}{vH_n^{(2)}(v)}\right] \right\} \frac{1}{D},$$

(4.108)

$$D = n^2 \cos^2 \vartheta \left(\frac{1}{v^2} - \frac{1}{u^2}\right)^2 - \left[\frac{H_n^{(2)'}(v)}{vH_n^{(2)}(v)} - \frac{\varepsilon_v J_n'(u)}{uJ_n(u)}\right]\left[\frac{H_n^{(2)'}(v)}{vH_n^{(2)}(v)} - \frac{J_n'(u)}{uJ_n(u)}\right],$$

(4.109)

$$v = k_0 a \sin \vartheta, \quad u = k_0 a \sqrt{\varepsilon_v - \cos^2 \vartheta} .$$

(4.110)

When the vector $\vec{H}_0$ of the incident field is in a plane passing through the cylinder axis, coefficients a_n and b_n should be substituted by c_n and d_n, respectively:

$$c_n = \frac{J_n(v)}{H_n^{(2)}(v)} \times$$

$$\times \left\{ n^2 \cos^2 \vartheta \left(\frac{1}{v^2} - \frac{1}{u^2}\right)^2 - \left[\frac{H_n^{(2)'}(v)}{vH_n^{(2)}(v)} - \frac{\varepsilon_v J_n'(u)}{uJ_n(u)}\right]\left[\frac{J_n'(v)}{vJ_n(v)} - \frac{J_n'(u)}{\varepsilon_v uJ_n(u)}\right] \right\} \frac{1}{D_1}$$

(4.111)

$$d_n = b_n \frac{D}{D_1},$$

(4.112)

$$D_1 = n^2 \cos^2 \vartheta \left(\frac{1}{v^2} - \frac{1}{u^2}\right)^2 -$$

$$-\left[\frac{H_n^{(2)'}(v)}{vH_n^{(2)}(v)} - \frac{\varepsilon_v J_n'(u)}{uJ_n(u)}\right]\left[\frac{H_n^{(2)'}(v)}{vH_n^{(2)}(v)} - \frac{J_n'(u)}{\varepsilon_v uJ_n(u)}\right]$$

(4.113)

Calculations performed for these cross sections and the corresponding experimental data show that, for cylinders, the resonance attenuation and scattering effects may be considerable (Ulaby *et al.*, 1986; Chukhlantsev, 1986)

4.4.2. Experimental Research

Actually, there are rather few papers where experimental data on the microwave scattering and absorption by plant elements are presented. Passive (radiometric) and active (radar) methods were used to measure the extinction, scattering, and backscattering cross sections of leaves, stalks, and branches.

Measurements of the attenuation loss for horizontally polarized and vertically polarized waves transmitted through a fully grown corn canopy, and of the phase difference between the two transmitted waves, were conducted in Ulaby *et al.* (1986) at frequencies of 1.62, 4.75, and 10.2 GHz. The measurements were made at incidence angles of 20°, 40°, 60°, and 90° relative to the normal incidence. Experimental data were compared with computed ones which were based on a model of infinite cylinders (stalks) and randomly oriented disks (leaves). The proposed model was suitable for corn-like canopies.

Measurements of backscattering from leaves were conducted in Senior *et al.* (1987) and Karam *et al.* (1988). In Karam *et al.* (1988), the disk and finite cylinder models based on the generalized Rayleigh-Gans approximation were compared with measurements of backscattering cross sections for an aspen leaf and a birch stick. There was a good agreement between the theory and the measurement except for small incident angles with respect to the stick axis. It was explained by possible diffraction from the stick ends. The diffraction problem at the end of a circular cylinder was discussed in Chukhlantsev (1981). An additional extinction due to the diffraction at the end of a cylinder was calculated.

Propagation in simulated canopies composed of bare deciduous twigs and leafy coniferous branches was investigated at 9 GHz (Mougin *et al.*, 1990) and interpreted (Lopes and Mougin, 1990) by the Foldy-Lax theory for coherent wave propagation through a sparsely discrete random medium. The infinite cylinder scattering function was applied. A good agreement was obtained between the model and attenuation observations.

A radiometric method was proposed in Mätzler and Sume (1989) to measure the transmissivity and the reflectivity of leaves at frequencies 21, 35, and 94 GHz. Different kinds of crops and trees leaves were investigated. The model of an infinite flat dielectric slab (a very large plane particle) was in good agreement with the measurement and was used to examine the dielectric behavior of leaves in the frequency band reported.

The radiometric method was also applied to measure the extinction and scattering cross sections of leaves and branches at frequencies of 1, 1.67, and 13.3 GHz (Chukhlantsev, 1981, 1986). The measurement configuration is presented in Fig. 4.4.

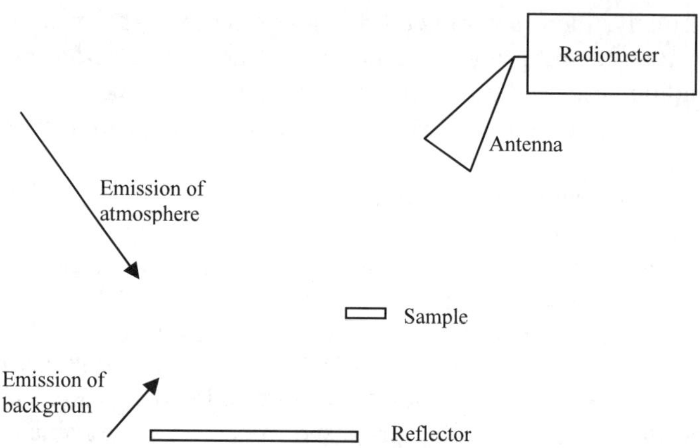

Fig. 4.4. Configuration of measurements of absorption and scattering by plant elements.

A reflector, which had dimensions much greater than the wavelength and served as a calibrator, was placed in the far-zone of a radiometer antenna to reflect microwave emission from the atmosphere. A sample was placed in the antenna beam but not above the reflector to avoid multiple reflections between the sample and the reflector. A metal plate (with reflectivity $R = 1$) and an absorbing material ($R = 0$) were used as the reflector. In the first case, the measured brightness temperature is given by

$$T_b^1 = \frac{S}{S_A}\left\{T_{bb}\left(\frac{S_A}{S}-1\right)+\frac{\sigma_a}{S\cos\vartheta}T_0+ \right.$$
$$\left. +\frac{\sigma_s}{S\cos\vartheta}T'_{bat}+\left(1-\frac{\sigma_a}{S\cos\vartheta}-\frac{\sigma_s}{S\cos\vartheta}\right)T_{bat}\right\}$$

(4.114)

where S_a is the area of the antenna footprint, S is the area of the reflector, T_{bb} is the brightness temperature of background, T_{bat} is the brightness temperature of the atmosphere, ϑ is the observation angle, T'_{bat} is the brightness temperature of the atmosphere averaged over the sample phase function, рассеяния исследуемого образца, T_0 is the temperature of the sample, and σ_a and σ_s are the absorption and scattering cross sections of the sample, respectively. In the second case ($R = 0$), the measured brightness temperature is given by

$$T_b^2 = \frac{S}{S_A}\left\{T_{bb}\left(\frac{S_A}{S}-1\right)+\frac{\sigma_a}{S\cos\vartheta}T_0 + \right.$$
$$\left. +\frac{\sigma_s}{S\cos\vartheta}T_{bat}' +\left(1-\frac{\sigma_a}{S\cos\vartheta}-\frac{\sigma_s}{S\cos\vartheta}\right)T_0\right\}$$

(4.115)

When the sample is absent, the measured brightness temperature is given by

$$T_b^1 = \frac{S}{S_a}\left[T_{bb}\left(\frac{S_a}{S}-1\right)+T_{bat}\right],$$

(4.116)

$$T_b^2 = \frac{S}{S_a}\left[T_{bb}\left(\frac{S_a}{S}-1\right)+T_0\right].$$

(4.117)

The absorption and scattering cross sections are found as

$$\sigma_a = \frac{\delta T_b^1}{\Delta T}S\cos\vartheta,$$

(4.118)

$$\sigma_s' = \frac{\delta T_b^2}{\Delta T}S\cos\vartheta$$

(4.119)

where $\delta T_b^{1,2}$ is the difference between $T_b^{1,2}$ in measurements with and with no samples, ΔT is the difference between T_b^2 and T_b^1 in measurements without samples. Limitations of the above measuring technique and the errors of measurements are considered in detail in Chukhlantsev (1986). Some results of measurements are presented in Fig. 4.5 – Fig. 4.7.

Dependencies of the normalized absorption cross section on the water surface density for disk-like leaves is presented in Fig.4.5 (a normalized cross section — or a relative cross section — is determined as the ratio of the cross section to the geometric cross section of a particle; a water surface density σ_w is the weight of water per the unit area of a leaf and represents the thickness of water layer in the leaf). These dependencies were obtained at the frequency 13.3 GHz (the centimeter wave band). Results of calculations for the models of very small particle and very large plane particle are also presented in Fig. 4.5. One can see that for very thin disks ($\sigma_w < 0.005$ g/cm^2) both models and experimental results coincide (this result follows from equations (4.84) and (4.87) at small d). For typical leaf values of $\sigma_w \sim 0.02\text{-}0.03$ g/cm^2, the normalized absorption cross section depends on the diameter of the leaves. When the diameter D is ten or more times greater than the wavelength, the absorption cross section tends to that of a very large particle. When D is less than the wavelength, the absorption cross section tends to that of a small particle. The scattering cross section of leaves is close to that of a very large particle for $D > \lambda$ ($k_0 a > 3$, see Fig. 4.4). In the decimeter wave band (1 and 1.67 GHz), the leaves are very thin compared to the wave length and, as the results of measurements showed, the model of plane thin particle works well in this frequency band. Results obtained for strip-like leaves were close to those for disk-like leaves.

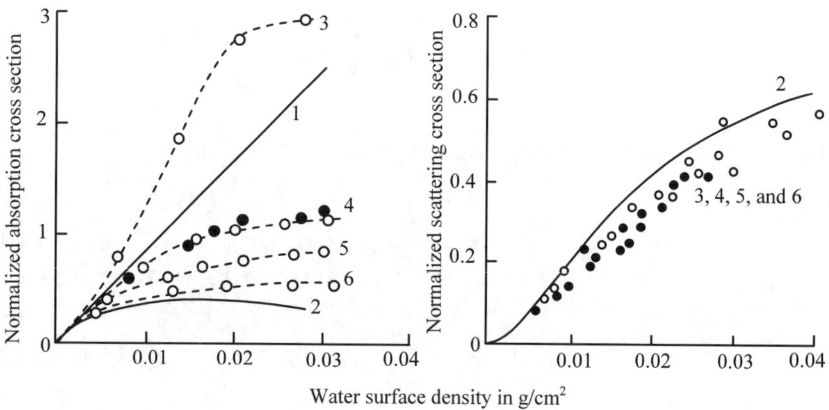

Water surface density in g/cm^2

Fig. 4.5. Absorption and scattering cross sections of disk-like leaves with a diameter of 3 cm (3), 7 cm (4), 11 cm (5), and 15 cm (6) versus their water surface density at 13.3 GHz. Results of calculation for small particle (1) and very large plane particle (2).

For plant elements in the form of a cylinder, dependencies of the normalized extinction cross section on the radius and the length of the elements were obtained (Fig. 4.6 and 4.7). These dependencies were compared with results of calculation by the model of a long dielectric cylinder. The model satisfactorily fitted experimental data for a cylinder with length more than three-five wavelengths.

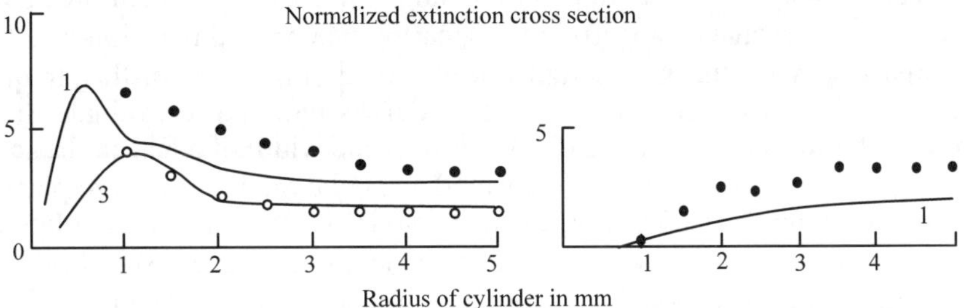

Fig. 4.6. Extinction cross section of sticks with a length of 15 cm versus their radius at 13.3 GHz. Vector E is in the plane passing through the axis of cylinders (on the left), and vector H is in the plane passing through the axis of cylinders (on the right). Angle between the wave vector and the cylinder axis is 90° (1) and 30° (2). Lines show results of calculation for a long cylinder and points represent experimental data.

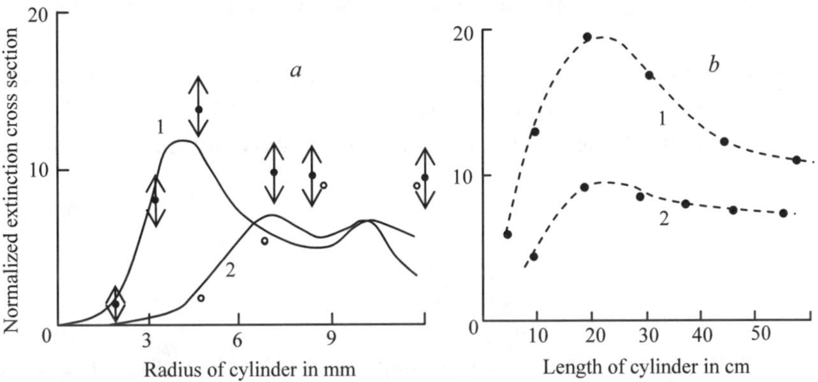

Fig. 4.7.a. Normalized extinction cross section of sticks with a length of 45 cm versus their radius at 1.67 GHz. Vector E is in the plane passing through the axis of cylinders. Angle between the wave vector and the cylinder axis is 90° (1) and 30° (2). Lines show results of calculation for a long cylinder and points represent experimental data.
Fig. 4.7.b. Normalized extinction cross section of sticks with a length of 45 cm versus their length at 1.67 GHz. Vector E is in the plane passing through the axis of cylinders. Angle between the wave vector and the cylinder axis is 90°. Radius of cylinder is 4.5-5 mm (1) and 3-3.5 mm (2).

An analysis of the results of the aforementioned studies enables one to make the following conclusions. The extinction and scattering of electromagnetic waves by leaves is adequately described by the thin plane particle model (the generalized Rayleigh-Gans approximation) at frequencies below the S-band. This approximation is successfully used in known models of backscattering from vegetation canopies, namely, the MIMICS and Santa

Barbara models (Ulaby *et al.*, 1990; Wang *et al.*, 1993). At centimeter wavelengths, the resonance absorption (appearing when the dimension of a leaf is comparable with the wavelength) is observed (Fig. 4.5). In this frequency range, a further development of the models is required to evaluate the absorption by leaves. For $a, D \geq (5...10)\lambda$ (a is the width of a leaf in the form of a strip, and D is the diameter of a round leaf), the model of a large plane particle can be applied. The long cylinder model is suitable for calculating σ_s and σ_e of stems and branches when their length is $l \geq (3...5)\lambda$. For $l \ll \lambda$, the small particles model can be used. When $l \sim \lambda$, resonance effects of extinction can be significant (Fig. 4.7b).

4.5. MICROWAVE PROPAGATION THROUGH A VEGETATION LAYER. RELATION OF ELECTRODYNAMIC PARAMETERS TO BIOMETRIC FEATURES OF VEGETATION

Propagation of the coherent field in a vegetation medium is described by the dispersion equation for the wave number (4.69) or (4.76). The extinction coefficient of the coherent field should be determined according to (4.71) as

$$\gamma = \sum_{i=1}^{n} \sigma_e^i . \tag{4.120}$$

An estimate of the extinction coefficient can be found for a given vegetation canopy by a determination of the number of plant elements with a given shape in a unit volume, by measuring the dimensions of each element, and by a determination of the orientation of each element with respect to the direction of propagation and polarization of the incident wave. Then, one needs to perform model calculations of the extinction cross sections for all plant elements in the unit volume and to sum these cross sections. It could be noted that the described procedure, to a greater or lesser extent, is used in some research of backscattering (Ulaby *et al.*, 1990) or emission (Ferrazzoli and Guerriero, 1996) from vegetation canopies. Plant elements in the unit volume are divided into groups of elements of close shape and dimensions. The extinction cross sections for each group are calculated for simple geometric forms (ellipsoids, cylinders, strips, and disks) (see previous Section) which are characterized by their dimensions and dielectric constant (see Chapter 2). It is clear, however, that this approach has a limited practical application because measurements of above-named vegetation parameters in every given case are hardly possible. Nevertheless, this approach is useful for an under-standing of the influence of different plant parameters on the electro-magnetic wave extinction and scattering in a canopy.

For practical needs, it is expedient to establish a relation of an average extinction coefficient to average statistical characteristics of a vegetation canopy, which are determined in the practice of biometrical measurements, e.g., vegetation biomass or water content per unit area, trunk volume per unit area, etc. A vegetation canopy should be considered at that as a statistical ensemble, which is characterized by distributions of its elements over dimensions and orientations. Besides, it is useful to consider a possibility to separate vegetation canopies into some types that have close frequency dependencies of microwave attenuation and scattering.

The derivation of expressions that relate the average coefficient of ex-tinction to the biometrical parameters is presented in detail in Chukhlantsev (1989a). For the simplest case of a canopy consisting of small chaotic ellip-soids, it was obtained (Kirdiashev *et al.*, 1979; Basharinov *et al.*, 1979) using (4.120), (4.84), and (4.85) that:

$$\gamma = \frac{u}{3} p k_0 \varepsilon_v'' = \frac{u}{3} k_0 \frac{w}{m_v \rho_w} \varepsilon_v'', \tag{4.121}$$

$$\tau = \gamma h / \cos \vartheta, \tag{4.122}$$

$$\tau = \frac{u}{3} p k_0 \varepsilon_v'' h / \cos \vartheta = \frac{u}{3} k_0 \frac{W}{m_v \rho_w} \varepsilon_v'' / \cos \vartheta \tag{4.123}$$

where τ is the optical thickness (depth) of the canopy for the coherent radia-tion, h is the height of the canopy, p is the vegetation volume density, w is the weight of water per unit volume of the canopy, W is the water content per unit area, m_v is the volumetric water content of plant elements, ρ_w is the density of water, k_0 is the free-space wave number, ε_v'' is the imaginary part of the dielectric permittivity of plant elements, ϑ is the observation angle, and $u = 1$ and $u = 2$ for small needles and small disks, respectively. One can see that for canopies with small, compared to wavelength, elements the opti-cal thickness is proportional to the water content per unit area. The propor-tionality coefficient is accepted to designate as b (Jackson and Schmugge, 1991) so the expression for the optical depth is reduced to the form:

$$\tau = bW / \cos \vartheta. \tag{4.124}$$

The above expression has very simple form and so is rather attractive for modeling of attenuation properties of vegetation. Therefore, it was proposed (Chukhlantsev, 1981; Chukhlantsev and Shutko, 1982; Shutko and Chukhlantsev, 1982; Mo *et al.*, 1982; Jackson and Schmugge, 1991) to use this model for all types of vegetation canopies but not obligatory for those with small elements. For the case of a canopy with arbitrary dimensions of elements, the model (4.124) is obtained as follows (Chukhlantsev, 1989a):

$$\tau = n_i\,\sigma_{ei}\,h\,/\cos\vartheta = n_i\,\frac{V_i m_{vi}\rho_w}{V_i m_{vi}\rho_w}\sigma_{ei}\,h\,/\cos\vartheta = \frac{w_i}{V_i m_{vi}\rho_w}\sigma_{ei}\,h\,/\cos\vartheta$$

$$= \frac{\sigma_{ei}}{V_i m_{vi}\rho_w}W_i\,/\cos\vartheta = W\frac{\sigma_{ei}}{V_i m_{vi}\rho_w}\frac{W_i}{W}\,/\cos\vartheta = bW\,/\cos\vartheta$$

(4.125)

$$b = \frac{\sigma_{ei}}{V_i m_{vi}\rho_w}\frac{W_i}{W} = \eta_i\frac{\sigma_{ei}}{V_i m_{vi}\rho_w}$$

(4.126)

where V is the volume of an element and η_i is the weight fraction of given shape elements in the canopy. The coefficient b can be calculated on the basis of equation (4.126), of the models for the extinction cross section of plant elements presented in the previous Section, and of statistical data on the distribution of plant elements over shapes, dimensions, and orientations. Some exemplary dependencies of the coefficient b are presented in Fig. 4.8 and Fig. 4.9 (Chukhlantsev, 1989a). Of course, due to the approximate character of the models for the extinction cross sections of plant elements, the calculated values of b represent only estimates of this coefficient. Therefore, it is more correct to specify the value of b in every given case from experimental data and most researchers prefer to do so. At the moment, a rather extensive experimental data base has been collected for the coefficient b for different crops at different frequencies (Van de Griend and Wigneron, 2004a).

The model (4.124) was extensively tested by numerous researchers both experimentally and theoretically (e.g., Le Vine and Karam, 1996). This model is widely used and, actually, is a basic model in the modeling of microwave emission from vegetation canopies (Pellarin *et al.*, 2003a, 2003b), in the interpretation of microwave radiometric data (e.g., Burke *et al.*, 2002b; Chanzy and Wigneron, 2000; Jackson *et al.*, 1992, 1995, 1997, 1999; Njoku *et al.*, 2003; Wang *et al.*, 1987), and the retrieval of vegetation and soil parameters from microwave radiometric data (Crow *et al.*, 2005; Njoku and Li, 1999; Wigneron *et al.*, 2003).

The single scattering albedo ω is found (4.72) as the ensemble averaged ratio of the scattering cross section to the extinction cross section. For small chaotic disks, one gets from (4.84)

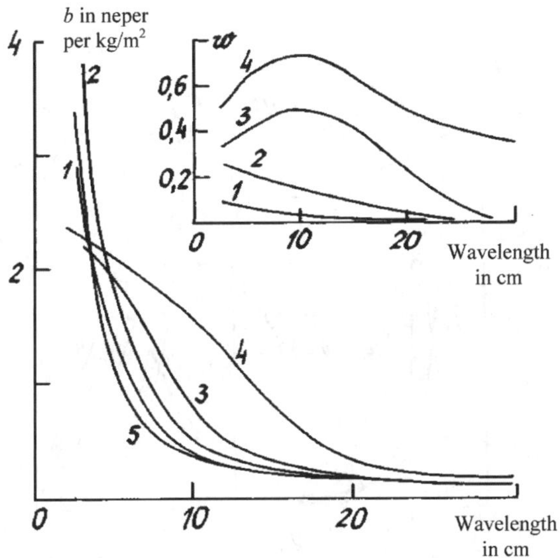

Fig. 4.8. Dependence of the coefficient b and the single scattering albedo ω on the wavelength for leaves of small grains (1), alfalfa (2), soybean and cotton (3), corn (4), and the model of small particles (5).

$$\omega = \frac{\dfrac{1}{12}(k_0 d)^2 |\varepsilon_v - 1|^2 (k_0 a)^2}{k_0 d \varepsilon_v'' + \dfrac{1}{12}(k_0 d)^2 |\varepsilon_v - 1|^2 (k_0 a)^2}. \tag{4.127}$$

To find the expressions of ω for vegetation with plane thin leaves, one has to average the extinction cross section (which is given for the normal incidence by (4.94)) over the leaf angle distribution. Corresponding numerical calculations have been performed in Ferrazzoli and Guerriero (1994) for some model distribution. For an estimate of albedo, one can use the approach $\omega = \omega_{ni}$ where ω_{ni} is the albedo of leaf in the case of normal incidence (Chukhlantsev, 1992). For leaves of disk shape, one obtains from (4.87a) and (4.94)

$$\omega = \frac{\frac{1}{12}(k_0 d)^2 |\varepsilon_v - 1|^2 (k_0 a)^2 (\Phi_1 + \Phi_2)}{k_0 d \varepsilon_v'' + \frac{1}{12}(k_0 d)^2 |\varepsilon_v - 1|^2 (k_0 a)^2 (\Phi_1 + \Phi_2)}. \tag{4.128}$$

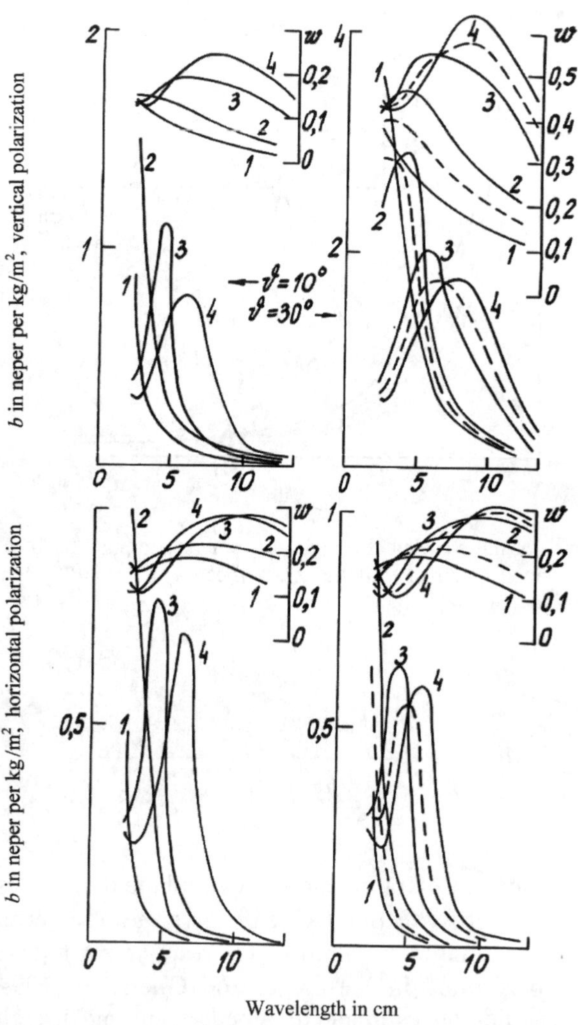

Fig. 4.9. Dependence of the coefficient b and the single scattering albedo ω on the wavelength for stems of small grains with a radius of 1 mm (1) and 2 mm (2), and stems of soybean with a radius of 2.5 mm (3) and 3 mm (4). Dashed lines represent dependencies averaged over the radius of stems.

It is important to emphasize that in the independent particles approach the single scattering albedo does not depend on the vegetation volume density but is a function of leaf shape, size, and orientation (type of vegetation) and water content of leaves (state of vegetation). Single scattering albedos were calculated in (Chukhlantsev, 1989a) for a number of crops under different leaf moisture conditions. Some results are presented in Fig. 4.8 and Fig. 4.9.

The phase function of vegetation unit volume is found by conventional manner by averaging the amplitude function over all leaf orientations. In simple models, isotropic scattering is assumed. More complicated models were considered in detail in Ferrazzoli and Guerriero (1994, 1996).

Results of calculations presented in Fig. 4.8 and Fig. 4.9 are based on rather exhausting computations. In some special cases, several simplified forms of equation (4.124) are used. In the decimeter wave band (1-3 GHz), the optical depth of vegetation layer is often estimated by equation (4.123) in the form (Kirdiashev *et al.*, 1979; Kerr and Wigneron, 1994):

$$\tau = A f \varepsilon_v'' W \qquad (4.129)$$

where A is the form factor that takes into account the structure of vegetation canopy, f is the frequency. Wegmüller *et al.* (1994) proposed to use equation (4.129) in the centimeter wave band too with a correction accounting for the attenuation of electromagnetic wave inside a leaf, i.e., actually, with the use of equation (4.93) for the field inside the leaf. A simple estimating formula for leafy vegetation can be then obtained:

$$\tau = A f \varepsilon_v'' W \frac{1}{1 + \frac{1}{4}(k_0 d)^2 |\varepsilon_v - 1|^2 + k_0 d \varepsilon_v''}. \qquad (4.130)$$

This expression reflects the discrepancy between the attenuation by crops with small leaves (curves 1 and 2 in Fig. 4.8) and the attenuation by crops with large leaves (curves 3 and 4 in Fig. 4.8).

The form factor A depends on the type of vegetation canopy and on the polarization of the microwave radiation. For vegetation with a quasi-chaotic orientation of plant elements (grass, small grains at early stages of growth, alfalfa), an estimate of the optical thickness in the decimeter wave band can be obtained by (Chukhlantsev, 1981)

$$\tau = bW / \cos\vartheta = \frac{1}{3} k_0 \varepsilon_s'' [2\eta_1 + (1 - \eta_1)] \frac{W}{m_v \rho_w} / \cos\vartheta \qquad (4.131)$$

where η_1 is the weight fraction of leaves in the form of disks ($\eta_1 = 0$ for small grains and $\eta_1 = 0.5$ for alfalfa). For vegetation canopies having a pronounced orientation of elements, the difference of attenuation at different polarizations is observed (Chukhlantsev, 1981). An estimate of the optical thickness can be obtained by

$$\tau_{v,h} = b_{v,h} W$$

$$b_v = k_0 \varepsilon_s'' [\frac{u}{3}\eta_1 + (1-\eta_1)\sin^2 \vartheta] \frac{1}{m_v \rho_w} / \cos \vartheta \qquad (4.132)$$

$$b_h = \frac{u}{3m_v \rho_w} k_0 \varepsilon_s'' \eta_1 / \cos \vartheta$$

where the weight fraction of leaves $\eta_1 \sim 0.5$ for alfalfa and $\eta_1 \sim 0.2\text{-}0.3$ for small grains; $u = 1$ and $u = 2$ for small grains and alfalfa, respectively.

In the general case, the extinction coefficient of a vegetation canopy for a part of a microwave band is expedient to determine in the form that follows from (4.121) (Chukhlantsev *et al.*, 2003c):

$$\gamma = Awf^\alpha \qquad (4.133)$$

where A is the coefficient depending on the type of vegetation, w is the water content per unit volume of the vegetation canopy, f is the frequency, and α is a coefficient, which determines the frequency dependence of attenuation. Coefficients A and α are found from the results of theoretical simulations and specified on the basis of experimental data.

Forest canopies are characterized as the most complex structure. Ferrazzoli and Guerriero (1996) computed the microwave attenuation by different components (leaves, branches, trunks) of deciduous and coniferous forests and by the crowns in total. The discrete model of vegetation was used. The contribution of trunks to the total attenuation was found to be negligible. At the same time, some researchers reported results of calculations that show a rather significant contribution of trunks to the total attenuation by forests. It follows from Fig. 4.7,a that the resonance extinction for thick trunks could be expected at low frequencies corresponding to the very high frequency (VHF) band. It is confirmed by results of calculations reported by Lang (2004). In the microwave band, one should expect that the extinction cross section of a trunk tends to its optical limit $\sigma_e = 2\sigma$ where σ is the geometric cross section of the trunk. Kruopis *et al.* (1999) studied experimentally the transmissivity of forests in the 6.8-18.7 GHz frequency range as a function of forest-stem volume. Measurements were conducted at both vertical and horizontal polarizations at 50° incidence angle. The transmissivity t can be related to the optical thickness as

$$t \approx \exp\{-\tau\}. \tag{4.134}$$

It was found that the transmissivity at the vertical polarization can be modeled as a decreasing function of frequency f [GHz] and forest-stem volume V_f [m^3/ha], which exponentially approaches some saturation limit:

$$t = t_{high} + (1 - t_{high})\exp\{0.035V_f\}, \tag{4.135}$$

$$t_{high} = 0.42 - [1 - 0.42]\exp\{0.028f\} \tag{4.136}$$

where t_{high} = frequency dependent transmissivity of a very dense forest, i.e., the saturation value of forest transmissivity at a certain frequency. According to the model, the forest transmissivity decreases with the frequency. The transmissivity reaches 99% of its saturation level at 132 m^3/ha stem volume. The forest transmissivity to horizontally polarized waves can be modeled by the expression of the same form as (4.135) but was found to be significantly lower than to vertically polarized ones, which was explained by the dominant horizontal orientation of branches and coniferous needles. The saturation character of the transmissivity versus stem volume dependence indicates that the contribution of stems to the total attenuation seems to be really small and the attenuation is mainly determined by the mass of leaves and branches. This mass correlates to the forest-stem volume and, above a certain value of stem volume in a forest, does not increase that causes the saturation of the transmissivity – stem volume dependence. Actually, it was noted in Kruopis *et al.* (1999) that at higher stem volumes forest transmissivity seems to increase slightly, rather than saturate, which was explained by sparser crowns of bigger and older pines. Since the empirical model (4.135) is based on a statistical link between the forest-stem volume and the mass (the water content) of leaves and branches, the coefficients of the model can be dependent on the type of forest and, moreover, on the region. Indeed, Pardé *et al.* (2005) indicated that for other forests in another region the model (4.135) can be applied but with the other coefficients, namely,

$$t = 0.39 + (1 - 0.39)\exp\{0.038V_f\} \quad \text{for 19 GHz}, \tag{4.137}$$

$$t = 0.26 + (1 - 0.26)\exp\{0.054V_f\} \quad \text{for 37 Ghz.} \tag{4.138}$$

Because a vegetation canopy is an inhomogeneous medium, its coherent attenuation should be treated as a random value. Calculating extinction in a canopy with given plant element dimensions, angular distributions, and number densities, one can obtain an estimate of the mean value of extinction. This estimate is important in the case of a low spatial resolution of the remote sensing antennas, which perform the spatial averaging of the object characteristics. Nevertheless, the sampling values of attenuation can change in rather wide limits for the canopy. For example, at a high spatial resolution, variations of attenuation for a corn canopy can reach 50% and more within the test site Ulaby *et al.* (1987). Spatial variations of attenuation for a forest can reach 20% and more (Pardé *et al.*, 2005). It shows that it is interesting to know the relative level of attenuation at different frequencies rather than absolute values of attenuation. The ratio of attenuation levels at different frequencies is used for the inversion of radiation models and for the retrieval of object characteristics in multi-frequency remote sensing. Therefore, the determination of the coefficients α and A in (4.131) is of a great importance. This could be done on the basis of experimental research of microwave attenuation by vegetation canopies. This problem is considered in the next Chapter.

Chapter 5

EXPERIMENTAL STUDIES OF MICROWAVE PROPAGATION IN VEGETATION CANOPIES

5.1. METHODS OF EXPERIMENTAL RESEARCH

Measurements of attenuation by vegetation were conducted under laboratory conditions, under ground-based field conditions, and from aircraft. Laboratory experiments have the advantage that parameters of investigated objects can be changed and controlled. They allow examining the influence of individual parameters on the microwave propagation characteristics of the medium.

Ground-based field measurements produced the largest amount of information on propagation properties of vegetation canopies. The natural conditions make it possible to obtain data during the whole period of canopy growth and to compare them with ground truth data.

Airborne sensors make it possible to gather quickly a large data set for different types of vegetation under various states and conditions.

Methods which were used to measure the attenuation can be divided into active and passive ones. In active methods, the attenuation over the path between transmitting and receiving antennas is measured. Airborne radar studies used corner reflectors and active calibrators placed on the ground at different locations within a canopy. Several typical configurations of active measurements are presented in Fig. 5.1.

Configuration Fig. 5.1A, was used in Allen and Ulaby (1989), Ulaby and Jedlicka (1984), Ulaby and Wilson (1985), and Ulaby *et al.* (1987) to measure attenuation in different crops such as wheat, corn, soybeans, etc. Ulaby *et al.* (1987) conducted several types of experiments to measure the transmission loss for a mature corn canopy at incidence angles of 20°, 40°, 60°, and 90°. The transmitters for the 20°, 40°, and 60° measurements were placed on a truck-mounted platform at a height of 11.5 m above the ground surface, and the receivers were placed underneath the canopy. For the 90°

119

measurements, the transmitter platform was placed on the truck bed and the receivers were placed on a wooden platform whose height above the ground was the same as that of the transmitters. The transmitters used three dual-polarized antennas with center frequency of 1.62, 4.74, and 10.2 GHz. To investigate the spatial variations of the canopy attenuation, the receiving platform was placed on a rail system on which it slid in synchronism with the motion of the truck through a pulley system. The canopy loss L is defined in dB as

$$L = 10\log(P_0 / P_r) \tag{5.1}$$

where P_r is the power received when the canopy is present and P_0 is the free-space level received under identical conditions but without an intervening canopy between the transmitter and the receiver. Besides, the phase measurements were conducted in Ulaby *et al.* (1987), which referred to the phase difference between the vertically polarized (V) and horizontally polarized (H) waves transmitted through the canopy.

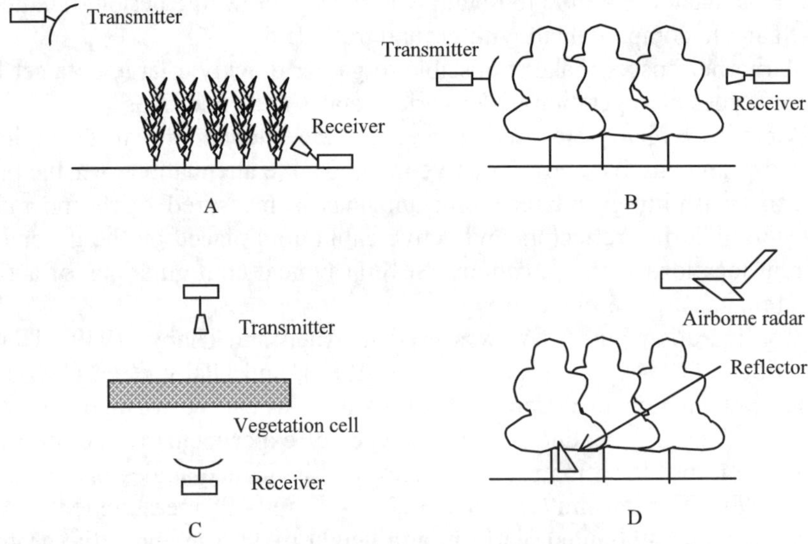

Fig. 5.1. Typical configurations of active measurements of attenuation.

A similar configuration of attenuation measurements was used by Ulaby and Wilson (1985). The attenuation experiments were conducted for canopies of winter wheat and soybeans in the late spring and summer period. Attenuation data were acquired at 1.55, 4.75, and 10.2 GHz for horizontal and vertical polarizations at incidence angles near 20° and 50°. In addition, wheat decapitation and soybean defoliation experiments were conducted to evaluate the relative importance of different canopy constituents. Ulaby and Jedlicka (1984) reported some attenuation data for corn and soybean canopies at 10.2 GHz that were obtained with the use of the same configuration.

Allen and Ulaby (1989) used the transmitter section of a 10.2 GHz truck-mounted radar as the source in a canopy transmission experiment to measure the attenuation properties of wheat stalks and heads. The transmitting antenna was at approximately 10 m above the soil. The receiver was a small X-band horn antenna, which was placed in the field within the main beam of the transmitting antenna. This antenna was mounted on a structure that enabled the operator to continuously vary its height above the soil from 132 cm down to 23 cm. The system was calibrated without canopy between the antennas.

Paris (1986) reported results of probing thick vegetation canopies with a narrow-beam field microwave scatterometer. The one-way canopy transmittance was estimated from a set of calibrated backscattering coefficient data over a span of slant ranges. O'Neill *et al.* (2003) used the multi-frequency (L, C, and X band) quad-polarized radar system mounted on a hydraulic boom truck with a 20 m boom and a dual-polarized L band radiometer mounted on a portable 18 m tower, to acquire a simultaneous active/passive data over a dense crop like a corn canopy. The vegetation scattering model was used to produce estimates of attenuation and canopy transmissivity data.

Configuration Fig. 5.1,B with different heights of antennas above the ground is usually used for investigation of electromagnetic waves propagation in forest environment (e.g., Li *et al.*, 1999; Herbstreit and Crichlow, 1964; Whale, 1968; Tewary *et al.*, 1990; Tamir, 1967; Nashashibi *et al.*, 2002; Savage *et al.*, 2003; Yakubov *et al.*, 2002; Grankov *et al.*, 2005). In work by Yakubov *et al.* (2002), transmitting antennas were mounted on a tower at a height of 21 m above the ground. A uniform site of 30-year larch forest with a height of trees of 16 m and a crown thickness of 5 m was used for measurements of attenuation. The mean diameter of trees was 13 cm, and the number of trees per unit area was 0.5 m^{-2}. Receiving antennas were placed at a height of 1.6 m above the ground beneath the forest crown. Measurements were conducted at frequencies 0.2 and 1.275 GHz. Special vector receiving antennas allowed measuring attenuation values at both vertical and horizontal polarizations and, besides, measuring intensities of cross polarized waves.

Savage *et al.* (2003) reported results of an extensive wide band channel sounding measurements campaign to investigate signal propagation through vegetation. The measurements were conducted at three frequencies, i.e., 1.3, 2, and 11.6 GHz, at sites with different measurement geometries and tree species. At each site, the transmitter was located in a clearing in front of the vegetation. The vegetation depth was measured from the front edge of the vegetation to the receiver antenna position.

A field measuring complex was used in Grankov *et al.* (2005) to measure attenuation spectra of trees. Measurements were conducted in the frequency range 0.47-2.1 GHz. Attenuation on the path between transmitting and receiving antennas was measured. A wide band noise generator was used as a transmitter and tunable radiometers were used as receivers. To obtain reliable data and to avoid the influence of the antenna's diagram effects on the measurements, the following procedure of measurements was applied. Antennas were placed at a height of 3 meters above the ground and at a distance of 6…10 meters from each other. First, nothing was positioned between the antennas. That provided the free space propagation reference. Secondly, trees were located between antennas and the attenuation at different frequencies was recorded. The main goal of measurements was obtaining the slope of frequency dependence of attenuation (that is a relative level of attenuation at different frequencies) and studying the spread of experimental data due to the different position of trees on the propagation path. Frequency dependencies of attenuation were obtained for single trees of spruce, apple, and chestnut at vertical and horizontal polarizations.

Configuration Fig. 5.1C, was used in laboratory measurements of attenuation by forest litter (De Roo *et al.*, 1991) and forest fragments (Mougin *et al.*, 1990). A vector network analyzer measured the power and phase of the signal passing through the cell (De Roo *et al.*, 1991). The attenuation constant of a layer was determined by comparing measurements of litter with that of the empty cell.

In work by Mougin *et al.* (1990), measurements were performed at 9 GHz. Vegetation samples were placed on a Styrofoam frame, in one row or several rows with different spacing. The frame was located on the path between transmitting and receiving antennas, and the P_r value was measured. To provide a quasi-constant power density of transmitted radiation along the frame, a lens was positioned in front of the transmitting antenna in such a way that its focus was at the phase center of the antenna. After removing the samples the reference power was then recorded to find the attenuation by equation (5.1). The measurements were performed on freshly cut vegetative samples with a typical volumetric water content of 0.5. In the first step, measurements were carried out only on bare twigs and needles. During the second step, measurements were done on different leafy coniferous branches.

To measure attenuation by vegetative samples in a wide frequency range, a super wideband waveguide system was used (Chukhlantsev *et al.*, 2003b, 2004a). The system consists of the wide band rectangular waveguide, two horn antennas matched with the waveguide, and the Vector Network Analyzer. The antennas serve as filters of spatial harmonics providing the single-mode propagation regime in the waveguide and a correct interpretation of attenuation measurements. A block diagram of the transition system is presented in Fig. 5.2.

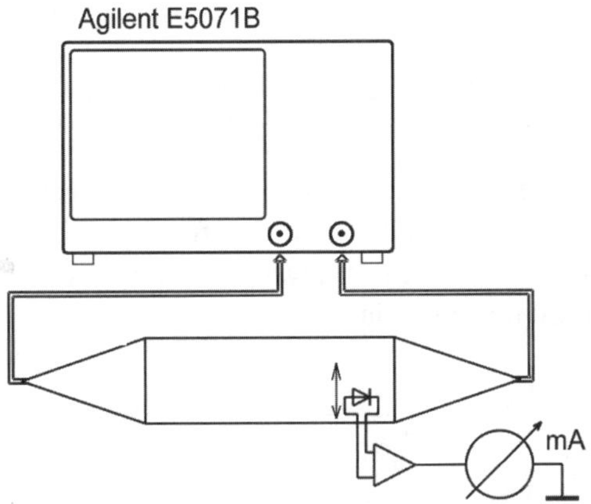

Fig. 5.2. Block diagram of the super wideband waveguide transition system.

The horn antenna transforms the TEM-wave of the coaxial cable into an H_{10}-wave of a rectangular waveguide. Since the divergence angle of the antenna is not large, an excitation level of high-order waves is small. When an H_{10}-wave passes through the camera it is attenuated by the investigated object. The second horn antenna, in turn, transforms the approaching wave into a TEM-wave of the coaxial cable. Therefore, the antennas serve as filters of spatial harmonics providing the single-mode propagation regime in the waveguide and a correct interpretation of attenuation measurements. Direct measurements of the electric field distribution in the waveguide confirm a realization of the single-mode propagation regime. When an H_{10}-wave of a rectangular waveguide propagates in the waveguide, the transition coefficient T of a layer with a thickness d is given by

$$T = \left| \frac{(1 - R^2)e^{j\gamma d}}{1 - R^2 e^{2j\gamma d}} \right|^2 \tag{5.2}$$

where R is the field reflection coefficient and γ is the propagation constant of the dielectric-filled waveguide. Because vegetation samples are very sparse, the effective dielectric constant of vegetation only slightly differs from unity, and the reflection coefficient is small. Equation (5.2) reduces in this case to

$$T \approx \left| e^{j\gamma d} \right|^2 . \tag{5.3}$$

The propagation constant of an H_{10}-wave is expressed by

$$\gamma = \sqrt{k_0^2 \varepsilon - \left(\frac{\pi}{a} \right)^2} = k_0 \sqrt{\varepsilon} \sqrt{1 - \frac{1}{\varepsilon}\left(\frac{\lambda}{2a} \right)^2} = \gamma_0 \sqrt{1 - \frac{1}{\varepsilon}\left(\frac{\lambda}{2a} \right)^2} \tag{5.4}$$

where γ_0 is the propagation constant in the free space, a is the waveguide width. Comparing the transition in the waveguide (5.2) with the transition in the free space given by the formula

$$T_0 \approx \left| e^{j\gamma_0 d} \right|^2 , \tag{5.5}$$

one can find that, due to the small values of the effective dielectric constant of vegetation close to unity, the measured attenuation values can be recalculated to the attenuation in the free space by the equation

$$T_0(dB) \cong \frac{T(dB)}{\sqrt{1 - \left(\dfrac{\lambda}{2a} \right)^2}} . \tag{5.6}$$

The transition coefficient T is found by subtraction of the transition coefficient of the empty waveguide from the transition coefficient of the vegetation-filled waveguide when the data are presented in dB. Attenuation is found to be equal to $1/T$.

The following measuring procedure was applied. A big branch of a tree (which could be a young tree) was cut. The branch was cut into smaller parts to put into the waveguide and to obtain the attenuation spectrum of the branch as a whole. Attenuation measurements were accompanied by measuring the weight and gravimetric moisture of samples. Finally, attenuation values were recalculated to find spectral dependencies of specific attenuation (the coefficient b (4.124)) for each branch, i.e., attenuation per kg/m^2 of the water content of the branch.

Configuration Fig. 5.1D, is used in airborne and space borne radar studies of attenuation and backscattering (Hoekman, 1987; Pitts *et al.*, 1988; Murata *et al.*, 1987; Shinohara *et al.*, 1992; Hallikainen *et al.*, 1990; Ulaby *et al.*, 1990; Durden *et al.*, 1991; Varekamp and Hoekman, 2002; Hoekman and Quinones, 2002; Koskinen *et al.*, 1999; Pulliainen *et al.*, 1999; Wagner *et al.*, 1999; Moghaddam and Saatchi, 1999; Magagi *et al.*, 2002). Attenuation is determined by comparing signals from reflectors placed beneath the canopy and placed on the open ground.

Passive methods use the measurements of thermal emission from the investigated object in the presence of atmosphere and space emission as the background. Typical configurations used in passive (radiometric) measurements of attenuation are shown in Fig. 5.3.

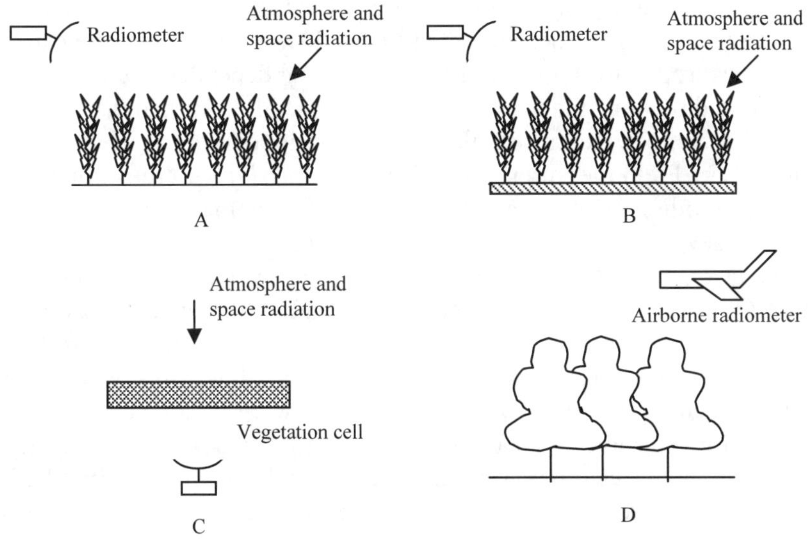

Fig. 5.3. Typical configurations of passive measurements of attenuation.

Configuration Fig. 5.4 A, was used by numerous researchers to retrieve the optical depth data from radiometric measurements. Mo *et al.* (1982), then, Jackson and Schmugge (1991), and, finally, Van de Griend and

Wigneron (2004a) gave an excellent overview of these investigations. Brunfeldt and Ulaby (1984, 1986) used this configuration for the field measurements of attenuation by corn and soybeans. The zero-order solution of the transfer equation for the emission from vegetated soil was used to expound measuring data:

$$T_b = \left(1 - r - t^2 R\right)T_0 + t^2 R_s T_{sky} \tag{5.7}$$

where T_0 is the vegetation and soil physical temperature, R_s is the reflectivity of the soil surface ($R_s = 1 - e_s$), and T_{sky} is the down welling sky (atmosphere and space) brightness temperature. The simplified equation (5.7) was used to determine the reflectivity r and the transmissivity t by measuring the brightness temperature T_b for a) bare soil, b) canopy over soil, c) bare absorbing material with an emissivity close to unity, and d) canopy over the absorbing material. To increase the accuracy of measurements, the soil was covered by screens which were wire meshes with an inter wire spacing of 1.6 mm and wire thickness of 0.3 mm. The brightness temperature was measured by a dual-frequency (2.7 and 5.1 GHz), dual-polarization radiometer mounted atop a truck-mounted boom. It was shown that the canopy is highly anisotropic, the emission exhibits a strong dependence on polarization and aiming direction, and the reflectivity is typically less than 0.1.

The technique described above ensures good accuracy but it is very cumbersome. Field measurements of the transmissivity were conducted with microwave radiometers mounted on a truck (Pampaloni and Paloscia, 1986; Chukhlantsev *et al.*, 1989; Golovachev *et al.*, 1989) and on a car (Vorobeichik *et al.*, 1988). Equation (5.7) was used by Pampaloni and Paloscia (1986) to retrieve the optical depth of corn and alfalfa canopies from measurements of the canopy brightness temperature in X (3 cm) and Ka (0.8 cm) bands. Radiometers were installed on a boom. In Chukhlantsev *et al.* (1989) and Golovachev *et al.* (1989), measurements were conducted in L (15-30 cm) band for vegetable crops and wheat. In Vorobeichik *et al.* (1988), attenuation properties of rice crops were investigated in L (18 cm) and Ku (2.25 cm) bands. The transmissivity was estimated by the relation that one can easy obtain from (5.7):

$$t = \left(\frac{1 - r - e_v}{1 - r - e_s}\right)^{\frac{1}{2}} \tag{5.8}$$

where e_v and e_s are the emissivity of vegetated and bare soil, respectively, with the same moisture content (and the reflectivity of the soil surface) within a site.

The technique proposed in Brunfeldt and Ulaby (1984, 1986) for field measurements can be easily performed under laboratory conditions (Chukhlantsev, 1981, 2002; Chukhlantsev and Golovachev, 1989; Kondratyev *et al.*, 1992) (configuration Fig .5.3,B)). The fresh-cut plants were placed on the reflector located in the far zone of the antenna of the radiometer. As the reflector, a "blackbody" and a metal plate were used. The transmissivity of the vegetation layer was estimated from the expression

$$t = \left(\Delta T_{bv} / \Delta T_b \right)^{1/2} \tag{5.9}$$

where ΔT_b is the brightness temperature contrast between the bare blackbody and the bare metal plate and ΔT_{bv} is the brightness temperature contrast between the metal plate covered by vegetation and the bare metal plate. The reflectivity of the vegetation layer was found from the expression:

$$r = \left(\Delta T'_{bv} / \Delta T_b \right)^{1/2} \tag{5.10}$$

where $\Delta T'_{bv}$ is the brightness temperature contrast between the black body and the black body covered by vegetation. The attenuation by different crops was studied in Chukhlantsev and Golovachev (1989), the transmissivity and the reflectivity of tree branches were investigated in Chukhlantsev (2002) and Kondratyev *et al.* (1992).

Configuration Fig. 5.4, C was used in Chukhlantsev and Golovachev (1989) to measure the transmissivity of a vegetation layer. To avoid the effect of side lobes of the receiving antenna, a configuration was used that is presented in Fig. 5.4. The antenna was placed in the lower base of a truncated pyramid made of an absorbing material. The upper base of the pyramid oriented to zenith was closed by a metal screen with a round aperture. A vegetative layer was placed over the aperture. The dimensions of the aperture were much greater than the wavelength but less than the footprint of the antenna. The transmissivity of the layer was found by

$$t = \left(T_{b2} - T_{b3} \right) / \left(T_{b2} - T_{b1} \right) \tag{5.11}$$

where T_{b1}, T_{b2}, and T_{b3} are measured brightness temperatures with open aperture, closed aperture, and aperture with a vegetative sample, respectively.

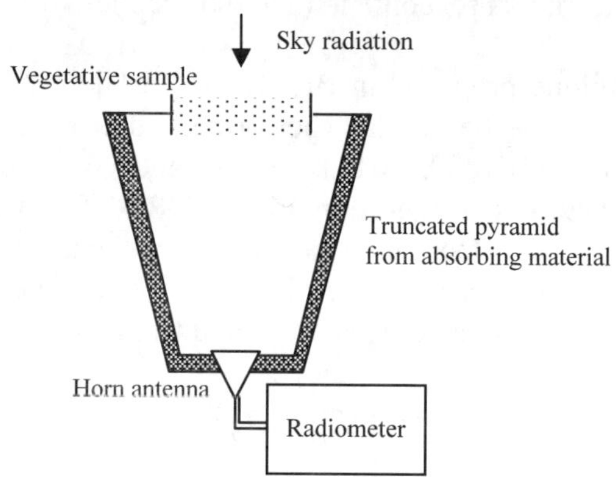

Fig. 5.4. Configuration for measurements of the transmissivity of vegetation layer.

In Vichev *et al.* (1995) and Milshin and Grankov (2000), configuration Fig. 5.3C, was used for field measurements of the transmissivity of trees. Antennas of radiometers were directed to the sky at a certain angle with respect to zenith, so that crowns of trees were on the path of the antenna beam. Attenuation values were estimated by comparing measurements of the brightness temperature with the presence of crowns and without crowns on the path.

Configuration Fig. 5.4D, was used for measurements of the forest transmissivity with the help of airborne radiometers (Kirdiashev *et al.*, 1979; Milshin *et al.*, 1999). Kruopis *et al.* (1999) estimated the transmissivity by the equation:

$$t = \sqrt{\frac{T_{bf} - T}{T_{bc} - T}} \qquad (5.12)$$

where T_{bf} and T_{bc} are brightness temperatures of forest and clear cut respectively and T is the arithmetic mean from air and snow physical temperature.

5.2. EXPERIMENTAL RESULTS

5.2.1. Dependence of the Optical Depth of a Vegetation Layer on the Vegetation Water Content

The linear relationship (4.124) between the optical depth τ and the canopy water content W was first experimentally revealed in (Kirdiashev *et al.*, 1979; Chukhlantsev, 1981) and then confirmed and reported by many researchers. Laboratory measurements conducted by the author (Chukhlantsev, 1981) for different crops under close control of biometric parameters distinctly displayed this linear dependence. Some examples are presented in Fig. 5.5. Jackson *et al.* (1982) and Mo *et al.* (1982) also found that the canopy optical thickness is directly proportional to the total amount of plant water per unit area. They presented some experimental data for soybeans, corn, and grass. Because of extreme importance of the discussed relationship in microwave radiometry of a soil-vegetation system, several special experimental studies relating to this problem were conducted. A comprehensive overview of these investigations was given by Jackson and Schmugge (1991) and Schmugge and Jackson (1992). All these investigations endorsed in general that the linear dependence of τ on W is quite appropriate in the decimeter wavelength frequency band. Pampaloni and Paloscia (1986) reported experimental data of τ versus W at 3.1 and 0.8 cm for corn and alfalfa. They found that for these high frequencies a slight deflection from the linear dependence of τ on W is observed. This fact made them approximate the τ versus W dependence by a logarithmic function as follows:

$$\tau = C \ln(1 + W) \tag{5.13}$$

where W is in kilograms per square meter and C is a dimensionless parameter that depends on wavelength and crop type. The authors found difficulty in the explanation of non-linear behavior of the optical depth. At the same time, expanding equation (5.13) into a power series, one obtains the following:

$$\tau = CW - \frac{1}{2}CW^2 + ... = CW(1 - \frac{1}{2}W + ...) = \tau_0(1 - \frac{1}{2C}\tau_0 + ...) \tag{5.14}$$

where τ_0 is the optical depth at low values of W. Equation (5.14) results directly from equation (4.74), which produces the following expression for the optical depth:

$$\tau = \tau_0\left(1 - c\gamma_0 / k_0\right).\tag{5.15}$$

From (5.14) and (5.15) a relation between coefficients C and c can be easily found:

$$c = \frac{k_0 h}{2C} = \frac{\pi h}{\lambda C}\tag{5.16}$$

where h is the height of vegetation. Comparing attenuation data for 3.1 and 0.8 cm, Pampaloni and Paloscia (1986) found that coefficient C can be presented in the form

$$C = k / \sqrt{\lambda}\tag{5.17}$$

where k in $m^{1/2}$, independent of wavelength, is 0.16 for corn and 0.25 for alfalfa.

The aforementioned results obtained by Pampaloni and Paloscia stimulated researchers to more detailed studies of the τ versus W dependence.

Chukhlantsev and Golovachev (1989) conducted radiometric measurements of the optical depth and the extinction coefficient with the use of configurations presented in Fig. 5.3 B, and Fig. 5.4. Some results for the optical depth are presented in Fig. 5.6. Similar dependencies were obtained for wheat, rye, and alfalfa. Variations of the water content were modeled by the change both of vegetation height and vegetation volume density p.

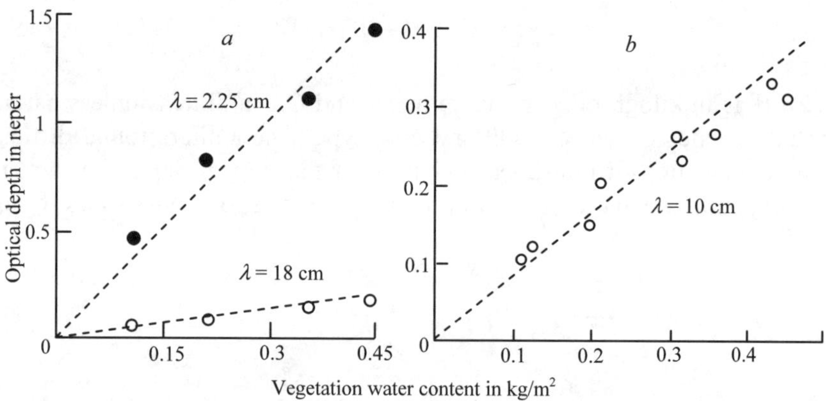

Fig. 5.5. Optical depth versus vegetation water content for barley canopy at bushing out stage of growth (a) and for alfalfa canopy with 30-40 cm stems (b).

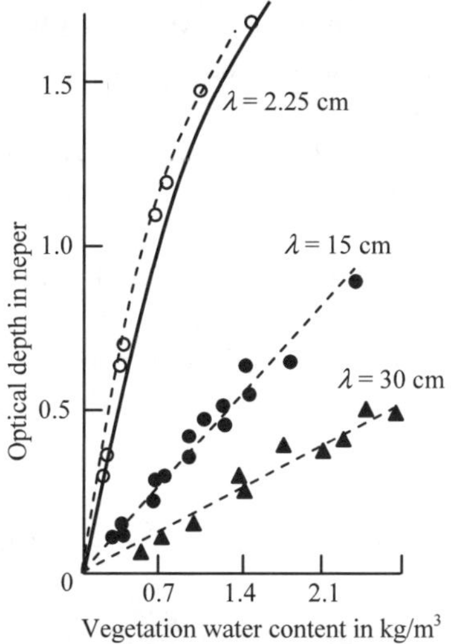

Fig. 5.6. Dependence of optical depth on water content of pea canopy. Solid curve is calculated by the model (5.18).

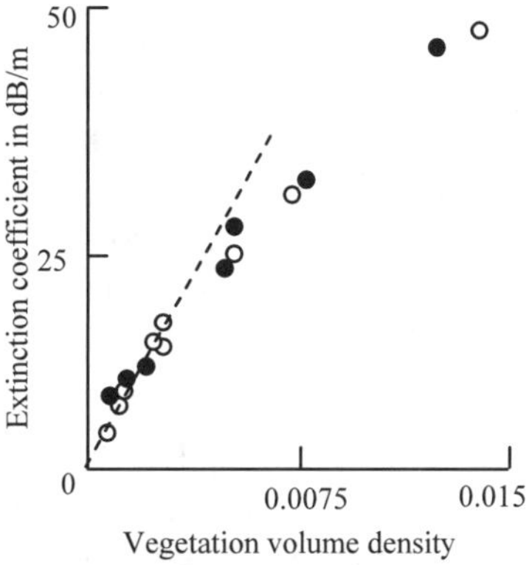

Fig. 5.7. Dependence of extinction coefficient on vegetation volume density.

The measurements have shown that at decimeter wavelengths the dependence of the optical depth on the water content is close to the linear one when W varies from 0 to 3-5 kg/m^2 . A change in the volume density in limits of 0-0.01 (that exceeds variations of p for cultural vegetation) with a given value of W does not influence appreciably the optical depth of investigated canopies in this frequency band. In the centimeter wavelength frequency band, it has been found that the optical depth for a given value of W depends on the vegetation volume density. That indicates a nonlinear dependence of τ on p and, as a consequence, a nonlinear dependence of τ versus W since $W = pm_v\rho_w h$ where m_v is the volumetric moisture of plant elements, h is the height of vegetation, and ρ_w is the density of water. At a constant volume density, the dependence of τ on W is linear. However, since in a real situation an increase of the water content in a crop takes place due to both the change of crop height and of crop volume density, the link between the optical depth and the water content becomes non-linear. A special study was conducted to examine the volume density influence on attenuation. The configuration presented in Fig. 5.4 was used. Two types of vegetation were under research: alfalfa (small disk-like leaves with chaotic orientation) and wheat (small strip-like leaves with chaotic orientation). The measurements have shown (Fig. 5.7) that the dependence of extinction coefficient γ on the vegetation volume density is close to linear only for small values of $p < 3 \cdot 10^{-3}$ which is in agreement with theoretical estimates (see previous chapter). For greater values of p, a deviation from the linear dependence is observed to lower values of attenuation. Coefficient c in equation (5.15) was found from experimental data and equals 1.34 dB^{-1} = 5.82 Np^{-1} for given wavelength and type of vegetation.

Comparison of results from Chukhlantsev and Golovachev (1989) with Pampaloni and Paloscia (1986) data (Fig. 5.6 in the text and Fig. 4 in the paper by Pampaloni and Paloscia (1986)) shows an excellent agreement between these two independent measurements. Using formulas (5.13) and (5.17), one can obtain for the optical depth at 2.25 cm:

$$\tau = (k / \sqrt{\lambda}) \ln(1 + W) = (0.25 / 0.15) \ln(1 + W). \qquad (5.18)$$

This dependence is plotted in Fig. 5.6 by the solid curve and coincides with experimental data within errors of measurements. It shows that Pampaloni and Paloscia's empirical model provides good estimates for the optical depth of agricultural vegetation in the centimeter wavelength frequency band.

A comprehensive study of attenuation properties of coniferous branches at 9 GHz was performed by Mougin *et al.* (1990). The measurements were done on simulated canopy composed of bare poplar and willow twigs with diameter ranging from 2 mm to 3.5 cm and the length between 10 and 30 cm. For a given diameter the results show that there is a linear relationship (in dB) between the attenuation and number density of twigs, at least for the number density values usually found in biological systems. However, a deviation from linear behavior was observed for higher densities or for larger diameters with relatively high densities. These results are expected since at high number densities the mutual influence of scatterers becomes noticeable (equation (5.74)). Mougin *et al.* (1990) also investigated the dependence of extinction coefficient on the volumetric water content of the slab of twigs. It was demonstrated that the extinction coefficient is a linear function of the volumetric water content but the proportionality coefficient strongly depends on the diameter of twigs. For example, coefficient b in (4.124) can be found from Mougin *et al.* (1990) data to be equal to 1.33 $dB/(kg/m^2)$ for a 1-cm diameter cylinder canopy and to 6.7 $dB/(kg/m^2)$ for a 0.2-cm diameter cylinder. It can be explained by a resonance character of attenuation by a cylinder and is consistent with the results of attenuation by a single twig (see Fig. 4.6 in Chapter 4).

Waveguide measurements conducted in Chukhlantsev *et al.* (2004a) also confirmed that the dependence of optical depth on the vegetation water content is close to a linear one. Some examples are presented in Fig. 5.8 and 5.9.

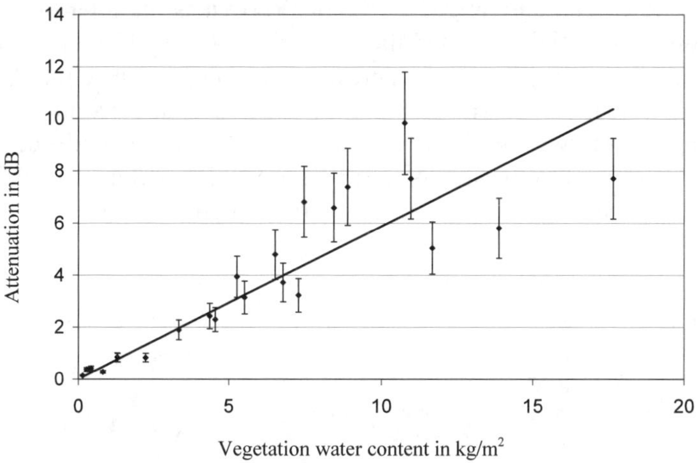

Fig. 5.8. Attenuation by a slab of pine branches versus their water content at 1 GHz.

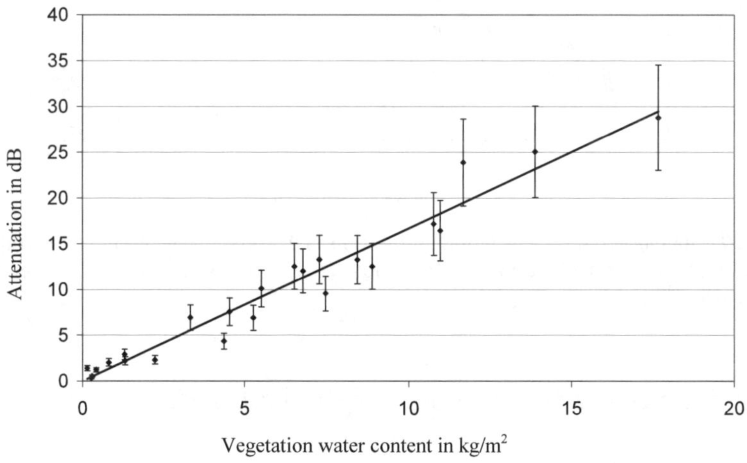

Fig. 5.9. Attenuation by a slab of pine branches versus their water content at 8 GHz.

At present, the linear dependence of the optical depth on the vegetation water content in the decimeter wavelength frequency band (and, hence, the linear dependence of the extinction coefficient on the volumetric water content (4.133)) is not called in question by most researchers. An excellent review by Van de Griend and Wigneron (2004a) provides the interested reader with data on the coefficient b (and the coefficient A in (4.133)) for different crops. As an example, Fig. 5.10 represents this dependence for different types of vegetation, which was obtained by the author at close to nadir observation angles. At centimetric waves, where the linear dependence is violated, Pampaloni and Paloscia's empirical model (Pampaloni and Paloscia, 1986) provides necessary corrections to the linear dependence.

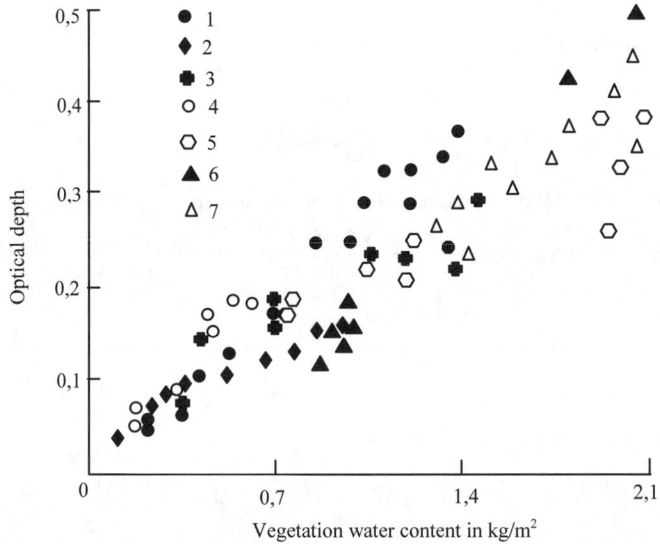

Fig. 5.10. Optical depth of vegetation at 1.4-1.66 GHz versus vegetation water content for alfalfa (1), wheat and rye (2), rice (3), grass (4), soybeans and cotton (5), corn (6), pine branches (7).

5.2.2. Frequency Dependence of the Extinction Coefficient

Equation (4.133) describes the frequency dependence of the extinction coefficient in a simple power form. Actually, it is difficult to expect so simple a dependence in the entire microwave band because of manifestations of the resonant character of attenuation by different plant elements (see Fig. 4.8 and 4.9 in Chapter 4). Nevertheless, this simple form of frequency dependence is rather attractive due to its ease and convenience. Thus it is no surprise that experimentalists use this form of frequency dependence at least for a certain part of the microwave band. Jackson and Schmugge (1991) were the first who tried to examine frequency dependencies of the coefficient b in simple power form by comparison with results of regression analysis of known experimental data. At that time, Kirdiashev *et al.* (1979) proposed a model that described the frequency dependence of b in the form

$$b = b' / \lambda \tag{5.19}$$

where λ is the wavelength. This approximation appeared to be good enough at decimetric waves and was successively used for estimates of vegetation screening effect in the decimeter wavelength frequency band. In their turn, Pampaloni and Paloscia (1986) found that at centimetric waves the frequency dependence is given by (see (5.18))

$$b = b' / \sqrt{\lambda} \ . \tag{5.20}$$

(Unfortunately, there was a misprint in the Jackson and Schmugge paper and the models (5.19) and (5.20) were interchanged. This misprint was reproduced in Van de Griend and Wigneron (2004a)). Jackson and Shmugge (1991) proposed to determine b in the form

$$b = b' / \lambda^x \tag{5.21}$$

where x should be found from analysis of experimental data ($x \equiv \alpha$ in equation (4.133)). Several data sets were examined and it was shown that parameter x is sensitive to vegetation type especially at centimetric waves. At longer wavelengths, the quantitative agreement with experimental data was good with $x = 1$. Van de Griend and Wigneron (2004a) collected a more representative data base on crop attenuation. They also confirmed that when a large frequency domain is considered, the coefficient b is inversely proportional to the power of the wavelength (5.21). They have found that different canopy types could be separated into different groups, each with a different combination of values for $\log(b')$ and x, which characterize the linearized equation (5.21):

$$\log(b) = \log(b') - x \cdot \log(\lambda) \ . \tag{5.22}$$

Based on visual interpretation, two groups/clusters and three individual canopy types were distinguished. The clusters were as follows.
Cluster A: short grass; short wheat; tall grass: $\log(b') \cong 1.3$; $x \cong 1.4$.

Cluster B: soybean; broadleaf; wheat; winter rye: $0 < \log(b') < 0.3;$, $0.7 < x < 1$.

The individual canopy types are as follows.

Oat: $\log(b') \cong 0.9;$ $x \cong 1.1$, alfalfa: $\log(b') \cong 0.2;$ $x \cong 0.6$,

corn: $\log(b') \cong -0.4;$ $x \cong 0.4$.

For all crop types, it was found that $b = 0.2 \pm 0.12$ at L-band. This result is consistent with the author's data presented in Fig. 5.10.

Though results presented in Van de Griend and Wigneron (2004a) are based on a big volume of measurements, an experimentalist should use these results with a certain care. First of all, one can see that measurements reviewed in Van de Griend and Wigneron (2004a) were conducted at some (one or two, and maximum three) frequency points of the microwave band. Since the spectral behavior of attenuation can be rather complex (see Figs. 4.8 and 4.9 in Chapter 4), it is difficult to make a conclusion on the frequency dependence of attenuation by its values at two or three frequencies. Secondly, in most cases attenuation values were estimated indirectly using inverse modeling. It was noted in Van de Griend and Wigneron (2004a) that values of optical depth presented in the literature have been derived using different theoretical approaches, different inversion models, and different measuring techniques either using passive or active systems. Therefore, the currently available figures are not very consistent and are difficult to compare. Thirdly, it is unreasonable to assume that the slope of frequency dependence of attenuation is constant over the entire microwave band. The limited number of wavelengths did not allow Van de Griend and Wigneron (2004a) to confirm an assumption that the relationship between b and λ is different for different parts of the microwave band (i.e., for $\lambda > \sim 5$ cm and $\lambda < \sim 5$ cm), as it was suggested by Jackson and Schmugge (1991). Experimental validation of this assumption requires data on attenuation to be gathered in a consistent manner and, desirably, at many frequency points of the microwave band for the same canopy.

To examine the frequency dependence of microwave attenuation more carefully, it is necessary to consider results of measurements obtained for a big number of different frequencies or for continuous frequency variations in a wide range of frequencies. Moreover, the data of direct or close to direct measurements of attenuation should be considered as preferable ones in comparison with the data obtained indirectly using inverse modeling.

Direct measurements of the transmissivity of a vegetation layer (and the optical depth linked to the transmissivity by equation (4.134) ($t \approx \exp\{-\tau\}$)) can be performed with the use of techniques that are realized with the configurations presented in Fig. 5.1 A, B, and C; Fig. 5.2, Fig. 5.3 B, and C; and Fig. 5.4. Chukhlantsev and Golovachev (1989) used configuration Fig. 5.3 B, and equation (5.9) to determine the transmissivity of vegetation. Tunable radiometers were used operating in the frequency range 1-6 GHz (wavelengths 5-30 cm). A radiometer was also used operating at the frequency 13.3 GHz (wavelength 2.25 cm). The step of frequency change was 0.5-1 GHz that allowed obtaining the frequency dependence of attenuation in a wide range at quasi-continuous frequency variations. Results of measurements for alfalfa are presented in Fig. 5.11.

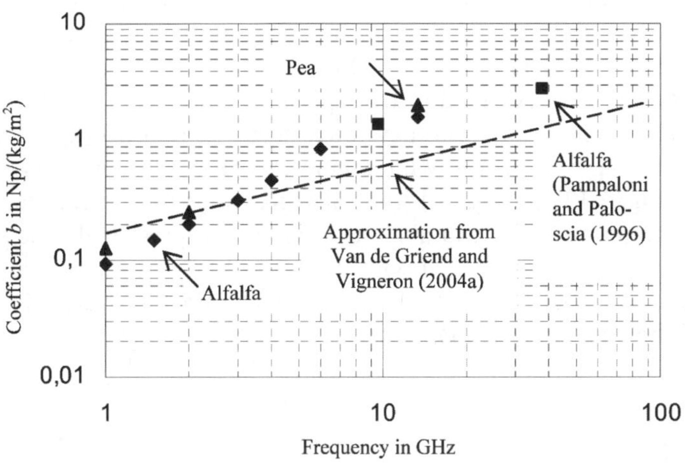

Fig. 5.11. Spectral dependence of coefficient b for alfalfa.

Data for pea canopy and data of Pampaloni and Paloscia (1996) for alfalfa are also presented in Fig. 5.11. Regression frequency dependence $b = 1.58 / \lambda^{0.6}$ obtained for alfalfa in Van de Griend and Wigneron (2004a) is depicted in Fig. 5.11 by a dashed line. One can see that the slope of the frequency dependence really changes: it is approximately unity ($x \sim 1$) for low frequencies (1-3 GHz); it is greater than unity ($x \sim 1.2$) for 3-10 GHz; and it decreases to approximately 0.5 or less for higher frequencies. This spectral behavior can be explained by the following. At low frequencies (L-band), the absorption is predominant and $\gamma \sim \lambda$. At centimetric waves, the scattering term becomes significant (see Fig. 4.2 in Chapter 4) and the slope of frequency dependence increases. At higher frequencies, the denominator in (4.93) becomes greater and greater which results in a reduction of the slope of frequency dependence. If a limited number of wavelengths (frequencies) are available for a statistical analysis, only a rather approximate value of the slope can be obtained. Experiments conducted in Chukhlantsev and Golovachev (1989) for vegetation with vertical stems and results of calculations (Le Vine and Karam, 1996) have shown that in this case the frequency dependence of b can be even more complex than for alfalfa.

Quite new possibilities in the research on attenuation spectra were achieved with the use of a super wideband waveguide transition system (Fig. 5.2) (Chukhlantsev *et al.*, 2004a). The system allows obtaining continuous attenuation spectra of vegetation fragments in the frequency range 0.8...8.5 GHz. Results of measurements for aspen samples are presented in Fig. 5.12. One can see that the spectral dependence of attenuation for branches of different size and leaves are quite different. It is explained by the resonant character of attenuation by tree branches in the microwave band. It could be noted that the data presented in Fig. 5.12 are consistent with the data of Mougin *et al.* (1990) obtained in free space measurements at 9 GHz. For branches with a diameter of 1 cm they found that $b = 1.33$ dB/(kg/m^2). This agrees with data for branches with diameters of 0.5-2 cm (curve 4 in Fig. 5.12). In Fig. 5.13 and Fig. 5.14, linear regressions for the coefficients b for each tree component and a branch as a whole are presented. The measurements of attenuation conducted with the use of the waveguide transition system have shown that the extinction coefficient of a forest crown can be estimated by the equation (4.133):

$$\gamma = Af^{\alpha} = 0.3 f^{0.44} w \qquad (5.23)$$

where γ is in Np/m, f is in GHz, and w is the crown volumetric water content in kg/m^3. This regression relationship is obtained for the entire frequency band of 0.8-8.5 GHz, in which measurements were conducted. However, it has been revealed that the slope of frequency dependence of attenuation changes within this frequency band. For example, Fig. 5.15 presents data on attenuation by aspen branches. It is seen that in the frequency range 0.8-2.5 GHz the slope of the frequency dependence is greater than that at higher frequencies. Averaged slope of frequency α dependence for a mixed crown (aspen, pine, maple) was found to be $\alpha = 0.6$ at frequencies below 3 GHz and $\alpha = 0.42$ for higher frequencies.

Mätzler and Wegmüller (1994) studied the transmissivity of spruce fir twigs with the use of configuration Fig. 5.3B. Their data are recalculated to the attenuation and presented in Fig. 5.16. The same configuration was used in Chukhlantsev (2002) for measurements of attenuation by different crown branches. Results are presented in Fig. 5.17.

Configuration Fig. 5.1B, was used in Grankov *et al.* (2005) to measure attenuation by crowns of single trees. Some results of measurements are presented in Fig. 5.18 and Fig. 5.19.

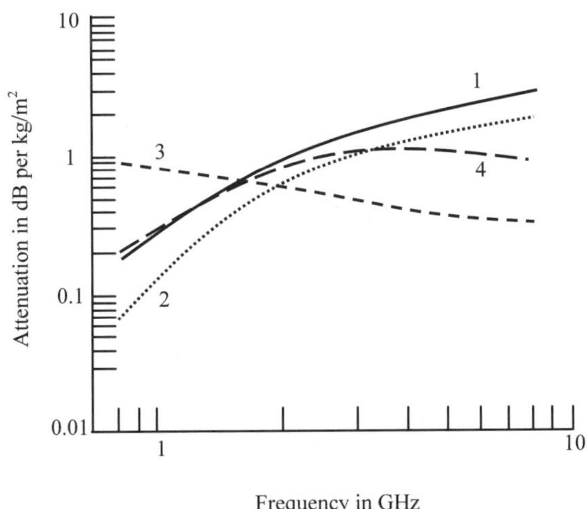

Fig. 5.12. Spectral dependence of specific attenuation by thin leafy branches with a diameter less than 0.5 cm (1), bare thin branches (2), thick branches with a diameter of 2-5 cm (3), and branches with a diameter of 0.5-2 cm (4).

Pardé *et al.* (2005) measured the transmissivity of natural crowns at 19 and 37 GHz. They did not find an essential difference between attenuation at vertical and horizontal polarizations. Their data produce a slope $\alpha \sim 0.4$. The empirical model of Kruopis *et al.* (1999) produces for 6-48 GHz a slope $\alpha \sim 0.8$. Brown and Curry (1982) collected some data on attenuation by forest crowns in the 0.1-1 GHz frequency range. These data produce a slope $\alpha \sim 0.7$-0.8. Close values of slope for this frequency band was obtained in Chukhlantsev *et al.* (2003c), where available experimental data on attenuation by forest are collected. Yakubov *et al.* (2002) measured attenuation in larch forest crowns at 0.2 and 1.275 GHz and also did not reveal an essential difference between attenuation at vertical and horizontal polarizations. Their data produce a slope $\alpha \sim 1$. The data of Savage *et al.* (2003) for the 1.3-11.6 frequency band produced different slopes $\alpha = 0.15$-0.5 with an average value $\alpha \sim 0.3$.

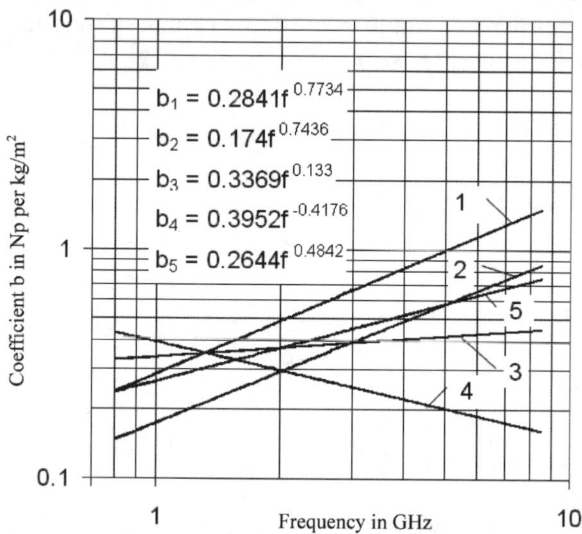

Fig. 5.13. Spectral dependence of specific attenuation by thin leafy aspen branches with a diameter less than 0.5 cm (1), bare thin branches (2), branches with a diameter of 0.5-2 cm (3), thick branches with a diameter of 2-5 cm (4), and the branch as a whole (5).

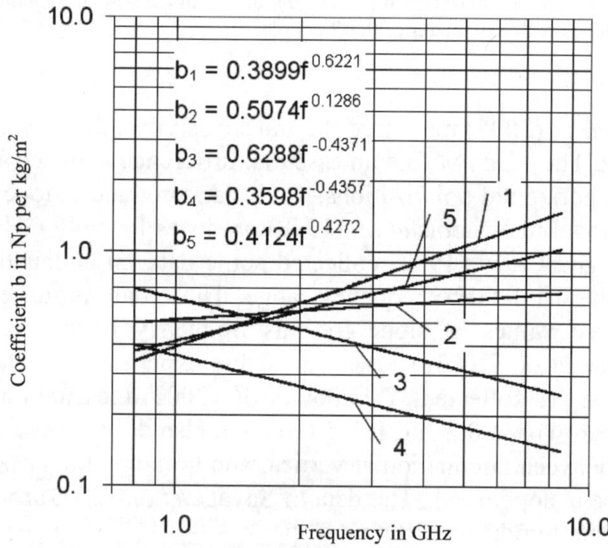

Fig. 5.14. Spectral dependence of specific attenuation by thin leafy maple branches with a diameter less than 0.5 cm (1), bare thin branches (2), branches with a diameter of 1-2 cm (3), thick branches with a diameter of 2-5 cm (4), and the branch as a whole (5).

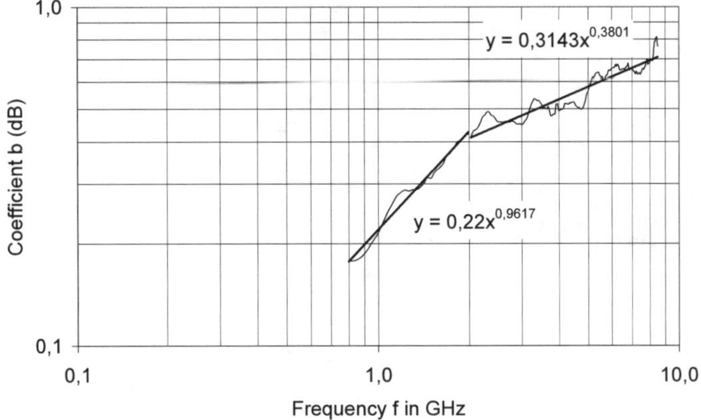

Fig. 5.15. Spectral dependence of specific attenuation by aspen branches.

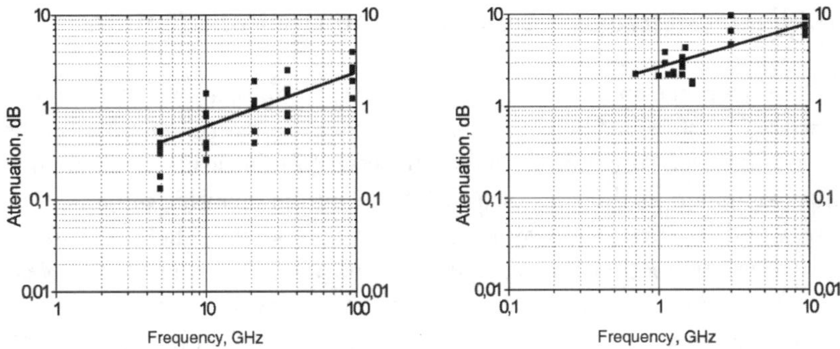

Fig. 5.16. Regression spectral dependence of microwave attenuation by the spruce fir twigs. $T(dB) = 0.1652 * F^{0.5805}$ (GHz).

Fig. 5.17. Regression spectral dependence of microwave attenuation by crown branches. $T(dB) = 2.66137 * F^{0.47492}$ (GHz).

Comparison of results of waveguide measurements with results of measurements for natural forest crowns allows one to conclude the following. Waveguide measurements provide reasonable values of slope for the frequency dependence of attenuation that are consistent with data obtained for natural forest stands. At the same time, the waveguide measurements enable one to find the dependence of attenuation on biometric parameters, i.e., to determine the coefficient A in (4.133) (or the coefficient b' in (5.22)).

The slope α for forest crowns changes with the increase in frequency from 0.7-1 at VHF to 0.4-0.5 for millimeter waves.

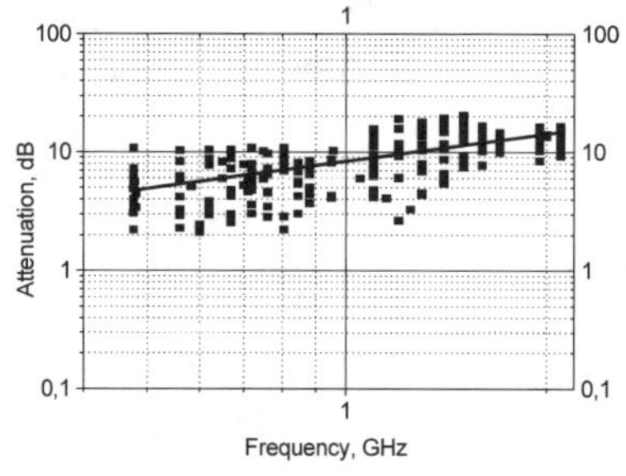
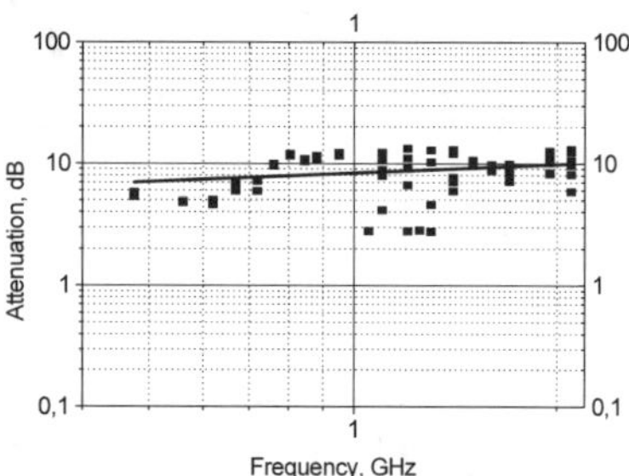

Fig. 5.18. Resultant regression spectral dependence of microwave attenuation by crowns of spruce, chestnut and apple-tree at VP, T (dB) = 8.34838*F$^{0.76256}$ (GHz).

Fig. 5.19. Resultant regression spectral dependence of microwave attenuation by crowns of spruce and chestnut at HP, T (dB) = 8.34947*F$^{0.2379}$ (GHz).

5.2.3. Dependence of Attenuation on the Polarization

The difference in attenuation for different polarizations of electromagnetic waves is observed when the vegetation canopy contains a component with a pronounced orientation (stalks, trunks). In this case, the attenuation is usually calculated as the sum of attenuation by the chaotic component, such as leaves and of attenuation by the oriented component, such as stalks. The difference in attenuation for vertically and horizontally polarized waves can be large for some crops at not close to nadir observation angles. For example, Fig. 5.20 presents data on the optical depth for alfalfa (Chukhlantsev, 1981). The alfalfa had stems with a length of 30-40 cm and a diameter of 0.4-0.6 cm. The weight fraction of stems in the crop was about 0.5. Ulaby and Wilson (1985) have observed a great difference in the attenuation at vertical and horizontal polarizations by wheat and soybeans for observation angles of 52-56°. This difference was not so big at 16-24° observation angles. Brunfeldt and Ulaby (1986) used passive measurements (configuration Fig. 5.3A,) of the transmissivity of wheat, soybean, and corn canopies at 2.7 and 5.1 GHz. Their data are recalculated to the optical depth and shown in Fig. 5.21.

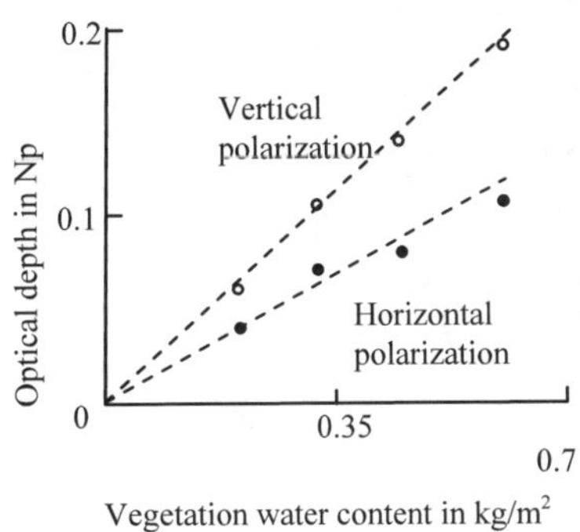

Fig. 5.20. Optical depth of alfalfa versus vegetation water content at 1.67 GHz. Observation angle is 30°.

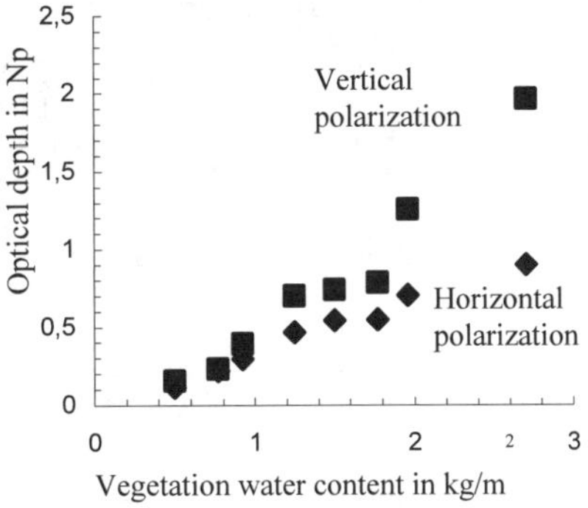

Fig. 5.21. Optical depth of soybeans versus vegetation water content at 2.7 GHz. Observation angle is 10°.

Van de Griend *et al.* (1996) have developed a measurement procedure to determine horizontal and vertical polarization radiative transfer properties, i.e., single scattering albedo and optical depth, of vegetation under field conditions. The procedure was applied to a wheat crop for a series of biomass densities. The measurements were done using two different radiometers (1.4 and 5 GHz) and for different view angles. The measurements indicated that the ratio of the optical depths at horizontal and vertical polarizations is slightly dependent on view angle. This ratio did not differ much from unity in the C-band. In the L-band, this ratio was about unity at small nadir view angles and diminished to 0.5-0.6 at 30-50° nadir view angles. Owe *et al.* (2001) retrieved values of optical depth from satellite radiometric observations and found that, for a variety of crops and for this particular large spatial scale of averaging, the ratio of the optical depths at horizontal and vertical polarizations is also close to unity in the C-band. An estimate of the ratio of the optical depths at horizontal and vertical polarizations can be obtained for the L-band from equation (4.132). For higher frequencies, the data of Fig. 4.8 and Fig. 4.9 from Chapter 4 can be used for the estimate. The calculated L-band ratio τ_h / τ_v as a function of the observation angle is presented in Fig. 5. 22 where it is compared with experimental data from Van de Griend *et al.* (1996). It is seen that the calculated data describe the general tendency of τ_h / τ_v angular dependence correctly.

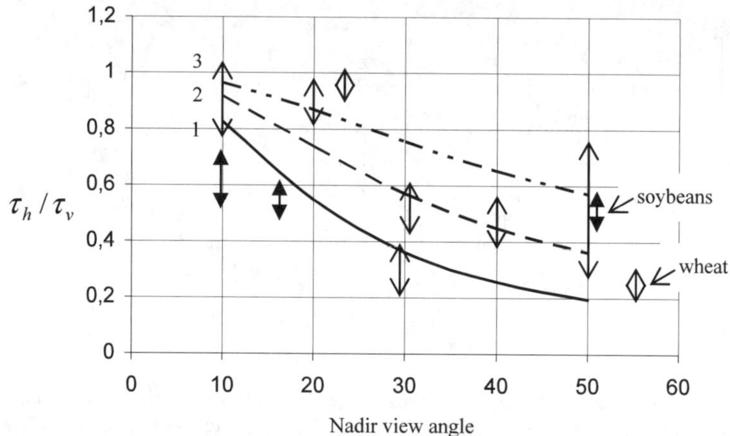

$$\tau_h / \tau_v$$

Nadir view angle

Fig. 5.22. Calculated L-band ratio τ_h / τ_v for different weight parts of chaotic component in the canopy: 0.3 (1) – wheat; 0.5 (2) – alfalfa; 0.7 (3) – soybeans.

It can be recognized that there are rather few data on direct measurements of τ_h / τ_v at different angles. This fact makes it difficult to choose an appropriate model for the description of canopy anisotropy. Due to this reason, some researchers propose semi-empirical models. Particularly, a simplified parameterization, derived from a physical analysis of the attenuation by vertical stems was developed in Wigneron *et al.* (2004b). The parameterization describes the dependence of τ on the measurement configuration (observation angle ϑ and polarization p) and is based on two parameters: a correction parameter [C_{pol}] and the optical depth at H polarization $\tau\,(h)$:

$$\tau_{h,v} = \frac{\tau(\vartheta, p)}{\cos \vartheta},$$

(5.24)

$$\tau(\vartheta, h) = \tau(h) = \text{ constant},$$

(5.25)

$$\tau(\vartheta, v) = \tau(h)(\cos^2 \vartheta + C_{pol} \sin^2 \vartheta)$$

(5.26)

where $\tau\,(h)$ is the optical depth at the nadir observation. One can see that this parameterization coincides with that given by (4.132) when

$$C_{pol} - 1 = \frac{3(1 - \eta_1)}{u \eta_1}.$$

(5.27)

In Fig. 5.22, curve 1 corresponds to $C_{\mathrm{pol}} = 8$, and curve 3 corresponds to $C_{\mathrm{pol}} = 2.3$. Several attempts were made to estimate C_{pol} by fitting the brightness temperature models to experimental data (Wigneron *et al.*, 2004b). The results showed that C_{pol} can change from 5-10 to ~1 for a canopy during the growth period. The use of the C_{pol} formulation, which was originally developed for wheat, was useful in soil moisture retrievals for other crop types. Besides, the retrieved values of C_{pol} may provide information on the time variations in the vegetation structure as sensed by the microwave radiometers.

Forest crowns demonstrate rather weak polarization dependence of attenuation except the case of perpendicular to trunks propagation (Figs. 5.18, 5.19).

5.2.4. Dependence of Attenuation on the Type of Vegetation

The architecture of vegetation canopies influence values of the coefficient b. This fact is displayed by different values of b' in (5.21) (or A in (4.133)) for different canopies. Available data on b' for different crops are collected in Van de Griend and Wigneron (2004a) and are given for some groups of crops in Section 5.2.2. For forest crowns, the form-factor A in (5.23) was found to be equal to 0.3-0.4. Besides, the slope of frequency dependence of the coefficient b is also influenced by the type of vegetation. Further research is needed to specify this influence.

Chapter 6

MODELING OF MICROWAVE EMISSION FROM VEGETATION CANOPIES

6.1. GENERAL APPROACH TO THE MODELING OF RADIATION PARAMETERS FOR VEGETATED SOILS

Characteristics of microwave emission from a surface are the brightness temperature T_b or the emissivity e. Microwave radiometry is based on retrieving algorithms, which solve the inverse problem of reconstructing the environmental parameters (such as soil moisture, vegetation biomass, etc.) from the measured radiation characteristics (brightness temperatures). Three types of retrieving algorithms are usually used. The first one is based on the use of experimental regression relations between geophysical or biophysical parameters and radiation characteristics. A disadvantage of this approach is its limited applicability (in many cases, the regression relations are valid only for test sites or regions where they were obtained). The second type of algorithms (which has been successfully developed in the last few years) is based on the use of neural networks. For this approach to provide good results, it is necessary to train the corresponding neural network by a statistically representative sampling. However, such training is not always possible. The third type of algorithms, which is most widely used, is based on an inversion of radiation models. The models relating the radiation parameters to environmental ones are developed on the basis of certain theoretical assumptions concerning an object to be modeled, e.g., vegetation canopies, bare soils, water surface, etc. Then, the models are tested, validated, and refined at test sites under a direct control of geophysical and biophysical parameters. After solving the forward problem, which establishes the relation between the radiation and environmental parameters, the inverse problem is stated, which is posed as the retrieval of the environmental parameters from remote sensing data through an inversion of the forward problem. Evidently, the model approach also has some disadvantages. One of them is the influence of the model's errors on the final result. Any modeling implies some simplification and idealization of a real object, and, hence, leads to a discrepancy between the calculated and measured values and, thus, to an error in the retrieval of environmental parameters on the basis of the forward model. However, the model approach is widely used because of its obvious advantages. With this approach, it is possible to reveal the parameters of a medium that are responsible for its thermal microwave emission, to estimate the sensitivity of the radiation parameters to changes in parameters of the medium, and to determine the dynamic range of variation for a radiation parameter as a function of one or another parameter of the medium. The model analysis of the spectral dependences of the radiation parameters allows one to determine the optimal frequencies for remote sensing. After these possibilities are realized, the model approach can be further used in combination with two other aforementioned algorithms to obtain the best possible results.

The brightness temperature is a measure of the intensity of radiation emitted by the Earth's surface. Therefore, modeling of the emission from a vegetation canopy is usually performed in terms of the theory of radiation transfer. The analytical solution of the radiative transfer equation is known only for certain particular forms of the unit volume phase function. Therefore, as a rule, this equation is solved numerically. However, this approach is rather complicated for experimentalists. In addition, in the model analysis, it is often difficult to separate the contributions of different parameters of the model to the final modeled result simulated. Hence, the usual practice is to use the models based on simple analytical expressions, which either are semi-empirical ones or which are approximate formulas for the numerical solutions to the transfer equation, or, else, which are approximate solutions to the transfer equation with some simplifying assumptions, e.g., single scattering, etc. These expressions are quite convenient for calculations and for the model analysis. Their accuracy is estimated by comparing with numerical solutions of the transfer equation or with experimental data.

On the basis of scalar transfer theory, Basharinov *et al.* (1968) developed an approach to the determination of the brightness temperature of a scattering layer and of a "rough surface – scattering layer" system. He started from equations describing the radiation transfer in plane-parallel layers, which are given by

$$\cos\vartheta \frac{dJ(z,\vartheta,\varphi)}{dz} = -\gamma J(z,\vartheta,\varphi) + u(z,\vartheta,\varphi), \qquad (6.1)$$

$$u(z,\vartheta,\varphi) = \frac{\omega\gamma}{4\pi} \iint J(z,\vartheta',\varphi')\zeta(\vartheta,\varphi,\vartheta',\varphi')\sin\vartheta' d\vartheta' d\varphi' + u_0 \qquad (6.2)$$

where $J(z,\vartheta,\varphi)$ is the spectral ray intensity of radiation flux, γ is the extinction coefficient, ω is the scattering albedo of unit volume (the single scattering albedo), z is the distance along the normal to the layer, ϑ is the angle between the normal and the direction of propagation, u_0 is the intensity of internal sources of the thermal emission per unit volume, ζ is the phase function of unit volume. In the isothermal case, radiative properties of the scattering layer are characterized by the emissivity, which according to Kirchhoff's law is expressed as

$$e = 1 - r - t \qquad (6.3)$$

where r is the reflectivity and t is the transmissivity of the layer. The last values are found from the solution of the transfer equation (6.1). This solution depends on the form of the phase function. Analytical solutions to this equation are known for two particular forms of the phase function, which are one-dimensional and isotropic ones. The reflectivity and transmissivity of the scattering layer in the case of one-dimensional scattering were found by Ambartsumian and are given by (Basharinov *et al.*, 1968)

$$t = \frac{(1-r_0^2)e^{-k^*\tau_0}}{1-r_0^2 e^{-2k^*\tau_0}}, \qquad (6.4)$$

$$r = r_0\frac{1-e^{-2k^*\tau_0}}{1+r_0^2 e^{-2k^*\tau_0}}, \qquad (6.5)$$

$$r_0 = \frac{1-\dfrac{\omega}{2}}{\dfrac{\omega}{2}} - \sqrt{\left(\frac{2-\omega}{\omega}\right)^2 - 1}, \qquad (6.6)$$

$$k^* = \frac{\omega}{2}(1-x)\frac{1-r_0^2}{r_0}, \qquad (6.7)$$

where x is the fraction of radiation scattered in the forward direction, r_0 is the reflectivity of an optically thick layer, $\tau_0 = \gamma h \sec \vartheta$ is the optical depth of the layer for the coherent radiation, h is the thickness of the layer.

The emissivity of a layer with the isotropic scattering is expressed in terms of Chandrasekhar's functions:

$$e(\tau_0, \cos \vartheta) = \frac{\Phi(\tau_0, \cos \vartheta) - \Psi(\tau_0, \cos \vartheta)}{1 - \dfrac{\omega}{2}[\alpha_0(\tau_0, \cos \vartheta) - \beta_0(\tau_0, \cos \vartheta)]}. \tag{6.8}$$

The functions Φ, Ψ and their zero moments α_0, β_0 are tabulated (Basharinov *et al.*, 1968). The emissivity of the "scattering layer – rough surface" system was obtained by Basharinov (Basharinov *et al.*, 1968) in the case of one-dimensional scattering as

$$e_\Sigma = e + (1 - R)\frac{t}{1 - rR} + eR\frac{t}{1 - rR} \tag{6.9}$$

where R is the surface reflectivity, $\dfrac{t}{1 - rR}$ is the effective transmissivity of the system accounting for multiple reflections of radiation fluxes. The first term in (6.9) represents the emission of the layer, the second term shows the emission of the surface passed through the layer, and the third term accounts for the emission of the layer reflected from the surface and attenuated by the layer. In the case of two-dimensional scattering, the emissivity of the system was found by Basharinov (Basharinov *et al.*, 1968) by introducing linear operators accounting for the transformation of angular distribution for scattered and reflected fluxes. The models developed were used to make clear the influence of the form of phase function on the emission properties of the system. Calculations of the emissivity were performed for two extreme cases: the one-dimensional model (6.9) and the model of a layer with isotropic scattering above a surface with Lambert scattering. It was shown that the form of the phase function does not influence essentially the emissivity. Particularly, in the second case considered, the emissivity is greater by 10-20% than that in the first case. This research gave the first grounds for the use of approximate analytical expressions for the emissivity instead of numerical solutions of the transfer equation.

It should be noted that the use of transfer theory and the aforementioned theoretical results for the description of radiation properties of vegetation canopies gave rise to sound doubts at first. Really, the transfer theory can be used if the extinction length $= 1/\gamma$ is much greater than the wavelength. But the first available data on the attenuation by vegetation were obtained in the centimeter wavelength frequency band and displayed rather great attenuation that resides in this particular frequency band. Therefore, a vegetation layer was considered first as an optically thick medium. Its reflectivity was first estimated in the single scattering approach. The emissivity of vegetated soils was estimated by Basharinov *et al.* (1974) with the use of equation (6.9) but without the third term:

$$e = (1 - R)(1 - \xi_v) + (1 - r)\xi_v (1 - e^{-\tau_0}) + (1 - R)(1 - r)\xi_v e^{-\tau_0} \qquad (6.10)$$

where ξ_v was introduced as the fraction of soil covered by vegetation. Model estimates of attenuation performed in Kirdiashev *et al.* (1979) and experimental data obtained in Kirdiashev *et al.* (1979), Basharinov *et al.* (1979), and Chukhlantsev (1981) showed that in the decimeter wavelength frequency band, vegetation canopies are semitransparent media. Hence, an improvement of models was required. Particularly, the question arose if the transfer theory can be used for a description of microwave propagation in vegetation canopies. Application of the transfer theory for a description of multiple scattering in a medium is approved when scattering particles are positioned in the far-field zone with respect to each other. If the positions of scatterers are not correlated, the following equations are used for the extinction and scattering coefficients:

$$\gamma = \sum_{i=1}^{n} \sigma_e^i, \qquad (6.11)$$

$$\omega\gamma = \sum_{i=1}^{n} \sigma_s^i \qquad (6.12)$$

where n is the number of scatterers in unit volume, σ_e^i and σ_s^i are the extinction and scattering cross sections of the i-th scatterer. There is no clear answer to the question if the positions of scatterers in a vegetation canopy are correlated. Qualitative analysis and direct measurements show that there are no priori reasons to consider the positions of plant elements as completely casual. However, an account for the correlation between their positions leads to rather serious mathematical difficulties. In this case, the extinction coefficient is calculated with the use of scattering amplitudes for systems of one, two, three, etc. particles (see (4.60)). If the size of particles is small compared to the wavelength, the pair distribution function of different forms is used to account for their correlation. That allows one to find the corrections to the extinction coefficient due to the correlation of scatterers.

Some plant elements in a canopy are positioned in the near-field with respect to each other. An idea of the effect of such positioning can be given by considering the extinction of a simplest system consisting of two elements. Measured values of the extinction by two cylindrical sticks (parts of freshly cut thin branches) are presented in Fig. 6.1 (Chukhlantsev, 1981). The geometry of measurements is depicted in the figure. The ratio of extinction cross section of the system σ_Σ to the sum of the individual extinction cross sections $\sigma_1 + \sigma_2$ was determined in the experiment. This ratio shows the mutual influence of the cylinders. It is possible to separate near-field and far-field zones in the dependence of $\sigma_\Sigma /(\sigma_1 + \sigma_2)$ on the distance between the sticks. In the near-field zone, the ratio $\sigma_\Sigma /(\sigma_1 + \sigma_2)$ essentially differs from unity that is caused by the mutual screening of the elements. For the geometry presented in Fig. 6.1,a, the size of the near-field zone is estimated by the greatest from values of λ and d_{ef}, where λ is the wavelength and d_{ef} is the effective diameter of one cylinder that is equal to the ratio of its extinction cross section to its length. For the geometry presented in Fig. 6.1,b, that size is estimated by the value of σ_e /λ, if $d_{ef} > \lambda$, $d_{ef} > d$ where d is the diameter of the cylinder, or the value of λ, if $\lambda > d, d_{ef}$. In the far-field zone, fading oscillations of the ratio $\sigma_\Sigma /(\sigma_1 + \sigma_2)$ around unity are observed. If the distance between scatterers in a canopy is larger than the size of the near-field zone and the scatterers are not correlated, statistical averaging of σ_Σ will give equation (6.11) for the extinction coefficient. The effect of near-field zone positioning of scatterers can be neglected if the extinction length $1/\gamma$ is much greater than the size of near-field zone R_0. To get an estimate, one can assume that $1/\gamma \approx 1/(n\sigma_e)$ and, then, the condition for the aforesaid neglect is written as

$$1/(n\sigma_e) \gg R_0 . \tag{6.13}$$

When $R_0 \sim \lambda$, $\sigma_e \sim a/\lambda$ (small particles, a is the polarizability of a particle), the condition (6.13) is written in the form: $na \ll 1$, and was obtained by Kravtsov (Rytov *et al.*, 1978). For chaotically oriented disks (Chapter 4), $\sigma_e = \frac{2}{3} k_0 \varepsilon_s'' V$, and the condition (6.13) puts a restriction on the volume density $p = nV$ of the scattering medium:

$$p << \frac{3}{4\pi\varepsilon_s''} \approx \frac{1}{4\varepsilon_s''}. \tag{6.14}$$

Taking into account that $\varepsilon_s'' \sim 10\text{-}15$ (Chapter 2), one obtains: $p << 0{,}015$. The density of natural vegetation does not exceed 0.003-0.005, and the condition (6.13) is valid for most canopies. It was noted in Chapter 4 that the small particles approach can be used for leafy vegetation at decimetric waves. Therefore, the use of the transfer theory for this case is approved.

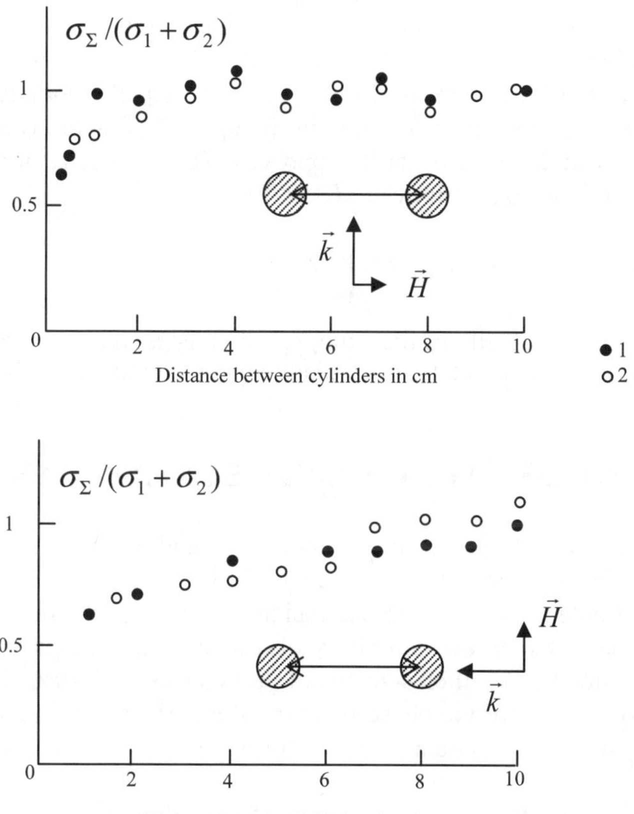

Fig. 6.1. Extinction by a system of two sticks at a wavelength of 2.25 cm. Diameter of sticks is 3 mm (1) and 6 mm (2).

If the scatterers are strong (stems at decimetric waves and leaves and stems at centimetric waves), the condition (6.13) is written as

$$p \ll \frac{\lambda V}{\sigma_e^2}. \tag{6.15}$$

Estimates show that this condition is satisfied at decimetric waves for canopies with $p \leq 0.003$, and at centimetric waves for canopies with the same density but with not very broad leaves (grass, tree crowns, grains, alfalfa, soybeans, etc.). The extinction length of microwave radiation for these types of canopies reaches to tens of centimeters at centimetric waves and some meters at decimetric waves. For larger values of the volume density, the mutual screening of elements is significant and the extinction coefficient (or the optical depth) is estimated by equations (4.74) (or (5.13)). For canopies with broad leaves (beets, cabbage, etc.), the extinction length at centimetric waves is comparable with the size (thickness) of a leaf. When there are no gaps between the leaves (the canopy is close enough), practically complete attenuation takes place at a single scattering on the upper leaf layer. The use of the transfer theory in this case is hardly approved. The emissivity of such a canopy is estimated by (Basharinov *et al.*, 1979),

$$e = 1 - \rho_c \tag{6.16}$$

where ρ_c is a total albedo of the canopy, which is estimated by the incoherent summation of fields scattered by plant elements in the upper half space.

6.2. THE EMISSIVITY OF A VEGETATION CANOPY

The emissivity of a vegetation canopy is found by the use of numerical solutions to the transfer equation or the use of known solutions to this equation for particular cases of one-dimensional and isotropic phase function. The difference between the emissivity of a layer with one-dimensional and isotropic scattering is rather significant and can exceed 10% (Basharinov *et al.*, 1968). Hence, for the phase function of an arbitrary form, a numerical solution to the transfer equation is performed to increase the accuracy of modeling.

The procedure for computing the emissivity of the vegetation is usually based on the complementary between emissivity and reflectivity (see (3.1) and (3.2)), so that the former can be evaluated once this latter is known. Details of the computing procedures have been given, particularly, in Ferrazzoli and Guerriero (1994, 1996) and Ferrazzoli *et al.* (1989, 1992a, 2000). Their brief outlines are the following. A) The vegetation canopy is modeled as a thick layer of elementary lossy scatterers of finite dimensions, i.e., disks and cylinders, over the rough soil surface. The layer is subdivided into elementary sub-layers, whose optical thickness is determined by the condition that the interactions among the scatterers within the same elementary layer are negligible. The scattering and transmission matrices of each sub-layer are then evaluated on the basis of the absorption and scattering properties of the individual vegetation elements. B). Absorption and scattering from an element of vegetation are expressed in terms of its absorption and bistatic scattering cross sections, which depend on the shape, dimensions, and permittivity of the element, as well as on the sensor parameters, i.e., frequency, observation angle, and polarization. The cross sections are computed with respect to local coordinate systems relative to the identified elements of vegetation; so that subsequent transformations in terms of the Eulerian angles and averaging over these angles are required to take the distributions of orientation of the scatterers into account (see Section 4.2.1 in Chapter 4 and paper by Eom and Fung (1984)). C). The terrain is analogously characterized by its scattering matrix, whose elements depend on the dielectric constant and on the roughness of the surface. For a given surface, the scattering model is selected according to the frequency and observation angle. D). The scattering and transmission matrices of adjacent layers are obtained in terms of the respective matrices by "doubling" (matrix doubling method is described in detail by Eom and Fung (1984)) and the process is reiterated to combine successively all the sub-layers into which the canopy was subdivided, until the scattering and transmission matrices of the whole layer are computed. Analogously, the soil is included through another "doubling" operation and, finally, the bistatic scattering coefficient of the terrain covered by vegetation is obtained. E). The emissivity of the vegetation canopy is computed by the energy conversation relation (3.1) and (3.2).

In recent years, described model simulations were successively compared with experimental emissivities measured over fields of different crops (e.g., Ferrazzoli *et al.*, 2000). The advantage of this approach is that the model is able to include multiple scattering effects and shows flexibility, since the dimensions, the orientation and the position of the scatterers may be properly selected in order to represent realistically a given crop geometry. In principle, the model is valid in the whole microwave spectrum, provided suitable approximations are adopted to compute the cross sections of single elements. However, the use of simple geometrical shapes, such as disks and cylinders, is acceptable at low frequencies but gradually loses validity when the frequency increases, since the microstructure of the elements becomes more and more important.

The use of simplified approximations for the cross sections of the plant elements and the limited applicability of the transfer theory itself for the description of microwave propagation in vegetation canopies lead to errors in modeling the vegetated terrains emission properties. The numerical approach described above provides an excellent fit of experimental data on an evolution of the emissivity during crop growth (Ferrazzoli *et al.*, 2000), on the dependence of polarization index on vegetation features (Ferrazzoli *et al.*, 1992a), on the emissivity of forests (Ferrazzoli and Guerriero, 1996). Nevertheless, a rather significant discrepancy between measured and computed data is often observed, which implies that further refinements of the model are needed.

The use of numerical simulations is not always convenient for practical applications. Due to this reason, experimentalists often use some simplified analytical models for the emissivity of vegetation canopies, which are based on known analytical solutions to the transfer equation and which approximate to the numerical solutions with an acceptable accuracy. These models are usually obtained on the basis of some assumptions that impose limitations on the considered scattering medium.

The emissivity of a vegetation layer is determined by the active losses of the plane wave in the layer (Levin and Rytov, 1967):

$$e_v = \frac{Q}{I_0} \tag{6.17}$$

where I_0 is the intensity of incident upon the layer plane wave, Q is the active (ohmic) losses of the wave in the layer. Under the single scattering assumption, the intensity of the coherent wave in the layer is determined by

$$J(z) = I_0 \exp(-\gamma z \sec \vartheta). \tag{6.18}$$

The ohmic losses in the layer are expressed as

$$Q = \int_0^h (1 - \omega)\gamma \sec \vartheta I_0 \exp(-\gamma z \sec \vartheta) dz$$

$$= (1 - \omega)(1 - e^{-\gamma h \sec \vartheta})I_0 = (1 - \omega)(1 - e^{-\tau_0})I_0. \tag{6.19}$$

The emissivity of the layer in the single scattering approach is expressed as

$$e_v = (1 - \omega)(1 - e^{-\tau \sec \vartheta}) = (1 - \omega)(1 - e^{-\tau_0})$$

$$= 1 - \omega(1 - e^{-\tau_0}) - e^{-\tau_0} = 1 - r - q. \tag{6.20}$$

The emissivity of the layer with one-dimensional scattering is given by (6.3)-(6.7). The emissivity of the layer with isotropic scattering is given by (6.8). Equation (6.20) is very simple and convenient for a model analysis and an inversion of the model. However, the effect of multiple scattering becomes significant when $\omega > 0,1$; $\tau_0 \sim 1$. The value of this effect has an order of ω. It reveals that multiple scattering effects should be taken into account for canopies with elements that are strong scatterers. Simplicity of equation (6.20) makes this expression very attractive for an interpretation of experimental data. Hence, it was proposed (Chukhlantsev, 1981) to use the form of this equation as a basic approximation for the emissivity of a vegetation layer. Parameters of this equation are to be chosen taking into account solutions of the transfer theory in general cases and available experimental data. Particularly, the substitution of ω in equation (6.20) for the reflectivity of the optically thick layer r_0 makes the model to be accurate under big values of the optical depth. Substitution of the term $\omega(1-e^{-\tau_0})$ in (6.20) for $r_0(1-e^{-2\tau_0})$ (this substitution follows from equation (6.5)) makes the model even more precise. A special investigation concerning an account for the multiple scattering effects was undertaken by Mätzler (2000). An analysis of different approximations and their accuracy was conducted in Chanzy and Wigneron (2000), Mätzler (2000), and Vinokurova *et al.* (1991) and will be given in brief below.

The emissivity of a vegetation canopy is usually found with the use of a three component model (6.9) (Kirdiashev *et al.*, 1979; Chukhlantsev, 1981, 1992). The brightness temperature of vegetated soil at a given polarization is found in this model as

$$T_b = T_v(1 - r - t) + e_s T_s t + T_v(1 - r - t)(1 - e_s)t \tag{6.21}$$

where T_v is the temperature of the vegetation, T_s is the temperature of the soil, r is the reflectivity, t is the transmissivity, e_s is the emissivity of the soil. For practical applications, the reflectivity and transmissivity of the vegetation layer are represented in simple analytical forms. In the single scattering approach, (6.21) reduces to the form:

$$T_b = (1 - \omega)(1 - e^{-\tau_0})T_v + e_s T_s e^{-\tau_0} + (1 - \omega)(1 - e^{-\tau_0})T_v(1 - e_s)e^{-\tau_0} \tag{6.22}$$

which is known and referred to as the τ-ω model (Van de Griend and Wigneron, 2004a) or the Kirdiashev *et al.* (1979) model (Mätzler, 2000). This form of the model today is generally accepted for the description of microwave emission from the Earth's surface in the presence of vegetation canopies. Numerous papers analyzed the applicability of this model. It was noted (Mätzler, 2000) that although the model gives a lower bound of the emissivity, by using effective parameters, the model becomes quite realistic. Several approaches were used for the effective choice of ω и τ_0. Mo *et al.* (1982) used approximations:

$$\tau_0^* = (1 - x\omega)\tau_0,\tag{6.23}$$

$$\omega^* = \frac{(1-x)\omega}{1-x\omega}\tag{6.24}$$

where x represents the fraction of radiation scattered by vegetation unit volume into the forward direction. Mätzler (2000) used somewhat more complicated approximations that take into account multiple reflections at the borders of vegetation layer. Vinokurova *et al.* (1991) proposed approximations based on the solution to the transfer equation for the layer with one-dimensional scattering. The following approximation formulas were obtained:

$$t \cong \{1 - r_0^2(\vartheta)\}\, e^{-k\tau\sec\vartheta} \cong e^{-k\tau_0},\tag{6.25}$$

$$r \cong r_0(\vartheta)(1 - e^{-2k\tau_0}),\tag{6.26}$$

$$r_0(\vartheta) \cong r_0(0)\{1 - \cos\vartheta\ln(1 + \sec\vartheta)\}/(1 - \ln 2),\tag{6.27}$$

$$r_0(0) = 0{,}6\,\omega(1-x)/(1 - \omega x + k),\tag{6.28}$$

$$k = \sqrt{1 - 2\omega x(1-\omega) - \omega^2}\tag{6.29}$$

where ϑ is the observation angle and x represents the fraction of radiation scattered by vegetation unit volume into the forward direction. For small values of ωx, these formulas and equations (6.23) and (6.24) produce practically the same transmissivity of the vegetation layer. But in the Mo *et al.* (1982) model, the reflectivity is found as

$$r \cong \omega^*(1 - e^{-k\tau_0}),\tag{6.30}$$

which differs from (6.26) by the index of exponent. In the isothermal case ($T_v = T_s = T$), the emissivity of a vegetation canopy is expressed by

$$e_c = \frac{T_b}{T} = e_s \{1 - (1 - t^2)(1 - r_0 t)\} + (1 - r_0 - r_0 t)(1 - t^2). \qquad (6.31)$$

Corrections to the emissivity for the non-isothermal case are easily obtained from (6.22) (Shutko, 1986). For $r_0 t \ll 1$, equation (6.31) reduces to the form

$$e_c = e_s t^2 + (1 - r_0)(1 - t^2). \qquad (6.32)$$

The approximation (6.31) and (6.32) were compared to the numerical solution of the transfer equation (Vinokurova *et al.*, 1991). It was shown that the difference between the numerically computed emissivity and approximation (6.32) does not exceed 0.01-0.02 for $\omega = 0\text{-}0.5$; $\tau = 0\text{-}4$; $\vartheta = 0\text{-}50°$; and for different forms of the phase function.

In experimental studies, most researchers concentrate on finding the parameters of the τ - ω model (6.22) from regression analysis of experimental data. The optical depth is found in these studies as $\tau = bW / \cos\vartheta$ (equation (4.124)), and the coefficient b and the single scattering albedo is estimated by fitting the model to the measured values of brightness temperatures. One can see that in this approach, as it was correctly noted in Van de Griend and Wigneron (2004b), the retrieved single scattering albedo represents the reflectivity of the optically thick vegetation layer r_0 (or the effective scattering albedo ω^*) rather than the scattering albedo of the unit volume of vegetation. With this semi-empirical approach and with the use of (6.32) for the isothermal case, the brightness temperature of a vegetation canopy is expressed in a simple form:

$$T_b = T\{e_s e^{-2\tau} + (1 - r_0)(1 - e^{-2\tau})\}. \qquad (6.33)$$

Kirdiashev *et al.* (1979) introduced the concept of the vegetation transfer coefficient (the slope reduction factor) as

$$\beta = t^2 = e^{-2\tau} = \frac{\Delta T_b}{\Delta T_{bs}} \qquad (6.34)$$

where $T_{bs} = e_s T$ is the brightness temperature of soil. This coefficient shows the decrease of the brightness temperature contrast ΔT_b, because of the presence of a vegetation canopy above a soil, comparing with that for the bare soil ΔT_{bs}. In other words, β-factor is the ratio of the slope of the emissivity-soil moisture function with a vegetation cover to the bare soil slope. A value of 1 indicates no effect while a small value means that most of radiometric sensitivity to soil moisture has been lost. In terms of the β-factor, the brightness temperature of vegetated soil is written as

$$T_b = T_{bs}\beta + (1 - r_0)(1 - \beta)T \,. \tag{6.35}$$

The models (6.22)-(6.35) have rather high accuracy that was confirmed by numerous theoretical and experimental studies. Their applicability was comprehensively grounded in Kirdiashev *et al.* (1979), Chukhlantsev (1981), Mo *et al.* (1982), Chukhlantsev and Shutko (1982), Pampaloni and Paloscia (1986), Brunfeldt and Ulaby (1986), Golovachev *et al.* (1989), Kerr and Njoku (1990), Vinokurova *et al.* (1991), Owe *et al.* (1992), Kerr and Wigneron (1994), Chanzy and Wigneron (2000), Mätzler (2000). At present, these models are basic ones for interpretation of microwave radiometric data (e.g., Shutko, 1986; Chukhlantsev and Shutko, 1987, 1988; Van de Griend *et al.*, 1996; Njoku and Li, 1999; Owe *et al.*, 2001; Jackson *et al.*, 2002; Burke *et al.*, 2002a; Njoku *et al.*, 2003; Crow *et al.*, (2001, 2005); Hornbuckle *et al.*, 2003; Liou *et al.*, 2001; Pellarin *et al.*, 2003; Davenport *et al.*, 2005). The models are very convenient for the sensitivity analysis (Davenport *et al.*, 2005) and for the development of inversion algorithms (Wigneron *et al.*, 2003; Crow *et al.*, 2005). Parameters of the models (the brightness temperature of soil, the optical depth, the reflectivity of optically thick vegetation layer, the physical temperature of vegetation and soil) are linked to geophysical parameters (soil moisture, soil roughness, vegetation water content, etc.). That makes it possible to retrieve the geophysical parameters from microwave radiometric measurements.

Relation of soil brightness temperature to soil parameters is considered in Chapter 3. Attenuation properties of vegetation canopies are discussed in Chapters 4 and 5. The reflectivity of vegetation layer r_0 (or the effective scattering albedo) is a function of vegetation type (dimensions of plant elements) and state (moisture of plant elements). Some calculated data for the single scattering albedo for different crops are presented in Chapter 4. These values allow one to calculate the reflectivity r_0 by using equations (6.24) or (6.27)-(6.28). The calculated reflectivity r_0 for typical crops are presented in Fig. 6.2.

Van de Griend and Wigneron (2004b) presented an overview of available experimental data of r_0. These data differ for different crops and different frequencies and vary typically between approximately 0.05 and 0.13, although also much higher values of up to 0.5 have been reported. Actually, there are very few experimental data on the reflectivity of vegetation. It is necessary to note that direct measurements of the reflectivity are practically absent. A few direct measurements of the reflectivity were performed at centimetric waves (Chukhlantsev, 1981) with the use of technique described in Chapter 5 (equation (5.10)). Some data of these measurements are presented in Fig. 6.2. As a rule, available data are retrieved by fitting the model to measured values of brightness temperatures. Since the values of reflectivity are rather small, one hardly can expect high accuracy of the retrieval. However, there is another possibility to estimate the reflectivity of vegetation layer. It is based on an existing relation between the emissivity and the backscattering coefficient.

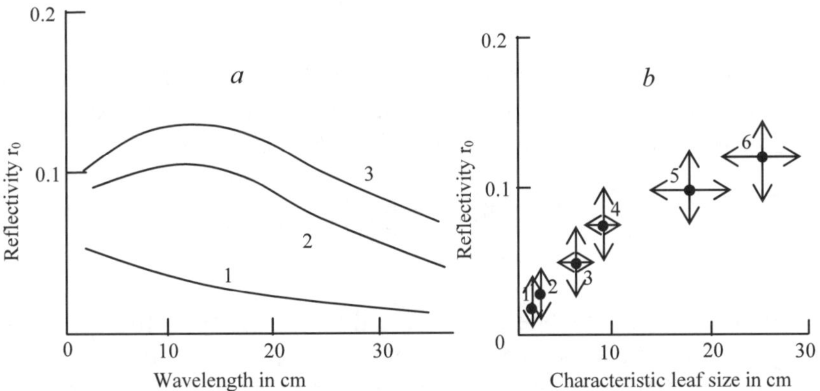

Fig. 6.2. a) Spectral dependence of reflectivity for small grains and alfalfa (1), soybeans (2), and corn (3). Observation angle is 0°; *b*) Experimental dependence of reflectivity on characteristic leaf size at 2.25 cm wavelength for wheat (1), alfalfa (2), cotton (3), corn (4), beets (5), and cabbage (6).

The backscattering coefficient, which is measured by a monostatic radar sensor, is a measure of intensity scattered by a surface into a backward direction. This value is a basic one in active microwave remote sensing of the Earth's surface (Ulaby *et al.*, 1981-1986). The backscattering coefficient $\sigma^0_{pq}(\vartheta_i,\varphi_i)$ is equal to the bistatic scattering coefficient, which is given by equation (3.4), upon the condition:

$$\sigma^0_{pq}(\vartheta_i,\varphi_i) = \sigma_{pq}(\vartheta_i,\varphi_i,\vartheta_r = \vartheta_i,\varphi_r = \varphi_i + \pi). \qquad (6.36)$$

From equations (6.36) and (3.1)-(3.2), one can see that there is an indirect link between the emissivity and the backscattering coefficient through the bistatic coefficient (Ferrazzoli *et al.*, 1989). Precisely, the emissivity is related to the latter by an integral relation, while the backscattering coefficient is the value the bistatic coefficient assumes when scattering is measured in the incidence direction.

The relation between the emissivity and backscatter coefficient was theoretically analyzed by Tsang *et al.* (1982) in the case of a uniform continuous medium embedded by isotropic scatterers. It was shown that when the layer of scatterers is sufficiently thick to behave as an infinite half space, a biunivocal relation holds which directly relates the emissivity to the backscattering coefficient. The relation is given by

$$\sigma^0(\vartheta) = \frac{[1 - e(\vartheta)]\cos\vartheta}{1 - \cos\vartheta \, \ln(1 + \sec\vartheta)} \, . \tag{6.37}$$

Taking into account that equation (6.27) represents the reflectivity of a layer with isotropic scattering, one obtains a relation between the backscattering coefficient and the reflectivity from (6.37), (6.27), and (3.1):

$$\sigma^0(\vartheta) = \frac{r_0(0)\cos\vartheta}{1 - \ln 2} \cong 3.27 \, r_0(0)\cos\vartheta \, . \tag{6.38}$$

Equation (6.38) allows one to estimate the reflectivity r_0 from radar measurements of the backscattering coefficient. It seems that the accuracy of this estimate is not worse than that of radiometric retrieval of r_0. Moreover, at present there are a lot of available data on measurements of backscattering coefficients at different frequencies, angles, polarizations, and for various crops. However, not all these data are suitable for estimation of reflectivity because in the real case of vegetated terrains the isotropic scattering conditions can not be satisfied, due to two major effects: a) the vegetation layer is not infinitely thick; b) the scatterers are not isotropic. Therefore, a selection of backscatter data should be gathered to provide the best estimate of the reflectivity. To satisfy the condition of an optically thick vegetation layer, one could choose the data of σ^0 that are obtained for well-developed crops at large nadir observation angles. Moreover, Ferrazzoli *et al.* (1989) have shown that, in this case, the assumption of isotropic scattering becomes more and more realistic although the isotropic backscatter still remains higher than that in a real situation. One can get some idea of the discrepancy between the isotropic scattering model and observed experimental data considering angular dependencies of the backscattering coefficient. For isotropic scattering, these dependencies obey the cosine law (6.38). An example of such angular dependence is presented in Fig. 6.3. Experimental data for the backscattering coefficient for soybeans from Eom and Fung (1984) are compared in the figure with calculated dependence (6.37) for different values of r_0. One can see that, for relatively high frequencies, calculated dependence fits experimental data rather well. In the *L*-band, the vegetation layer is, probably, not optically thick, and a discrepancy between calculated and experimental values is observed. Nevertheless, some estimates of the reflectivity can be obtained from the approach proposed. The reflectivity of soybeans is not less than 0.01-0.02 in the *L*-band, not less than 0.03-0.04 in the *C*-band, and is more than 0.05-0.06 in the *X*-band. The estimate can be improved taking into account results presented by Ferrazzoli *et al.* (1989). It was shown (Ferrazzoli *et al.*, 1989) that the isotropic model produces a backscattering coefficient that is 2-4 dB more than that in a real situation (Fig. 6.4). That leads to an underestimate of reflectivity from equation (6.37). Estimates of reflectivity obtained from (6.37) should be multiplied by a factor of 1.6-2. In this case, values of reflectivity retrieved from backscatter measurements are consistent with those retrieved from radiometric measurements. Now, a big volume of data on backscattering from vegetation canopies is available, which allows one to gather statistics on the vegetation reflectivity. This work was partially done in Chukhlantsev (1989a) where some data on the backscattering coefficient of different crops were collected. These data are added to those from Ferrazzoli *et al.* (1989) in Fig. 6.4. One can see that magnitudes of the reflectivity r_0 for natural crops hardly can exceed a value of 0.15. This is confirmed by data collected in Van de Griend and Wigneron (2004b) and some other data collected by Ferrazzoli and Guerriero (1994). Greater magnitudes of this value retrieved from radiometric measurements are connected, probably, with a great error of the retrieval. Ferrazzoli and Guerriero (1994) also reported a decreasing character of frequency dependence of r_0 for "wide leaf" crops at frequencies above 10 GHz, in general agreement with data presented in Fig. 6.2.

Data on angular dependence of the reflectivity r_0 are sparse. Calculated angular dependence of the ratio $r_0(\vartheta)/r_0(\vartheta=0)$ is presented in Fig. 6.5. These data are consistent with those retrieved from radiometric measurements (Van de Griend and Wigneron, 2004b).

The reflectivity of a vegetation canopy is a function of not only the type of vegetation but also depends on the moisture status of the vegetation. Dependence of the single scattering albedo (and the reflectivity that is linked to the single scattering albedo) on the leaf moisture content was calculated in several papers (e.g., Chukhlantsev, 1989a; Ferrazzoli and Guerriero, 1994). It has been shown (Chukhlantsev, 1989a) that when the gravimetric moisture (in wet weight basis) of a leaf varies from 20% to 60%, the single scattering albedo becomes approximately twice as large. That is confirmed by the measurements of scattering and absorption cross sections reported in Chapter 4. This fact points to a potential for microwave radiometric measurements in remote control of vegetation moisture status.

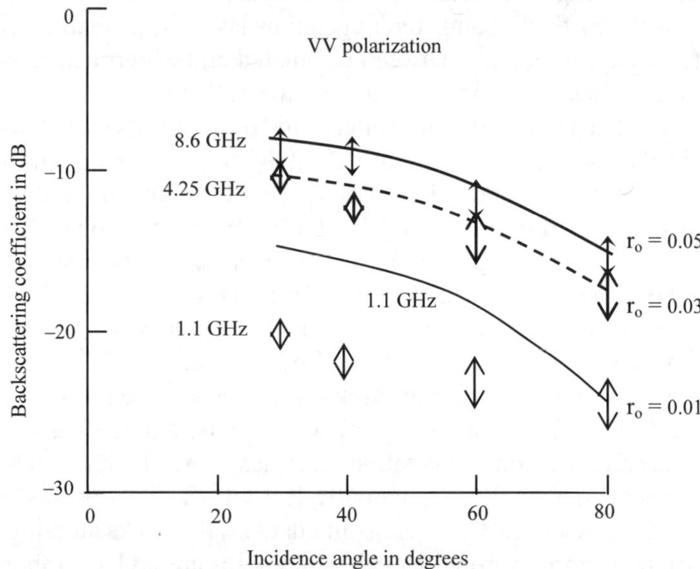

Fig. 6.3. Angular dependence of backscattering coefficient for soybeans.

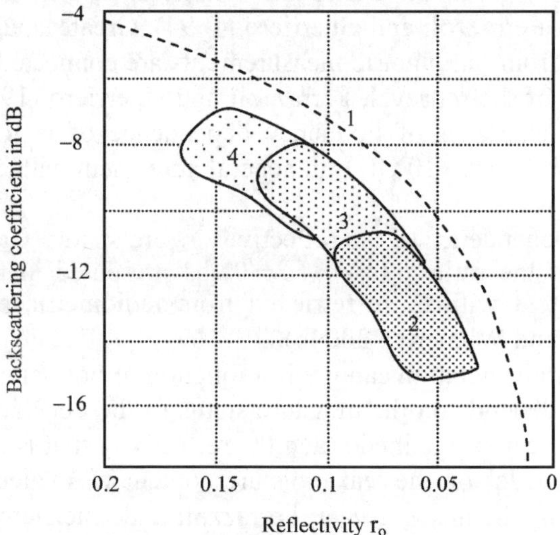

Fig. 6.4. Relation between reflectivity r_0 and backscattering coefficient for a half-space of isotropic scatterers (1) and in a real situation for wheat and alfalfa (2), corn and soybeans (3), and beets (4). Observation angle is 45°. Experimental data are for L-, C-, and X-band.

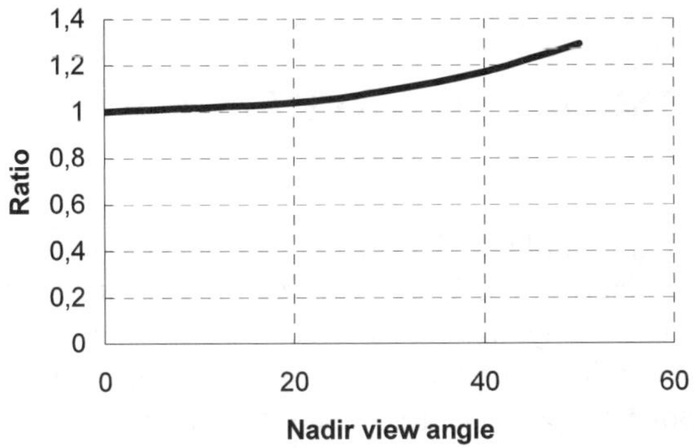

Fig. 6.5. Angular dependence of the ratio $r_0(\vartheta)/r_0(0°)$.

Sometimes vegetation has a mosaic structure and does not completely cover the Earth's surface. This is a condition inherited especially by forests. The presence of "gaps" in the vegetation cover can be taken into account by the introduction of factor ξ_v (see Basharinov *et al.* (1974) model (6.10)) that is the fraction of soil covered by vegetation which is determined by

$$\xi_v = 1 - \frac{\sum S_{gap}}{S_A} \tag{6.39}$$

where $\sum S_{gap}$ is the total area of the gaps within the area of the antenna footprint at the surface S_A. With incomplete coverage of the surface, the brightness temperature is found from modified equation (5.35):

$$T_b = [T_{bs}\beta + (1 - r_0)(1 - \beta)T]\xi_v + T_{bs}(1 - \xi_v). \tag{6.40}$$

The question of the gaps' effects was investigated in Chukhlantsev (1981). It has been shown that the gaps, with geometrical dimensions that satisfy the condition

$$D_{gap} = \frac{S_{gap}}{\lambda h} > 6...10 \tag{6.41}$$

where h is the canopy height, can be considered as an open area. The canopy with gaps satisfying the condition $D < 2$ can be considered as a continuous one.

6.3. MODELING MICROWAVE EMISSION FROM FORESTS

Forests have a complicated structure that makes modeling their microwave emission properties difficult. Several models are available now that can be grouped into physical ones based on calculating electromagnetic wave interactions with vegetation, and empirical models derived from radiometric observations by means of regression techniques.

The model (6.40) accounting for incomplete coverage of surface that inherited to forest stands was used by Kirdiashev *et al.* (1979) to interpret experimental data and to retrieve the forest transfer coefficient values from radiometric measurements. The model is physically grounded (see above) and has the advantage that it is very simple, and model parameters (β and r_0) have a clear physical sense and can be specified by fitting the model to experimental data. These parameters relate to biometric features of vegetation (crown water content and shape and size of leaves) that can be directly measured.

A numerical (Tor Fergata) model based on the radiative transfer theory and the matrix-doubling algorithm was implemented by Ferrazzoli and Guerriero (1996). The forest medium was subdivided into three main regions: crown, trunks, and soil. The crown is filled with scatterers representing leaves, needles, twigs, and branches. The scatterers were assumed uniformly located within the crown. The Rayleigh-Gans approximation was used to model the electromagnetic properties of needles and flat leaves. At higher frequencies, the physical optics approximation (the model of large plane particle) was used, which substitutes the field inside a thin disk with the one of a slab with the same thickness. The model of length cylinder (infinite length approximation) was used for cylindrical objects like twigs, branches, and trunks. The soil was described as a homogeneous half-space with known permittivity. The computational procedure applied by Ferrazzoli and Guerriero (1996) is already described in the beginning of Section 6.2. The following results were obtained from the calculations. Crown attenuation in dB practically linearly depends on the forest dry biomass. At *L*-band and at 45° incidence angle, the attenuation varies from ~3 to ~11 dB at horizontal polarization and from ~4 dB to ~14 dB at vertical polarization in the deciduous forest dry biomass range 45-240 tons/ha. Contribution of trunks to the total attenuation of crowns is negligible. The forest emissivity monotonically increases with the increase of forest dry biomass from 0 to 200 tons/ha. For a given biomass, the emissivity increases with an increase in frequency. At high frequencies, deciduous leaves made the emissivity lower, while the presence of coniferous needles made the emissivity higher. At low frequencies, even old forests are semitransparent, and underlying soil might play an important role in the emission.

The relation between L-band emission and the woody volume of trees was analyzed in detail by Macelloni *et al.* (2001). A first-order discrete element radiative transfer model was used, which describes a forest as a two-layer medium (trunk and crown) over a rough interface (soil). Trunks were approximated by cylinders and crowns by ensembles of cylinders and disks. An analysis of various contributions to total emission from trees showed that, at L-band, the emission is mainly due to crowns and that contribution for double reflection from soil is negligible. The main contribution to the crown emission is due to primary and medium branches, and the contribution of trunks and leaves in this particular frequency range is small. That is consistent with results presented by Ferrazzoli and Guerriero (1996).

Consistent results were also obtained by Ferrazzoli *et al.* (2002), who used Tor Fergata model (Ferrazzoli and Guerriero, 1996) to characterize the performance of an L-band space borne radiometer when flying over forests. The model was first refined on the basis of more recent experimental studies. The refinement concerned mostly the tree geometrical parameters that are needed as input to the model. In preliminary studies, particularly in Ferrazzoli and Guerriero (1996), morphological parameters of trees were assigned by making reasonable assumptions based on the scarce data available in literature at that time. Recent studies and measurements carried out in the coniferous Landes forest in France (Le Toan *et al.*, 1992), and the Duke forest in North Carolina, USA (Kasischke *et al.*, 1994) allowed the authors to develop a growth parameterization able to fulfill the whole input requirement of the Tor Fergata model. The measurements performed in the maritime pines of the French forest included biomass, density, diameter at breast height (*dbh*), basal area, trunk height, and tree height: they were synthesized by means of allometric equations that are functions of tree age. The Loblolly Pines of the American forest were studied in detail, including the size and density of primary and secondary branches: a set of allometric equations was defined to connect each tree parameter to *dbh*. In order to simulate emissivity measurements over the Landes forest, the authors used the allometric equations reported by Le Toan *et al.*, (1992) integrated with some equations developed for the Duke forest. In the modeling, tree age is considered as an independent parameter. Dependence of some model parameters on the tree age is presented in Ferrazzoli *et al.* (2002). Forest (Landes pine) dry biomass approximately linearly increases from ~50 tons/ha at 10 years of age to ~150 tons/ha at 50 years of age. Maximum branch radius approximately linearly increases from 2 cm at 10 years of age to 4 cm at 50 years of age. Branch volume fraction does not change significantly after 10 years of age and is about 10% of wood volume for the Landes forest and about 20% for the Duke forests. (That is consistent with known morphological data reported by Bazilevich and Rodin (1967). According to these data the weight fraction of trunks in a forest varies within 70-80%.) Results of model simulations for the Landes forest emissivity at C-band were compared in the paper with experimental results obtained from PORTOS radiometer (Wigneron *et al.*, 1997). It was found that experimental values are better reproduced by simulating a forest with an understory (with 0.8 kg/m^2 water content). Model simulations were conducted to reveal the contribution of different tree components to forest emissivity and transmissivity. It was found that the main contributor to both the emissivity and transmissivity is the branch component. The model developed is then applied to calibrate the τ - ω model (5.22) to be used for soil moisture retrieval on a global scale. It was found that the equivalent optical depth in the τ - ω model is proportional to the branch water content. From the simulated dependence of the optical depth on the branch water content presented in Ferrazzoli *et al.* (2002), it follows that at 1.4 GHz $\tau \approx bW_{branches}$, with the coefficient $b \sim 0.4$. A big spread of simulated equivalent single scattering albedo (from 0 to 0.2) was observed. The model approach developed in Ferrazzoli *et al.* (2002) was later used by Pellarin *et al.* (2003) for global simulation of L-band brightness temperature over land.

In Lang *et al.* (2001), the pine stands were modeled by using a discrete scattering approach. The model was used to investigate the radiometric sensitivity to soil moisture of the underlying ground in the case of the 18-year-old stand at L-band. The brightness temperature varied only about 10 K over the range from dry to wet soil.

Kruopis *et al.* (1999) used the model (5.35) with $r_0 = 0$ to estimate forest transmissivity $t = \sqrt{\beta}$. The model (5.35) was also used by Milshin *et al.* (1999) and Milshin and Grankov (2000) for interpretation of airborne microwave radiometric measurements.

It can be concluded that most researches more and more incline to the use of simple τ - ω model (5.22) or its modification (5.35) in modeling microwave emission properties of forests.

One special problem of microwave radiometry of forests is the use of microwave radiometric measurements for mapping forest fires (Borodin *et al.*, 1976; Borodin, 2001). In the case of a stable ground fire, the flame temperature over fire contour achieves a value of 800-900 K, while an increase in the temperature of soil and burned materials at burned-out parts of the fire is about 10-25 K and 30-80 K, respectively. The flame height is about 1-2 m at the fire front and about 0.3-0.6 m at its back and flanks. The width of intensively burning edge over the fire contour varies from 1 to 3 m. The speed of fire spreading is 1-2 m/min at its front and 0.6-1 m/min and 0.2-0.6 m/min at its flanks and back, respectively. Forest fire is usually accompanied by a significant smoke screen. In the presence of extensive smoke, it is not always possible to contour the forest fire visually from a plane or by the use of infrared sensors. Microwave radiometric sensors can be useful in such a situation since the smoke is transparent for microwaves. Radiative models of forest fires were developed by Borodin *et al.* (1976) and Stakankin (1986). It was shown that flame particles with dimensions of 100-300 µm produce the main contribution to the flame microwave emission. The intensity of forest fire emission is maximum in a wavelength range of 0.8-1-5 cm. An increase of brightness temperature over a front of forest fire with respect to the level of background forest emission reaches up to 200 K at 0.8 cm and 100 K at 3.4 cm even in the presence of a heavy smoke screen. Forest crowns decrease the aforesaid brightness temperature contrast to 40-60 K, but the contrast is still big enough to be confidently detected. These brightness temperature contrasts were observed in numerous experiments (Borodin *et al.*, 1976; Borodin, 2001).

6.4. POLARIZATION PROPERTIES OF MICROWAVE EMISSION FROM VEGETATION CANOPIES

Two approaches (the numerical model and semi-empirical analytical model) described in the previous section are generally accepted today by most experimentalists for modeling emission properties of vegetated soils. These approaches provide close results and a good fit of experimental data and can supplement each other. Both of them are used in modeling polarization properties of emission from soil covered with vegetation.

The first approach was used to study polarization features of microwave emission from vegetation canopies in papers by Ferrazzoli *et al.* (1992a), Ferrazzoli and Guerriero (1994, 1996), Ferrazzoli *et al.* (2000), Macelloni *et al.* (2001), Ferrazzoli *et al.* (2002). The numerical model allows one to calculate the emissivity at both vertical and horizontal polarizations, to obtain angular dependencies of the vegetation optical depth and of the vegetation half-space reflectivity. Once calculated and retouched by comparing with experimental data, the numerical results can be used to calibrate the semi-empirical analytical model (5.22). In the isothermal case, the latter can be represented in the form (6.33):

$$T_b(p,\vartheta) = T\{e_s e^{-2\tau} + (1-r_0)(1-e^{-2\tau})\} \tag{6.42}$$

where the soil emissivity e_s, the optical depth τ, and the reflectivity r_0 are functions of the polarization p and observation angle ϑ.

An increasing interest in polarization radiometric measurements is conditioned by the following reasons. In the lower range of the microwave band, where radiation is also contributed by subsurface layers, the emission from natural terrains is mainly controlled by soil water content. In addition to surface soil moisture, other surface parameters also play a role in the low frequency microwave emission: vegetation optical depth, surface temperature, vegetation reflectivity, soil surface roughness, etc. But at low frequencies the vegetation reflectivity is very small, and the following approximation is usually used for vegetated soil emissivity (Wang *et al.*, 1989):

$$e(p,\vartheta) \approx 1 - r_s(p,\vartheta)\cdot\exp\{-h_s - 2\tau\}; \quad \tau = b(p,\vartheta)\cdot W \tag{6.43}$$

where h_s is the soil roughness factor. Multi-configuration measurements (at different polarizations and observation angles) could thus be useful in soil moisture retrievals for decoupling between the effects of soil moisture and surface roughness parameters (Wigneron *et al.*, 2003; Crow *et al.*, 2005). These measurements, in principle, allow one to retrieve simultaneously soil moisture and vegetation optical depth (vegetation water content), as well as other parameters.

On the other hand, in the high frequency range of the microwave spectrum, the role of vegetation becomes considerable. Well-developed vegetation practically completely screens emission from soils at high frequencies and the vegetation emissivity is approximated by

$$e(p,\vartheta) \approx [1 - r_0(p,\vartheta)]\xi_v + (1 - \xi_v)e_s(p,\vartheta) \tag{6.44}$$

where ξ_v is the fraction of surface covered with vegetation that is close to unity for well-developed crops. Upon this condition, polarization properties of microwave emission from a vegetation canopy are determined by the vegetation reflectivity that is a function of type of vegetation, its leaf area index (the total area of leaves per unit area), and the moisture of plant elements. Therefore, polarization measurements at high frequencies have a potential in monitoring the vegetation state.

6.5. SPATIAL VARIATIONS OF MICROWAVE EMISSION FROM THE EARTH'S SURFACE

The brightness temperature of the Earth's surface is a random function of coordinate $\vec{r}$ and time t, and depends on the wavelength λ, observation angle ϑ_i, and polarization p:

$$T_b = T_b(\vec{r}, t, \lambda, \vartheta_i, p). \tag{6.45}$$

Assuming statistical uniformity of the brightness temperature random field within a landscape, statistical properties of this random field are determined by the brightness temperature probability density $P_{T_b}(T_b)dT_b$ and autocorrelation function

$$\psi(\vec{r}_1 - \vec{r}_2) = \langle T_b(\vec{r}_1) \cdot T_b(\vec{r}_2) \rangle - \langle T_b(\vec{r}_1) \rangle \langle T_b(\vec{r}_2) \rangle \tag{6.46}$$

where brackets denote statistical averaging. The form of autocorrelation function depends on the averaging procedure. Particularly, if the brightness temperature is measured along a given direction (profiling measurements) with an antenna having a certain finite directional pattern, an averaging of the brightness temperature takes place that is equivalent to the signal spatial filtration. For a simple case, when the antenna directional pattern is uniform within a given angle, the field of brightness temperature is transformed by the antenna as

$$T_A(x) = \frac{1}{R_A} \int_{x-\frac{R_A}{2}}^{x+\frac{R_A}{2}} T_b(x') dx' \tag{6.47}$$

where R_A is the linear size of antenna footprint, and x is the coordinate along the direction of measurements. With this transform, the autocorrelation function is expressed as

$$\psi_A(\tau) = \int_{-\infty}^{+\infty} \left(\frac{\sin[\kappa R_A / 2]}{\kappa R_A / 2} \right)^2 e^{j\kappa\tau} dG(\kappa), \quad \tau = x_1 - x_2 \tag{6.48}$$

where $G(\kappa)$ is the Fourier transform of the correlation function (6.46). Experimental research, which was carried out with airborne radiometers at low altitudes of flights, has shown (Shutko, 1983) that the correlation function (6.46) is satisfactory described by the approximation

$$\psi(\tau) = \sigma^2 e^{-\tau/\rho}$$
(6.49)

where ρ is the correlation length and σ^2 is the dispersion of brightness temperature variations. The Fourier transform of (6.49) is

$$G(\kappa) = \frac{\sigma^2 \rho}{\pi(1 + \kappa^2 \rho^2)}.$$
(6.50)

Integrating (6.48), one obtains:

$$\psi_A(\tau) = \frac{\sigma^2 \rho^2}{R_A^2}\left[\frac{2(R_A - \tau)}{\rho} + e^{-(R_A+\tau)/\rho} + e^{-(R_A-\tau)/\rho} - 2e^{-\tau/\rho}\right], \quad \tau \leq R_A$$

$$\psi_A(\tau) = \frac{\sigma^2 \rho^2}{R_A^2} e^{-\tau/\rho}\left[e^{-R_A/\rho} + e^{R_A/\rho} - 2\right], \quad \tau \geq R_A$$
(6.51)

When $R_A < (3-5)\rho$, the form of correlation function $\psi_A(\tau)$ is determined mainly by the brightness temperature spatial structure, and $\psi_A(\tau)$ tends to that given by (6.49). When $R_A > 10\rho$, the correlation length of $\psi_A(\tau)$ tends to R_A. The dispersion of brightness temperature decreases with an increase in ratio R_A/ρ. When $R_A/\rho \gg 1$,

$$\sigma_A^2 = \sigma^2 \cdot 2\frac{R_A}{\rho}.$$
(6.52)

Parameters σ^2 and ρ are determined by both landscape type and landscape hydrological situation. Measurements of these parameters are used, particularly, for an estimation of fire risk in forests (Grankov *et al.*, 1999; Yakimov, 1996).

6.6. GLOBAL SIMULATION OF MICROWAVE EMISSION FROM LAND

To evaluate potentials of projects based on L-band microwave radiometers on satellite platforms, such as the Soil Moisture and Ocean Salinity (SMOS) Mission (Kerr *et al.*, 2001; Silvestrin *et al.*, 2001), simulated brightness temperature maps at satellite scale resolutions were built by Pellarin *et al.* (2003a). The main objective of the study was to obtain a global synthetic L-band brightness temperature dataset, corresponding to a realistic range of surface state variables, and addressing seasonal and inter-annual variability. Such a dataset can be used to assess the feasibility of retrieving geophysical quantities, using different approaches (use of indices, different cost functions for methods based on forward modeling inversion). The simulated brightness temperatures were obtained at the half-degree resolution (corresponding to the SMOS spatial resolution, which ranges from 20 to 50 km).

The simulation process accounted for subpixel heterogeneity by considering several surface types. The brightness temperature of the mixed pixel was written as

$$T_b = f_B T_{bB} + f_F T_{bF} + f_H T_{bH} + f_W T_{bW} \qquad (6.53)$$

where f is the cover fraction of different surface types (B, F, H, W for bare soil, forests, herbaceous vegetation-covered surface, open water surface, respectively), and T_b is the brightness temperature at the incidence angle ϑ and at a given polarization. Within each vegetation tile, only one vegetation type was considered, corresponding to the largest cover fraction derived from a global land-cover database (Masson *et al.*, 2003), derived from land-cover and climatic maps at a resolution of 1 km. Within the herbaceous vegetation tile, either grassland or crops were considered; within the forest tile, either coniferous, broadleaf, or tropical forests were considered. For all vegetation tiles, the same τ - ω approach (6.22) was used to calculate the brightness temperature. Equation (4.124) $\tau = bW / \cos\vartheta$ was used to relate the optical depth to the vegetation water content. Vegetation parameters used in the modeling are presented in Table 6.1. Over forests, the value of τ was assumed to be constant and was related to the branch water content only.

Table 6.1. Vegetation parameters used in the global modeling of L-band brightness temperature.

Surface Type	ω	b (m^2 kg^{-1})	W (kg m^{-2})
Grassland	0.05	0.20	0.5 LAI
Crops	0.05	0.15	0.5 LAI
Rainforests	0.15	0.33	6
Broadleaf forests	0.15	0.33	4
Coniferous forests	0.15	0.33	3

A sensitivity study was made and showed that the uncertainty on W, b, and h_s (soil roughness parameter) produced the largest error. Global maps of synthetic L-band brightness temperatures were obtained. The maps show roughly an expected dynamic range of brightness temperature due to the change of geophysical quantities. Over rainforests the annual amplitude of brightness temperature is small and L-band radiometric measurements are unlikely to be useful. Over steppes this amplitude can achieve 60 K and more. Annual amplitude of brightness temperature over boreal forests is about 30 K. The authors emphasize that the main objective of the work was not to simulate the real world, but, rather obtain a dataset presenting a realistic range of surface states and the brightness temperature. In particular, the variability of brightness temperature of a real scenario can be only partially explained by a simulated database. Pellarin *et al.* (2003b) used the global simulation of L-band brightness temperature described above to test simple regression models for retrieval surface soil moisture from radiometric data.

As seen from Table 6.1, the model uses rather big values of vegetation water content for forest branches that provide the worst estimate of forest opacity ($\tau \sim 1$-1.2). In a real situation, forests can be rather sparse and an average value of forest optical depth at L-band can be less (see next chapter). Therefore, annual amplitudes of brightness temperature in a real scenario can be greater than those produced by the model.

The τ - ω approach (6.22) was used by Grankov *et al.* (2004) to calculate the brightness temperature of forests in satellite observations. It was noted that the equivalent single scattering albedo in the τ - ω model is, by definition, the reflectivity of an optically thick vegetation layer. It can be calculated with the use of a numerical approach (Ferrazzoli *et al.*, 2002). But it is better to use the values of the forest's single scattering albedo obtained from experimental data (Wigneron *et al.*, 1997; Macelloni *et al.*, 2001; Milshin and Grankov, 2000). As it follows from Wigneron *et al.* (1997) data, at C-band, the value of old forest reflectivity (equivalent ω) is about 0.05. One hardly can expect a great increase in this value (to 0.15 (see Table 6.1)) with a decrease in frequency (to L-band), since the reflectivity usually decreases with a decrease in frequency. Moreover, from experiments by Macelloni *et al.* (2001) and Milshin and Grankov (2000), it follows that at *L*-band the forest emissivity achieves 0.913-0.945. This implies that the equivalent single scattering albedo is at least less than 0.087. The value $\omega = 0.04\text{-}0.05$ is usually used (Grankov *et al.*, 2004) in modeling emission properties of forests at *L*-band. Another key problem in the modeling is an appropriate choice of model for the forest optical depth. The model (4.124) $\tau = bW / \cos\vartheta$ is very convenient but it requires prior knowledge of forest crown water content. In Pellarin *et al.* (2003a) the coefficient *b* equals 0.33 for all forests and only three values of crown (branch) water content are assigned for all forest types with a constant value of water content within a separate class (see Table 6.1). It is clear that the model can be improved much by assigning more accurate values of forest crown water content for each pixel, but it requires more detailed information on forest biomass distribution over the globe. The forest optical depth in global emission models can also be estimated from a semi-empirical model for the extinction rate of microwave radiation in forests (Chukhlantsev and Golovachev, 2002; Chukhlantsev *et al.*, 2003c):

$$\gamma(dB / m) = 8 \cdot 10^{-4} \cdot f^{0.8} \tag{6.54}$$

where *f* is the frequency in MHz. The optical depth in this approach is

$$\tau = \frac{1}{4.34} 8 \cdot 10^{-4} \cdot f^{0.8} \cdot h_f \sec\vartheta \tag{6.55}$$

where h_f is the height of forest crown. However, the use of this approach requires knowledge of forest height. It seems to be reasonable to combine both approaches in the modeling (Grankov *et al.*, 2004).

Chapter 7

EXPERIMENTAL RESEARCH ON MICROWAVE EMISSION FROM VEGETATION CANOPIES

7.1. RESEARCH ON MICROWAVE EMISSION FROM VEGETATED FIELDS

To develop, validate, and test theoretical models, experimental research needs to be performed. The research should be conducted with a well-adapted methodology over the whole range of possible observation conditions (frequency, polarization, observation angles) and for a large variety of surfaces (various crops, forests, natural vegetation, types of soils and soil moisture). All measurements can be separated into two main groups: measurements done over well-controlled sites, and large scale measurements mainly for remote sensing applications. Several papers are available now that present reviews of experimental studies of microwave emission from vegetation canopies (e.g., Chukhlantsev, 1992; Kerr and Wigneron, 1994; Chanzy and Wigneron, 2000; Pampaloni, 2004). A complete description of the various field experiments is too large. Therefore, only a brief historical review of some experiments will be given below.

Kirdiashev *et al.* (1979) presented some results of experiments conducted at the Institute of Radioengineering and Electronics of the USSR Academy of Sciences during the period from 1974 to 1976. The experiments were basically performed with the use of onboard radiometers (a four-engine plane Iliushin-18 was used) operated in the wavelength range from 0.8 to 30 cm. The brightness temperature and emissivity values were estimated with the calibration by internal standards of the radiometers as well as by the brightness temperature of a smooth water surface with known water temperature and dense forest areas. The emissivity of the water surface was determined by the Fresnel formulas; the emissivity of the calibration forest areas was assumed to be equal to 0.98. The estimated error of the brightness temperature measurements was 5-15 K. The study of microwave emission from vegetated areas showed that the effect of vegetation on the emission from the earth's surface is determined by the type of vegetative cover and its state. It was found that the basic parameters, determining the emission level, are the following: the emissivity of the soil surface, the transmissivity and reflectivity of the vegetation layer, and the degree to which the soil is covered with vegetation (the surface cover fraction). Spectral peculiarities of microwave emission that are inherent to various vegetation types arise because of their inherent different values of numbered parameters. Three groups of vegetation canopies were delineated. The first group included cultivated crops and natural vegetation with small leaves (the reflectivity is small and the transmissivity increases monotonically with an increase in wavelength). This vegetation was found to be practically transparent for radiation at low frequencies of the microwave band and semi-transparent at higher frequencies. The second group included "broad-leaf" crops like corn, sunflower, sugar beets, etc. The reflectivity of this kind of vegetation canopies was found to be significant. The third group included forest vegetation that is characterized by low values of transmissivity even at low frequencies of microwave band. Estimates of vegetation transmissivity and reflectivity were obtained comparing the emissivity contrasts observed between very wet and very dry soil conditions for the vegetated soil and the bare soil. The slope reduction factor β (6.34) was determined for each vegetation group (Fig. 7.1). Fig. 7.1 was later reproduced in numerous papers and is still used for estimates of vegetation screening effect.

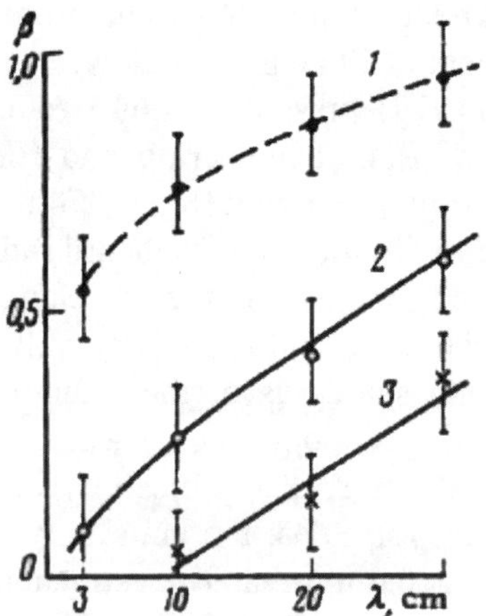

Fig. 7.1. The slope reduction factor of emissivity-soil moisture function versus the wavelength for small cereals (1), corn (2), and forest (3).

It was proposed in Kirdiashev *et al.* (1979) that the optical depth of vegetation layer is proportional to the vegetation water content and the proportionality coefficient depends on the type and state (plant moisture content) of vegetation.

In a paper by Basharinov *et al.* (1979), experimental data on the optical depth for corn canopies are presented. The data were obtained with the use of airborne radiometer (a light biplane Antonov-2 was used) that operated at a wavelength of 18 cm. The linear relationship between the optical depth and the vegetation wet biomass was observed. In 1997-1998, two more radiometers were installed on the board and measurements were conducted at 2.25, 18, and 30 cm wavelengths (Grankov *et al.*, 1979). Besides, an infrared radiometer was used for determining surface temperature. Measurements were performed over cotton fields from June to September. An infrared sensor allowed obtaining estimates of the surface cover fraction. During this period the height of cotton increased from 40 cm to 100 cm and the surface cover fraction increased from 0.5 to 1. At the same time, the slope reduction factor decreased from 0.95 to 0.85 at 30 cm, from 0.9 to 0.75 at 18 cm, and from 0.6 to 0.15 at 2.25 cm.

Burke and Schmugge (1982) reported some results of a soil moisture experiment conducted by NASA in 1975. Measurements of emissions at various microwave wavelengths were made over furrowed bare fields as well as fields with various vegetation canopies. Onboard radiometers operated in the wavelength range 0.8 to 21 cm. In addition to the microwave instruments the scientific package included an infrared radiometer for measuring surface temperature. Ground measurements were made in 46 fields, among which 28 were bare and 18 had vegetative covers of either alfalfa or wheat. The data analysis was carried out for measurements made at 21, 2.8, and 1.67 cm. At 21 cm, for vegetated fields, the sensitivity to soil moisture was still strong. For moist fields, the vegetation canopy produced an increase in brightness temperature of about 15 K. It was noted that the increase could be even more, but it was further compensated by the smoother surface under the vegetation of the furrow bare fields. As a result, for drier fields the brightness temperature responses over vegetated fields are similar to, or in some cases lower than, those over bare fields. (The slope reduction factor of this particular vegetation at 21 cm can be estimated from the data in Fig. 7.1. It was at least greater than 0.6 for nadir observations.) For vegetated fields, the 2.8 cm and 1.67 cm data showed no correlation to the background soil moisture. It indicated that the vegetation canopy becomes "opaque" at these wavelengths. Furthermore, brightness temperature at 2.8 cm was generally 20-30 K lower than the physical temperature that can be attributed to the scattering effect (r_0) of the vegetation layer. At 1.67 cm, the difference between brightness and physical temperatures was even more. It was concluded in the paper that, at longer wavelengths (e.g., *L*-, *S*-, and *C*-band), the radiation from soil can still penetrate through the vegetation layer providing sufficient surface moisture information. At shorter wavelengths, the radiation from soil cannot penetrate through a vegetation canopy. At these wavelengths, the brightness temperature response is, in general, 20-30K cooler than the physical temperature.

Barton (1978) used a 2.8 cm microwave radiometer mounted on an aircraft to measure bare soil and uniform grass sites. Strong relationships between soil moisture and emissivity were observed for the bare soil sites; however, as expected in the vegetated sites, no pattern could be detected. Newton and Rouse (1980) compared data collected using truck-mounted radiometers operating at 2.8 and 21 cm for bare soil and sorghum at two plant heights. At the shorter wavelength, they observed the same results obtained by Barton. At 21 cm they found that even with the highest sorghum cover the radiometer was still sensitive to the soil moisture variations. No attempt was made to quantify the vegetation effect.

Jackson *et al.* (1982) and Wang *et al.* (1982a) reported about a series of experiments conducted by NASA and USDA during October, 1979 and the summer of 1980 on plots located on Beltsville Agricultural Research Center in Maryland. Plot land covers included bare soil, corn, soybeans, and orchard grass. Wet biomass and vegetation water content measurements were made approximately every week during the course of the experiments. Plots of vegetation wet biomass versus time are presented in the paper. *C*- (6 cm) and *L*-band (21.4 cm) microwave radiometers were installed in a mobile truck system with the antennas mounted on an extendable boom. At the operating position the nominal area measured by the sensor was 1.5 m in diameter. Linear regressions for the emissivity-soil moisture function were obtained for both bare and vegetated fields (Jackson *et al.*, 1982). The slope reduction factor β (6.34) was then estimated for all vegetation types. The coefficient *b* (5.124) was found to be 0.11-0.12 in the *L*-band and 0.15-0.18 in the *C*-band. It was found that the *C*-band system alone is of little value in estimating the soil moisture under vegetation canopies because of a rather great error of soil moisture retrieval. The paper by Wang *et al.* (1982a) presents detailed data on the emissivity and polarization factor (at 40° observation angle) of bare and vegetated soils as a function of the soil moisture. The presence of vegetation cover reduced the measured polarization factor (the polarization index). While this factor increased monotonically from 0.1-0.16 up to 0.5-0.6 with increase in soil moisture content for bare soil at 1.4 and 5 GHz frequencies, the factor at 5 GHz was limited to <0.14 for vegetated fields over a wide range of moisture content. It is noted in the paper that a detailed study of the polarization factor could help delineate field type and determine field surface soil moisture. Some more data on polarization properties of microwave emission from investigated canopies were reported by Wang *et al.* (1982b, 1984).

Ulaby *et al.* (1983a, 1983b) reported data of radiometric measurements at 1.4 and 5 GHz performed in 1978 with airborne sensors. The canopy loss factor in dB (the optical depth) of mature corn was found comparing data for vegetated and bare soils. For 215 cm corn ($W = 4$ kg/m^2), these data (Ulaby *et al.*, 1983b) produced the slope reduction factor $\beta = 0.46$ and $\beta = 0.25$ at 1.4 and 5 GHz, respectively (compare with data in Fig. 7.1).

The described above first experiments on microwave radiometry of vegetated fields have shown that the presence of a vegetation cover reduces microwave sensitivity to variations in the underlying soil moisture. The reduction in sensitivity for a given cover type, as compared to a bare soil relationship, increased as both frequency and the amount of vegetation increased. It was clear that in order for remotely sensed data to be used effectively in developing algorithms for extracting soil moisture information from observations of a vegetation-soil complex, the effect of vegetation on these data must be well understood. However, developing experiments to isolate the individual effects of vegetation biomass and structure seemed to be difficult under typical crop or plant conditions because both factors vary simultaneously. In addition, there was no general agreement on how best to parameterize the influence of vegetation. It gave rise a series of special research projects to reveal the effect of vegetation parameters on microwave emission from vegetated soils.

In order to obtain information about the effect of vegetation parameters, NASA and USDA conducted a series of field experiments during the summer of 1982 with truck-mounted radiometers at frequencies of 1.4 and 5 GHz (O'Neill *et al.*, 1984). The radiometers had a comparable 3 dB beam width of 13° and a calibration accuracy of ± 3 K. Vertically and horizontally polarized measurements at both frequencies were obtained concurrent with ground observations of soil moisture and vegetation parameters. The data obtained at a viewing angle of 20° were discussed in the paper. The aim of the study was to isolate the influence of vegetation water content using cut corn stalks. Radiometer measurements were made with the sensors looking down the crop rows. Plant components were oriented parallel and perpendicular to the rows. Random orientation of the stalks was achieved by throwing the stalks in the air from different points around and within the perimeter of the sample site. The radiometer's plane of incidence was defined by the antenna line of sight and the normal to the surface, and contained the vertically polarized electrical field vector. The horizontally polarized field vector is at right angles to this plane, parallel to the ground surface. During most of the experiment, radiometer observations were obtained in sequence over full crop canopies, standing stalks from which leaves and cobs have been stripped, cut stalks oriented parallel, perpendicular, and random relative to the radiometer line of sight, and the stubble background remaining after the crop had been cut down and removed. Observations of brightness temperature, soil moisture, and vegetation water content for different measurement configurations were reported in the paper. One series of measurements involved observations of full canopy brightness temperature, stripped standing stalks brightness temperature, and stubble brightness temperature. An interesting outcome of the data presented from this series is evidence of a frequency-dependent response to removal of different portions of the corn canopy. The removal of leaves and cobs produced 69-72% of the total reduction in the brightness temperature due to the entire canopy removal at 5 GHz, and 32-38% of that at 1.4 GHz. These data have indicated that the leaves (and possibly cobs) within the canopy have a greater effect on the C-band, while the stalks are the major contributor to the total L-band response. Other series of measurements have indicated that the orientation of the stalks and their water content also affect the microwave emission from a vegetation/soil scene. The magnitude of the effect varied with polarization and frequency, disappearing at very low levels of vegetation water content.

To isolate the soil surface from the vegetation canopy, Brunfeldt and Ulaby (1984, 1986) used metal screens put on the ground beneath the canopy and then proceeded to make radiometric observations of the canopy as a function of several parameters. The screens were wire meshes. Temporal measurements were made at 2.7 and 5.1 GHz for soybean, wheat, and corn canopies in 1982-1983. The following questions were to be investigated: Is the brightness temperature of a vegetation canopy polarization-independent? Is it a reasonable assumption to regard the canopy as being symmetrical in azimuth? Does the brightness temperature depend on row direction for row crops? Is it possible to relate the brightness temperature of a canopy to its biophysical properties? The three component model (6.22) was used to examine the emission from a vegetation canopy. For the bare screen (measured following removal of the soybean plants), the observed brightness temperature was about 55-85 K. That provided a good brightness temperature contrast between the bare screen and the canopy over the screen. A row-direction experiment was conducted to evaluate the response of the brightness temperature to row direction and polarization. The experiment has shown that the canopy is highly anisotropic; the emission exhibits a strong dependence on polarization and viewing direction. The significant (23 K) difference between horizontal and vertical brightness temperatures was observed even for nadir observations of soybean canopy. To determine the single scattering albedo and the optical depth (from (6.22)), the brightness temperature was measured for bare screens, canopy over screens, bare absorbing material with an emissivity close to unity, and canopy over absorbing material. It has been found that the optical depth is greater at 5.1 than at 2.7. The albedo is generally smaller at 5.1 GHz than at 2.7 GHz (see Fig. 6.2,a in Chapter 6) and typically less than 0.1. The optical depth may be related to the integrated water content W through a simple formula based on linear regression.

Chukhlantsev (1981) conducted a series of laboratory radiometric measurements at 2.25, 10, 18, and 30 cm to test the relation between the optical depth and the vegetation wet biomass. Some data are presented in Chapter 5. It was found that the optical depth is related to the vegetation water content through a linear dependence. Later, Chukhlantsev and Golovachev (1989) performed more extensive laboratory measurements (see Chapter 5) that confirmed the linear relation between the optical depth and vegetation water content.

The influence of soil on microwave emission can be completely cancelled for such a specific crop like rice. For this crop underlying surface is water with well-known microwave radiation properties. Vorobeichik *et al.* (1988) performed measurements of rice crops emissivity with onboard and car-mounted radiometers during 1980-83 in Krasnodar Region. The radiometers operated at the wavelengths 2.25 and 18 cm. Spatial resolution at the earth surface was 20-100 m and 0.5-1 for airplane and car measurements, respectively. Radiometers were calibrated with internal calibrators and by the radiation level of open water surface. A black body and metal plate were used as external calibrators in measurements from the car. Measured emissivities were compared with averages over the antenna footprint values of rice above the water wet biomass. In airplane measurements, 10-20 samples of biomass were taken over a rice field. In car measurements, the plants were removed directly from the area of the antenna footprint. Biometric measurements were conducted by specialists of the Krasnodar Institute of Rice. Wet rice biomass, gravimetric moisture, relative weight of different plant components, and conductivity of vegetation sap were determined for different rice types during growth period. A dependence of the emissivity on the rice wet biomass is presented in Fig. 7.2. This dependence was a stable one for three complete vegetation cycles for the types of rice investigated. It implies that the dependence can be used for operational control of rice biomass (Vorobeichik *et al.*, 1986).

Chukhlantsev *et al.* (1989) and Golovachev *et al.* (1989) used a truck-mounted radiometer to study microwave emission from vegetable crops. The radiometer operated in the wavelength range 15-30 cm. The receiving antenna was located at a height of 2.5 m above soil surface. The linear size of antenna footprint on the soil surface was equal to 1-2 m, depending on the wavelength. The nadir view angle was about 15° to avoid reflections between antenna and soil surface. The radiometer sensitivity was about 2 K for a time constant of about 1 s. The calibration of the radiometer was conducted before and after measurements with the use of two external references with known emissivities (absorbing and metal plates). Besides, calibration levels were controlled by the emissivity of open water surface (a small pond was used) and the sky brightness temperature (the antenna was directed to zenith). The measurements were conducted at experimental agricultural fields, consisting of 13 thirty meters long sites with different soil moisture conditions. The soil moisture was uniform within each site. One part of each site was covered by vegetation, the other part not.

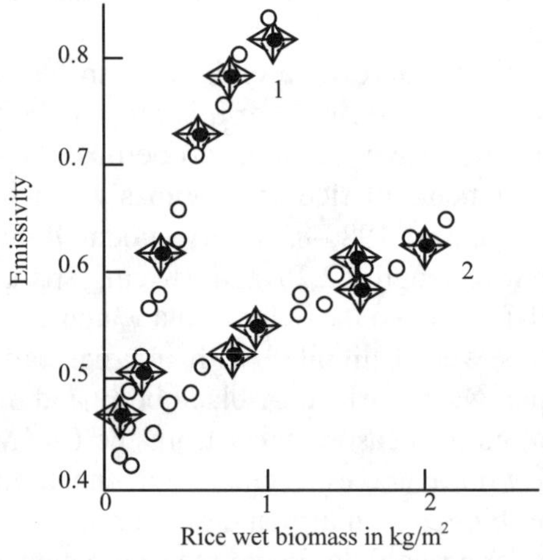

Fig. 7.2. Emissivity of rice crop versus its above water wet biomass at 2.25 (1) and 18 (2) cm. Points are car measurements. Crosses are airplane measurements and show a statistical spread of remote and ground truth data.

So, the canopy and bare soil emission could be measured by crossing every site. The measurements were conducted in 1982 every 2-3 days for a whole period of growth. The in situ measurements included: determining the soil moisture content at different depth by thermo-gravimetric method; determining the canopy wet biomass and water content by measuring the weight of the "mean" plant and the number of plants within every site. Experiments have shown that, without taking into account vegetation screening, the error of soil moisture retrieval may reach considerable values – about 0.1-0.12 g/cm^3 or 30-40% of the whole range of productive water (for the given soil type). The vegetation screening effect was described by the slope reduction factor β that was found by measuring the canopy emissivity and the bare soil emissivity after removing the plants (see equation (5.8)). Since this technique requires cutting a part of the canopy, this factor was also found by comparing the emissivities of vegetated and bare soil at the same soil moisture content within a site. Experimental data of the slope reduction factor versus the vegetation wet biomass were obtained. Biometric measurements on accumulation of biomass in a canopy enabled understanding of the influence of different plant bodies on the microwave emission attenuation of the canopy. It has been shown in particular that leaves and side stalks of tomato plants play the main role in screening. The averaged dependence of β on the biomass (stage of growth) was used to determine the emissivity of soil under vegetation and to retrieve the soil moisture content. The error of retrieved soil moisture data did not exceed 0.06 g/cm^3 when the canopy screening effect was taken into account.

Results of the experiments described above have distinctly shown that, in the low-frequency part of the microwave band (*L*- and *S*-band), vegetation canopies are transparent that allows the retrieval of soil moisture from radiometric measurements at these frequencies. In the high-frequency part of the microwave band (from *X*- to *Ka*-band), well-developed crops become opaque for microwave radiation and these frequencies can hardly be used at all for soil moisture retrieval in this case. On the other hand, at these high frequencies, emission from vegetated fields is basically determined by emission from the vegetation itself. This result provided motivation to study the efficiency of high-frequency microwave radiometry in remote sensing of crop biophysical parameters. A series of investigations in this direction was conducted by Pampaloni and Paloscia.

Vegetation emission measurements were carried out on different test sites through 1980-84 by means of truck-mounted microwave and infrared radiometers (Paloscia and Pampaloni, 1984; Pampaloni and Paloscia, 1985a, 1985b). The IR sensor was a commercial type operating on the 8-14 μm band, with a temperature accuracy of ± 0.5 K. The *X* (10 GHz) and *Ka* (36 GHz) band were Dicke type with two simultaneous vertical and horizontal polarization outputs and an internal calibration system which included a variable temperature load and a precision noise source. In addition, an external black body was used to obtain an accuracy of 1 K or better. Relative accuracy between two polarizations was more than 0.2 K. Ground truth data were simultaneously collected and recorded together with microwave data. Crops measured were corn, alfalfa, and wheat. It was found that a contribution from soil under vegetation into the total canopy emissivity is observed even in the *Ka*-band. Measurements, carried out before and after flooding of a corn field (plant height = 120 cm, density = ~ 6 plants/m^2), showed that emission is sensitive to soil moisture content and soil temperature. The emissivity difference between wet and dry conditions achieved 0.1 at 0-20° nadir view angles and diminished to 0.02 at 40-50° nadir view angles in the

X-band. In the *Ka*-band, the same values were 0.04 and 0.01. (For near to nadir observation angles, these data produce the slope reduction factor $\beta \approx 0.15$ (*X*-band) and $\beta \approx 0.06$ (*Ka*-band) that is consistent with data presented in Fig. 7.1.) The observations generally confirmed that, at frequencies equal to or higher than *X*-band and at incidence angles greater than 30-40°, the soil contribution to the observed microwave emission of a canopy is negligible for most crops. Under these conditions, the emission is mainly determined by the reflectivity and physical temperature of the vegetation layer. Since reflectivity is a function of leaf size and moisture, different crops are distinguished in the two-frequency diagram for emissivity at 10 and 36 GHz. The emissivity of corn was typically 0.88-0.96 at 10 GHz and 0.9-0.96 at 36 GHz (the 36 GHz emissivity was typically higher than the 10 GHz emissivity that implies a greater reflectivity at 10 GHz than that at 36 GHz (see Fig. 6.2,a in Chapter 6)). The wheat canopies had the highest emissivity close to unity at both frequencies. A diurnal change in the emissivity was observed that could be explained by a change in the physical conditions (leaf moisture) of the crops. Data from the *Ka*-band taken during the growth cycle of corn showed that the polarization index measured at an incidence angle of 50° has very low values when vegetation is in stress conditions and increases when the plants take on water.

The connection of microwave signatures to the physical conditions of plants was further considered in Paloscia and Pampaloni (1984). It was found that there exists a relationship between the *Ka*-band emissivity and the air vapor pressure at the top of vegetation. This relationship was approximated by linear regressions for corn and wheat canopies. Moreover, the crop water stress index (*CWSI*), which was previously introduced as a function of the difference between air and crop temperatures and was found to be useful for yield prediction, appeared to be correlated with the microwave polarization index (*PI* = 2*NPI*). The analysis of experimental data at 50° observation angle for 22 days has produced the equation

$$PI = 1.8(CWSI)^{-0.54} . \tag{7.1}$$

The correlation coefficient was greater than 0.9. It has indicated that the polarization index is sensitive to plant stress.

Further, Pampaloni and Paloscia (1985a, 1985b) have found that the polarization index can be expressed as a linear function of the leaf water potential which changes from zero (full hydration) to its lowest negative values the more the plant loses water and becomes stressed.

A relation of the vegetation optical depth to the vegetation water content was considered in Pampaloni and Paloscia (1986). The results were discussed in Chapter 6. The use of microwave polarization index for monitoring vegetation growth was revealed in Paloscia and Pampaloni (1987, 1988). The comparison between the model (6.22) and the radiometric data at 10 and 36 GHz has shown that, on the whole, the model is able to predict leaf area index and plant water content during the first phase of the growing cycle. However, the predictive accuracy changes according to observation frequency and crop type. The best results have been obtained for corn at 10 GHz, where different levels of the Leaf Area Index can be estimated as it changes between 0 and 5. Some results of their research were summarized in Paloscia and Pampaloni (1992).

Thus, general ideas on microwave emission from the Earth's surface in the presence of vegetation canopies were formulated up to the middle of the 1980s. Further investigations were basically directed to an accumulation of experimental data on vegetation screening effect and to an application of microwave radiometric measurements to monitoring soil moisture and vegetation features.

Wang *et al.* (1987) reported some results of radiometric measurements obtained with an airborne 4-beam pushbroom 1.4 GHz radiometer in 1983-84. Two types of vegetated fields, alfalfa and lettuce, were extensively sampled for soil moisture. The alfalfa plants ranged from 5 to 15 cm in height, while the lettuce plants were mature and ready to be harvested. The emissivity of the fields versus the volumetric soil moisture was plotted. (These data allow one to estimate the slope reduction factor: $\beta \approx 0.6$ and $\beta \approx 0.8$ for the lettuce and alfalfa fields, respectively.) It was found that the effect of vegetation cover on the emissivity is more pronounced than that of soil roughness.

In a paper by Wang (1985), the microwave radiometric response to surface soil moisture variation was explored. The data obtained from the Skylab 1.4 GHz radiometer and from the 6.6 GHz and 10.7 GHz channels of the nimbus-7 Scanning Multichannel Microwave Radiometer were used for analyses. It has been shown that the presence of a vegetation cover reduces the slopes of regression between the brightness temperature and soil moisture. The decrease of sensitivity of moisture sensing can be expressed in terms of the slope reduction factor. The analyses based on the Skylab and SMMR observations clearly supported the strong frequency dependence of sensitivity reduction due to vegetation cover claimed earlier (Fig. 7.1). It was asserted that the radiometric measurements at 6.6 and 10.7 GHz over densely vegetated regions have little use because of the substantial loss of sensitivity to soil moisture variations at these frequencies. At 1.4 GHz useful measurements of soil moisture can be made if the vegetation biomass is properly estimated. It was also pointed out that the polarization factor (PI) defined from the SSMR dual polarized radiometric measurements is sensitive to soil moisture variations when vegetation cover is not dense. Correlation of this factor with horizontally polarized brightness temperature could be useful in isolating the effects of vegetation cover and surface roughness from soil moisture variation.

Wang *et al.* (1989) reported results of radiometric measurements that were made over two small watersheds with a four-beam pushbroom aboard a microwave *L*-band radiometer, during a dry-down period following a heavy rainfall in May and June 1987. The two watersheds were in the tall grass prairie region of Kansas. The slope of the emissivity-soil moisture dependence for the data from the first watershed was about 73% of that expected for the bare-smooth soil case and for the second one it was 51%. This reduction in sensitivity was related to surface and vegetation factor by the equation (Schmugge *et al.*, 1986)

$$\beta_\Sigma = \beta_r \beta = \exp\{-h_s - 2\tau\} \tag{7.2}$$

where the optical depth was estimated as $\tau = 0.115W$. The results reported confirmed that vegetation can have a very significant effect on the retrieved soil moisture values and the magnitude of the effect is generally proportional to the biomass of overlying vegetation. It was noted that the green vegetation biomass can be estimated from visible and near infrared observations.

Mätzler (1990) performed microwave radiometric observations of emissivity temporal evolution for an oat field from seeding to harvest. The radiometers used were part of the Passive and Active Microwave and Infrared Radiometer (PAMIR). The linearly polarized Dicke instruments operated at frequencies of 4.9, 10.4, 21, 35, and 94 GHz. The PAMIR box was installed on a trailer along the southern border of the oat field. The key measurement in the experiment parameter was the reflectivity, which was defined from measured brightness temperatures T_b as

$$r = \frac{T - T_b}{T - T_s} \tag{7.3}$$

where T is the physical object temperature and T_s is the effective brightness temperature of the sky (a mean brightness temperature over the sky hemisphere). The measurement accuracy for small r values was mainly limited by the ability to measure the physical object temperature. For homogeneous object temperature distributions, which were observed during cloud-covered sky in the evenings, the error in r did not exceed 0.004. Volumetric soil moisture of the top 3 cm layer was determined from the product of the measured gravimetric soil moisture and bulk density. In addition to canopy height, plant density, and plant water content, the plant dry matter fraction and the leaf thickness were measured. Seasonal variations of these soil and vegetation parameters are presented and discussed in the paper. Temporal evolution of reflectivity at both vertical and horizontal polarizations, angular variations of bare soil reflectivity, spectral dependence of reflectivity for bare and vegetated soils, and evolution of the polarization difference ($r_h - r_v$) are also plotted in the paper. The following results can be noted. At 10-94 GHz frequencies, the reflectivity of vegetated soil was basically determined by that of the vegetation. At 65-70° observation angles, the reflectivity varied in this frequency band from 0.02 to 0.1. The maximum reflectivity is observed at 10 GHz. A clear polarization inversion ($r_v > r_h$) appeared during the ripening phase at 10 GHz. This inversion was earlier observed by Pampaloni and Paloscia (1985a). The polarization difference is less than 0.01 for the well-developed crop. At 4.9 GHz it was found that the main effect of vegetation is one of attenuation, since the scattering at low frequencies is small. The polarization difference on a logarithmic scale as a function of the crop biomass demonstrates close to a linear behavior. That was treated as the law of attenuation in a stratified dielectric medium:

$$r_h - r_v = \Delta r_0 \cdot \exp(-2\tau \sec \vartheta) \tag{7.4}$$

where Δr_0 is the polarization difference of the soil surface for zero crop attenuation. The nadir optical depth was found as a linear function of vegetation water content $\tau = bW$. By fitting experimental data, it was found that $b = 0.716$ m^2/ kg, $b = 3.7$ m^2/ kg, and $b = 6$ m^2/ kg at 4.9, 10.4, and 35 GHz, respectively. It was concluded in the paper that the different behavior of the radiation at different frequencies and different polarizations during the various stages of crop development demonstrated the high information content on vegetation parameters in passive microwave data. Further experiments with PAMIR and with a set of new portable radiometers were summarized by Mätzler and Wegmüller (1994).

In Spring/Summer 1988 an airborne scatterometer campaign was organized by ESA over several sites in Europe. Radiometric data were almost simultaneously collected by using the helicopter-borne Multi-band Sensor Package. Ferrazzoli *et al.* (1992b) presented a comparative evaluation of the potentials of active and passive microwave sensors in estimating vegetation biomass and soil moisture content. Microwave radiometers operated in *L*-, *X*-, and *Ka*-band. Active sensors operated in *L*-, *S*-, *C*-, *X*-, *Ku*1-, and *Ku*2 -band. In situ measurements of soil moisture, surface roughness, and plant parameters (including plant water content, leaf area index, and dimensions of stems and leaves) were carried out at the same time as the overflights. Corn, sunflower, alfalfa, wheat, and barley crops were observed.

According to theory and previous measurements, the authors expected micro-wave emission to be generally mostly influenced by soil parameters at the lower of nadir observation angles, especially for the lower frequencies, while vegetation effects were dominant at the higher observation angles. Since the collected data confirmed this trend, the authors used measurements at $\vartheta \approx 40°$ to assess vegetation biomass, while they discussed the relations between microwave measurements and soil moisture on the basis of data collected at a steeper (10°) observation angle. The quantity of biomass over-laying the ground is usually represented by the vegetation water content in kg/m^2. However, since during the growth phase this quantity was well corre-lated to leaf area index, the last parameter was used in the analysis. The measurements with passive sensors between 1.4 GHz and 36 GHz have shown a marked frequency-dependent behavior of the emissivity. Measure-ments carried out on moist soils showed that, at 1.4 GHz, H pol., the emis-sivity of wide leaf crops increased with the increase in LAI as the vegetation cover expanded. Small leaf crops exhibited at 10 GHz a trend in the emissi-vity – LAI dependence similar to that observed at 1.4 GHz for wide leaf crops. It has demonstrated the importance of the leaf dimension to the wave-length ratio in the emission from vegetation layer. The L-band radiometric data at 10° observation angle have been used to check the performance of passive microwave sensors in estimating soil moisture content of agricultural fields. Radiometric data were spread over a regression emissivity-soil mois-ture dependence that was due to the different vegetation cover and soil roughness. For this dependence, the authors have found a sensitivity equal to 0.7 K/0.01 g/cm^3 which was slightly lower than the values (1-2 K/0.01 g/cm^3) generally found in literature for medium rough and vegetated terrains. Nevertheless, comparing the abilities of active and passive sensors for soil moisture measurements, the authors concluded that, if the ground resolution were ignored, passive sensors are more preferable.

A set of experimental data was collected at the INRA Research Center near Avignon in 1991 and 1993 (Chanzy *et al.*, 2000; Wigneron *et al.*, 1993, 1995). Microwave emissivity values for bare soil, soybean, wheat, and sor-ghum canopies were estimated from the brightness temperatures measured by a PORTOS radiometer. That was a dual-polarized radiometer operated at 1.4, 5.05, 10.65, 23.8, 36.5, and 90 GHz. The radiometer was installed on a 20 m crane boom, and measurements were obtained over a range of inci-dence angles from 0° to 60° with a 10° increment. Regular external calibra-tions were performed: over liquid water at 45° incidence angle and over absorbing slabs either at ambient temperature or immersed into liquid nitro-gen. The estimated absolute accuracy of the microwave data was about ±3 K. These measurements stimulated appearance of a number of papers.

Wigneron *et al.* (1993b) presented a model analysis of leaf-characteristics effect on the microwave emission of land surfaces. The radiative transfer model and the discrete approach were used in the paper to study the leaf effect on the microwave emission of a mature soybean canopy at *L*-, *C*-, and *X*-bands. The main results of the model simulation were: the trend of the brightness temperature versus frequency is largely conditioned by soil moisture; the leaf gravimetric moisture content appears to be a key factor of the vegetation emissivity especially at *C*-, and *X*-bands; for a well-developed vegetation canopy the effects of the leaf volume fraction are limited to the low frequencies; the leaf size effect increases with frequency; and the effects of the distribution of leaf inclination are significant at high frequencies and weaker at low frequencies. Results of the measurements with the radiometer PORTOS were consistent with the computations from the model.

Wigneron (1994) investigated the temporal variation of a soybean field microwave emission using a continuous model. The microwave data (at 1.4, 5.05, and 36.5 GHz) were acquired with the PORTOS radiometer. Two correlation lengths were used to describe volume scattering inside the canopy. A two-step method of the model inversion is described. It consists in calibrating first these two lengths and then in retrieving the surface parameters: the soil moisture and the vegetation volume fraction. A good agreement between measured and inversed surface parameters was observed.

A composite discrete-continuous approach was used by Wigneron *et al.* (1995a) to model the microwave emission of vegetation. Model simulations have been compared to radiometric data observed at 23.8 GHz, during the vegetation development of a wheat field. Based on simulations and radiometric measurements, the interest of medium-frequency microwave channels (roughly from 10 to 40 GHz) to monitor a wheat crop development was analyzed. It was concluded that the biomass increase can be monitored at the beginning of the crop development, until the vegetation wet biomass goes beyond 1.5-2 kg/m^2. These results are consistent with those reported before.

Ferrazzoli *et al.* (2000) reported some more data on the multifrequency emission of wheat. The emissivities measured in *L*-, *C*-, *X*-, and *K*-band at two polarizations and several angles were compared with those simulated by a discrete multiple scattering model. It was emphasized that the use of model simulations helps someone to identify the radiometric configuration most sensitive to ground variables. Moreover, it aids to develop inversion algorithms that perform well under different soil and vegetation conditions.

Chauhan *et al.* (1994) examined the discrete scatter model for microwave radar and radiometric response to corn. The experiment was conducted during July, 1990. At that time, the ground was covered with a variety of agricultural crops as well as some pasture and forest. Radar data were acquired at *P*-, *L*-, and *C*-band at HH, VV and HV polarizations. The pushbroom microwave radiometer provided data at *L*-band. Several different types of measurements were made to determine plant parameters. Canopy height and plant density were determined. Measurements were made of the leaf inclination angle, the leaf and the stalk thickness, and the stalk length. The gravimetric moisture of plants was also measured. The gravimetric soil moisture was determined together with the soil bulk density. Soil roughness was estimated from photographs of the surface. It was found that the contrast between the brightness temperature of dry and wet soils decreased at 10° observation angle from ~110 K (bare soil) to ~50 K (corn canopy). It produces a slope reduction factor equal to 0.45 that is consistent with data in Fig. 7.1. The model data was found to be in reasonably good agreement with the experiments. It was noted that this result is important because a single set of vegetation parameters and single model was used to predict both active and passive sensor responses.

Van de Griend *et al.* (1996) performed microwave measurements with a radiometer operating at frequencies of 1.4 and 5.0 GHz. The radiometers were mounted on an aluminum field-movable pedestal with interconnected turntables allowing for easy change of the polarization angles for both radiometers simultaneously. View angles were easily adjusted by lifting the framework holding the radiometers. The instruments were regularly calibrated by measuring echosorb and clear sky brightness temperatures. The measurements were done over two surfaces with wheat of 3.5×6 m^2 each, selected to have near identical vegetation conditions. One surface was flooded to create a significantly different surface emissivity below the canopy. The other was kept at its original state. Different biomass densities were created by systematic thinning of the vegetation cover after each measurement series until complete unvegetated conditions (bare soil or water) occurred. The harvested vegetation was used to determine wet and dry biomass and LAI. All measurements on the dry plot were carried out during the first day. The flooded plot was measured during the second day. The measured data of brightness temperature were used to estimate the reflectivity and transmissivity of the wheat. The data on the transmissivity can be recalculated to the slope reduction factor values that are consistent with those presented in Fig. 7.1.

Jackson *et al.* (1997) described results of field experiments which were conducted during 1994 using the *L*- and *S*-band truck-based system through extended diurnal observations over bare soil and corn. The *S* and *L* microwave radiometer was a dual frequency passive sensor system operating at 2.65 and 1.413 GHz. The system was mounted on a truck hydraulic boom, which permitted deployment of the sensor package to a height of approximately 10 m above the ground. The instrument platform at the end of the boom could be moved to vary incidence angle from nadir to sky. The antennas were mounted to observe horizontal polarization. Calibration of the radiometers was based on external reference targets. The external targets typically used were the sky (~5 K) and a microwave absorber at ambient temperature (~300 K). The vegetation effect was taken into account by introducing the slope reduction factor with the optical depth calculated by (4.124). To determine the coefficient b, a canopy removal experiment was conducted. In this type of experiment, the brightness temperature was first observed for the soil-canopy complex. Then, as quickly as possible, the canopy was removed and the brightness temperature of the soil was observed. By using the appropriate model, an estimate of the optical depth of the canopy was obtained. Using these data, the estimated b values were S-band$=0.1$ and L-band $= 0.082$. A vegetation correction was applied to each corn canopy observation using these b values and an estimate of the vegetation water content. The observed emissivity plotted versus soil moisture was compared with the emissivity model based on the Fresnel approach for the soil reflectivity and the vegetation correction for vegetation water content of 2.5 kg/m^2. It was found that the vegetation attenuation effect on the sensitivity of the relationship between emissivity and soil moisture was obvious (change of slope) and more significant at S-band. The slope reduction factor was approximately 0.5 at L-band that is consistent with results of previous research.

Macelloni *et al.* (1998) reported results of experiments aiming at evaluating in detail the contribution of leaves, stalks, and fruits to total emission from the canopy. Dual polarized microwave radiometers at 10 and 36.6 GHz and a thermal IR sensor were used in the measurements. The microwave sensors were portable, battery-operated, self-calibrated system with a horn antenna for each polarization channel and an internal calibration based on two loads at different temperatures. Calibration checks were carried out during the field experiments by means of reference targets of known emissivity and temperature, i.e., absorbing panels, a smooth water surface, and clear sky. The achieved measurement accuracy was better than ± 1.0 K, with an integration time of 1 s. The experiments included measurements of emission from well-developed wheat fields in natural conditions and after sequential cuts of fruits, leaves, and stems. The fresh and dry biomass as well as the geometrical characteristics of plant constituents was measured on samples of 20 to 30 plants. The measurements were carried out on May 20, 1994, when crops were in the flowering stage, and on June 8, during early ripening stage. It has been shown that, for well-developed vegetation, with plant water content higher than 2.5 kg/m^2, vertical stalks gave the highest contribution to the total emission, since, when leaves and fruits were cut, the brightness temperature of the canopy remained almost the same as full crop, whereas it changed conspicuously when the stems too were cut. It was also found that the reflectivity of the canopy was linearly correlated to plant water content.

Hornbuckle *et al.* (2003) investigated anisotropy in 1.4 GHz brightness induced by a field corn vegetation canopy. Experiments were conducted in 2001. Two radiometers, oriented to record *H-* and *V*-polarized 1.4 GHz brightness, were mounted on the hydraulic arm of a truck. Brightness temperatures were measured at incidence angles of 15°, 35°, and 55° shortly after dawn when soil and canopy temperatures were nearly uniform. The radiometers were calibrated at least two times during each experiment. The sky was used as one calibration point. The internal reference loads were used for the other calibration point. The accuracy of brightness temperature measurements was estimated to be within ± 2.0 K. Leaf area index as well as water column densities were measured periodically throughout the summer. Soil moisture was estimated with the use of a buried time-domain reflectometry instrument. A laser profiler was used to measure soil surface height variations. It was found that both polarizations of brightness were isotropic in azimuth during most of the growing season. When the canopy was senescent, the brightness was a strong function of row direction. On the other hand, the 1.4 brightness was anisotropic in elevation. A new zero-order parameterization was formulated by allowing the volume scattering coefficient to be a function of incidence angle and polarization. It was noted that the small magnitudes of the scattering coefficients allowed the zero-order model to retain its physical significance for the case considered.

Numerous papers were devoted to large scale microwave radiometric measurements for remote sensing applications. Jackson *et al.* (1999) evaluated surface soil moisture retrieval algorithms, developed and verified at high spatial resolution, in a regional scale experiment. Using previous investigations as a base, the Southern Great Plains Hydrology Experiment (1997) was designed and conducted to extend the algorithm to coarser resolutions, larger regions with more diverse conditions, and longer time periods. The *L*-band electronically scanned thinned array radiometer (ESTAR) was used for daily mapping of surface soil moisture over an area greater than 10 000 km^2 for a one month period. The radiometer was a synthetic aperture instrument operating at a center frequency of 1.413 GHz and bandwidth of 20 MHz. For this experiment it was installed on a P3B aircraft board to provide horizontally polarized data. This instrument was the most efficient soil moisture mapping device available at that time. It provided for obtaining a radiometric image with the effective swath of about $\pm 45°$ wide with the width of a synthesized beam of 8-10°. Calibration of the ESTAR was achieved by viewing two scenes of known brightness temperature. These scenes included a black body and water. Flights were conducted at an altitude of 7.5 km in four parallel lines and a water calibration line (at 200 m over a lake). Soil moisture sampling was done on sites that were approximately a quarter sections (0.8 km $\times$ 0.8 km) in size. Vegetation type was used to define the functional form of the vegetation attenuation. For each land cover/vegetation category a vegetation parameter, the coefficient *b*, utilized in the retrieval algorithm was assigned based on published data. For all vegetation besides the grass categories a fixed value of the vegetation water content was used. For the grass and shrub categories, ground based measurements were used to obtain a regression equation, connected the vegetation water content to the Normalized Difference Vegetation Index (NDVI). This index will be described in Chapter 8. The NDVI was computed using satellite observations data (Thematic Mapper). Results of the experiment showed that the soil moisture retrieval algorithm performed the same as in previous investigations, demonstrating consistency of both the retrieval and the instrument. The results matched with the smaller scale 1992 and 1994 experiments (Jackson et al., 1993, 1995). This fact allowed a conclusion that the algorithms can be extrapolated from higher resolution ground experiments to satellite scales.

The ESTAR was also utilized for soil moisture mapping during the next Southern Great Plain Experiment (1999) (Guha *et al.*, 2003). The results indicated a good correlation between observed and predicted soil moisture in values and were consistent with results obtained from the same instrument previous experiments.

The Special Issue of the *IEEE Transactions on Geoscience and Remote Sensing* (Vol. 39, No.8, August 2001) was devoted to "large scale passive microwave remote sensing of soil moisture". Jackson and Hsu (2001) compared satellite data, collected by the Tropical Rainfall Measuring Mission microwave imager (TMI) and the special sensor microwave/imager (SSM/I), to soil moisture observations as a part of the Southern Great Plains 1999 Experiment. Satellite, aircraft, and ground based data collection were conducted during a sequence of meteorological events. The analysis of the satellite data and the resulting soil moisture maps have shown that consistent satellite-based soil moisture retrieval at coarse resolutions was possible.

Judge *et al.* (2001) compared ground-based and the SSM/I brightness temperatures at 19 and 37 GHz in the Northern and the Southern Great Plains. It was found that these brightness temperatures are moderately correlated. They matched well during winter. During summer, ground-based brightness temperatures at the bare soil were on average 10 K cooler than satellite-measured ones. In effect, the ground-based measurements bracketed the satellite observations with the SSM/I brightness temperatures lying closest to those of the bare soil.

Kerr *et al.* (2001) discussed the main aspects of the planed Soil Moisture and Ocean Salinity (SMOS) mission. They also described how soil moisture will be retrieved from SMOS data.

Owe *et al.* (2001) proposed a methodology for retrieving surface soil moisture and vegetation optical depth from satellite microwave radiometer data. The procedure was tested with historical 6.6 GHz H and V polarized brightness temperature observations from the scanning multi-channel microwave radiometer (SMMR) over several test sites in Illinois. The methodology used a radiative transfer model (6.22) to solve for surface soil moisture and vegetation optical depth simultaneously using a nonlinear iterative optimization procedure. It assumed known constant values for the scattering albedo and soil roughness, and that vegetation optical depth for H-polarization is the same as for V-polarization. Surface temperature was derived by a procedure using high frequency V-polarized brightness temperatures. The methodology does not require any prior field observations of soil moisture or canopy biophysical properties for calibration purposes and may be applied to other wavelengths. Results compared well with field observations of soil moisture and satellite-derived vegetation index data from optical sensors.

Crow *et al.* (2001) performed an observing system simulation experiment to assess the impact of land surface heterogeneity on large-scale retrieval and validation of soil moisture products. Vegetation effect was taken into account following equation (6.22). It was found that a large error of soil moisture retrieval occurred for the heavily vegetated areas. The error depends on the vegetation water content and the vegetation fractional coverage. For the particular considered frequency of 6.9 GHz (AMSR, see below), obtaining an acceptable error establishes a threshold for the vegetation water content and the fractional coverage. At $W = 0.5$ kg/m^2, the fractional coverage must not exceed 25%, at $W = 1$ kg/m^2, the fractional coverage must be less than 10%. It seems that at L-band, these estimates will be less restrictive.

The Priroda remote sensing module on the Mir Spase Station offered many new sensors for monitoring surface soil moisture. Jackson *et al.* (2002a) evaluated the Priroda observations and found that the 13 GHz radiometric sensor could provide some information on soil moisture even at the spacecraft resolution.

Several papers were devoted to the utilization of the Advanced Microwave Scanning Radiometer (AMSR and AMSR-E) for soil moisture retrieval. The radiometer was developed by National Space Development Agency of Japan and provided to NASA for launch on its Aqua satellite. The AMSR-E instrument measures radiation at six frequencies in the range 6.9-89 89 GHz, all dual polarized. The AMSR 6.9 GHz channel held promise for retrieving soil moisture in regions with low levels of vegetation. Jackson *et al.* (2002b) reported on the experiment which was designed to provide C-band datasets for AMSR algorithm development and validation. Ground observations of soil moisture and related variables were collected in conjunction with aircraft measurements using a C-band radiometer similar to the AMSR sensor (6.9 GHz), the Polarimetric Scanning Radiometer with its

C-band scanhead. Flights were conducted under a wide range of soil moisture conditions, thus providing a robust dataset for validation. The dynamic range of the *C*-band radiometric observations indicated that the AMSR instrument can provide useful soil moisture information. Nevertheless, it was noted that vegetation is a very significant factor when interpreting *C*-band brightness temperatures.

Njoku *et al.* (2003) described the AMSR-E soil moisture retrieval approach and its implementation after AMSR-E was launched on the Earth Observing System Aqua satellite on May 4, 2002. Examples of AMSR-E brightness temperature observations over land were shown from the first few months of instrument operation, indicating general features of global vegetation and soil moisture variability.

Some other results on microwave radiometry of vegetation canopies are reported in Jackson and Le Vine (1996), Jackson *et al.* (1992), Owe *et al.* (1992), Wang *et al.* (1990), Laymon *et al.* (2001).

One can see that microwave radiometry of vegetation canopies has come a long way from first experiments, which established relationships between radiative characteristics and geophysical quantities, to global projects providing information on the global soil moisture distribution and vegetation state.

7.2. RESEARCH ON MICROWAVE EMISSION FROM FORESTS

A review of experimental research related to the microwave emission of forests was given by Pampaloni (2004). Unfortunately, some radiometric observations, conducted in the Institute of Radioengineering and Electronics of the Russian Academy of Sciences and other Russian research centers, were not considered in the review. Below, a brief description of experiments on the microwave radiometry of forests is given.

The first experimental data related to microwave emission from forests were probably those presented by Kirdiashev *et al.* (1979). The data were obtained from microwave radiometers installed onboard an IL-18 airplane. The radiometers operated in the wavelength range 0.8-30 cm. Flights were performed over the same forested area with a length of several tens of kilometers in the Krasnoyarsk region during several days. The area included coniferous and mixed forests underlain with dry and wet soils, swampy areas, and partially flooded forest sites. It was found that the emissivity of a dense forest over dry soil is close to unity within the whole of the wavelength band. An idea on the forest screening effect can be obtained examining the emissivity spectrum of a partially flooded pine forest. According to the data of ground observations and aerial photography, the surface beneath the trees was 60-80% water-covered. The emissivity of forest in these two particular cases versus the wavelength is presented in Fig. 7. 3. It is seen from the data in Fig. 7.3 that in the centimeter wavelength band the forest practically completely screens the emission from the soil surface. The deviation of the emissivity from unity in this band (and a possibility of soil emission detection) is mainly determined by the presence of "gaps" in the forest canopy. In the decimeter wavelength band, the forest is semitransparent. The data presented in Fig. 7.3 were used to estimate the slope reduction factor for the forest (Fig. 7.1).

A group from the Helsinki University of Technology (HUT) has carried out several campaigns using the profiling (non-imaging) airborne multi-frequency radiometer operating at 6.8, 10.65, 18.7, 23.8, 36.5, and 94 GHz (Kurvonen and Hallikainen, 1997; Kurvonen *et al.*, 1998; Kruopis *et al.*, 1999; Hallikainen *et al.*, 2000). The measurements were performed on several sites in Finland since 1991 and were mostly addressed in estimating boreal forest transmissivity to evaluate the effect of forest vegetation on sensing the underlying surface.

Profiling airborne measurements were carried out by the Institute of Applied Physics of the Italian National Research Council (IFAC-CNR) in summer 1999 and winter 2002 in Italy at frequencies of 1.4, 6.8, 10, 19, and 37 GHz (Macelloni *et al.*, 2001; Macelloni *et al.*, 2003). Microwave emission spectra of broadleaved forests were obtained.

Multi-frequency, dual polarized measurements between 5 and 90 GHz were carried out in May and August 1994, by using the airborne radiometer PORTOS over the coniferous Les Landes forest (southwest France). The sites consisted of large homogeneous stands of maritime pines with three ages covering a wide range from seedlings to 40-50 years, corresponding to a standing biomass from zero to about 180 tons of dry matter per hectare (Wigneron *et al.*, 1997). The ground data collection included measurements of tree characteristics and of the undergrowth vegetation. The same year, a winter campaign was held in Canada during the BOREAS (boreal ecosystem atmosphere study) to collect data at 18 and 37 GHz (Goita *et al.*, 1997).

Milshin *et al.* (1999) reported results of complex experiments on mapping forest temperature and underlying soil moisture. Measurements were performed in May 15, 1993 over the Rzhev test site of the Russian Institute of Agricultural Aero-Photo-Geodesy Research. Data were collected from a profiling airborne microwave *L*-band (1.4 GHz) radiometer, a profiling IR radiometer (7.5-13 μm), and an aero-photo camera. Navigation TransPack equipment was used to bind the data to the area. The length of the site was about 60 km and its width was about 7-12 km. The site has hilly relief and consists of forests, swamped forests, parts of open soil, and meadows and agricultural fields. Forests cover about 60% of the site. There are birch, aspen, and alder forests, as well as coniferous forests (fir and pine) in the site.

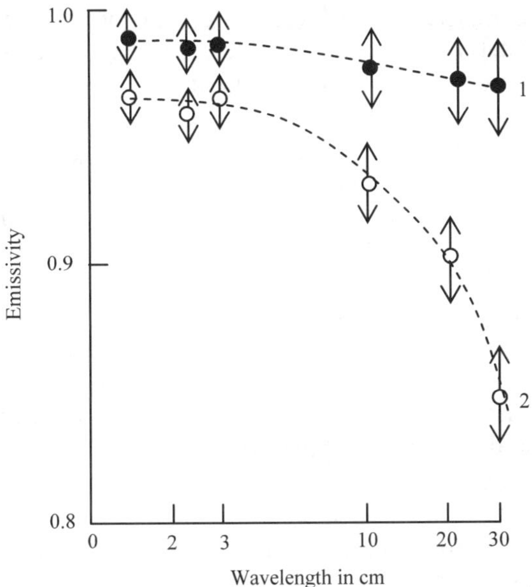

Fig. 7.3. Emissivity of forest over dry soil (1) and of flooded forest (2).

The IR radiometer was used to synthesize a map of surface temperature. Within the site, the temperature varied from 14.5°C to 23.5°C. The spatial gradient of temperature at the site changed from 0.15 to 3°C/km. The spatial resolution of the microwave radiometer at the surface was about 2 kilometers. At this resolution, a mosaic structure of forest cover within the antenna footprint had to be taken into account. To estimate the forest cover density (fraction of surface covered with forest) within the pixel, aero-photo pictures were used. It was determined that the density varied from 0.3 to 0.7 with an average value of 0.55. Absolute calibration of the microwave radiometer was performed by the emission of smooth water surface and very dense old forest at low altitudes of flight. The brightness temperature of the latter was estimated to be equal to $0.95\,T$, where T is the physical temperature. The map of brightness temperature distribution over the test site was synthesized. It was found that, at this particular spatial resolution of 2 km, the brightness temperature of the forested areas varied from 226 K to 260 K with an average value of 247 K. Equation (6.35) was then used to determine the equivalent optical depth of forests. With $r_0 = 0$, the optical depth is

$$\tau = -0.5 \ln[(T - T_b)/(T - T_{bs})]$$ (7.5)

where $T_b = 247$ K and $T = 293$ K are average values for pixels with "dense" forest (with no other types of surfaces within the pixel). To estimate the T_{bs} value the open parts of the test site were used and a histogram of the brightness temperature was built for these parts. The distribution of T_{bs} was a symmetrical one with an average of 226 K. In 90% of events, the T_{bs} value varied from 215 K to 235 K. The use of this range in equation (7.5) yielded a value of forest transfer coefficient β of 0.6-0.8 and a value of optical depth τ of 0.115-0.265. Equation (4.123) was used to simulate optical depth data. According to the ground measurements, the crown biomass of forests varied from 14.5 tons/ha for a coniferous forest to 30.4 tons/ha for a mixed forest containing 40% of deciduous trees. Spatial variability of the crown biomass amounted 4.6 − 10.3 tons/ha. At these values of crown biomass, the simulated optical depth is 0.1 − 0.25 that is consistent with the experiment. It was noted in Milshin *et al.* (1999) that these are "effective" values of the optical depth averaged over the area of 2×2 km^2 (pixel dimensions). At a higher spatial resolution, the values of β and τ decrease and increase, respectively. This is an important conclusion that points to a dependence of equivalent forest optical depth on the size of spatial resolution. The average values of the transfer function were used in Milshin *et al.* (1999) to synthesize a map of soil moisture within the test site. It was found that an additional error of soil moisture retrieval due to the presence of the forest canopies does not exceed 0.06 g/cm^3.

The paper by Milshin and Grankov (2000) provides a review of microwave radiometric measurements over forests performed in the former Soviet Union (Kirdiashev *et al.*, 1979; Baturin *et al.*, 1985; Liberman *et al.*, 1995, 1996; Milshin *et al.*, 1999; Kosolapov and Kozoderov, 1994; Ambarnikov and Elagin, 1989; Grankov *et al.*, 1992). The experimental data on the forest emissivity and brightness temperature are presented in Table 7.1. The data obtained by Macelloni *et al.* (2001) and Wigneron *et al.* (1997) are also included. One can see a consistency of data reported by different research groups.

Statistical characteristics of forest brightness temperature were considered by Milshin and Grankov (2000) in more detail for different dimensions of the antenna footprint. In Moscow region (Milshin and Grankov, 2000), measurements were performed at a spatial resolution of ~150 m, while in Tver region (Milshin *et al.*, 1999), the spatial resolution was about 2 km. The histograms of brightness temperature for these experiments are presented in Fig. 7.4. One can see that an increase in the footprint size yields a shift of the histogram to the range of lower frequencies. That can be explained by a decrease of average forest density with a worsening of spatial resolution, as well as a decrease in the effective optical depth of forest.

Table 7.1. Measured microwave emission characteristics of forests.

Region, forest type, carrier	Date	Wave-length, cm	Spatial resolution, m	Emissivity	Brightness tempera-ture, K
Krasnoyarsk region, coniferous forests, Iliushin-18	1974-1976	10, 20, 30	100-200	0.93; 0.9; 0.84	
Vladimir region, deciduous forest, Antonov-2	July 4, 1984	30	35-100	0.85-0.92	248-259
Kaluga region, mixed forest, helicopter KA-26	Oct. 10, 1990	27	140	0.81-0.91	224-251
Belarus, mixed forest, helicopter MI-2	Sept. 5, 1991	27	70	0.87-0.92	250-264
Bavaria, nixed forests, D-228	Nov. 6, 1992	21	200	0.83-0.9	235-260
Tver region, mixed forests, Antonov-2	May 14, 1993	21	2000	0.81-0.9	235-260
Moscow region, mixed forest, Antonov-2	Sept. 23, 1994	21	150	0.83-0.93	240-273
Southwest France, the Landes forest	May, August 1994	6		0.94-0.98	
Tuscany, broad-leaved forests, ARAT (Fokker-27)	June 15-16 and 24-25, 1999	21	100	0.913-0.945	

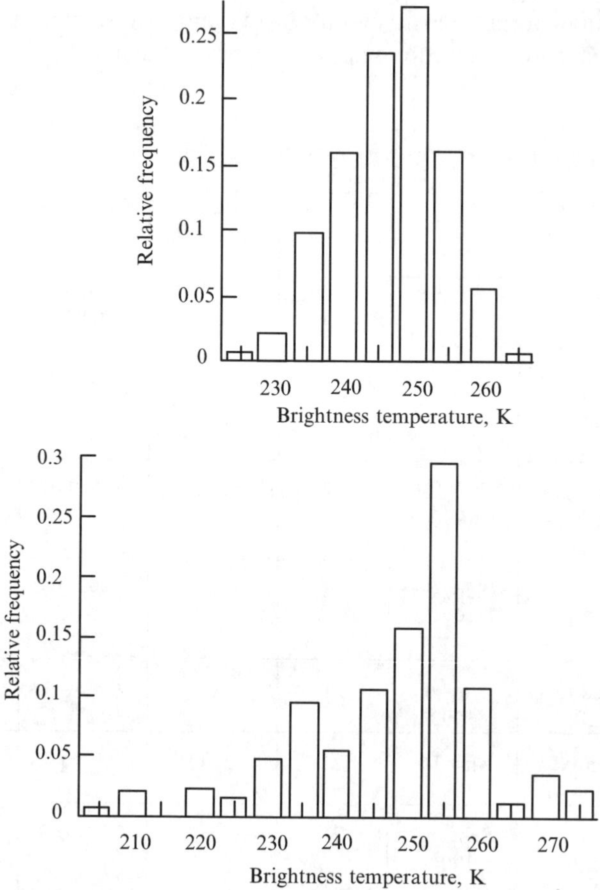

Fig. 7.4. A histogram of 1.4 GHz brightness temperature of forested area for low spatial resolution (upper diagram) and high spatial resolution (lower diagram).

An interesting dependence of the effective optical depth on the forest canopy density was established by Baturin *et al.* (1985). Measurements were carried out with an *L*-band (1 GHz) radiometer mounted onboard a light bi-plane Antonov- 2 at a spatial resolution of 35 and 100 m. Ground measurements of soil moisture were performed at nine small areas along the test site with a length of 4.5 km. The forest density at the areas varied from 0 to 0.7 whereas soil moisture at the areas varied from 0.14 to 0.35 g/cm^3. The optical depth was retrieved from the measurements by fitting the model (6.35) to the measured brightness temperature values. The dependence of the optical depth on the forest density is presented in Fig. 7.5. This dependence was approximated by a linear regression:

$$\tau_{eff} = 0.06606 + 1.28052\xi_v \qquad (7.6)$$

with a correlation coefficient of 0.94. For forest densities of 0.4-0.7, corresponding to measurements conditions reported in Milshin *et al.* (1999), equation (7.6) yields an optical depth of 0.4-1 that results in the transfer coefficient of forest β = 0.16-0.45. This range of β is consistent with the data presented in Fig. 7.1. Data for β reported by Milshin *et al.* (1999) (β = 0.6-0.8) contradict those in Fig. 7.1. It is probably explained by assigning zero value to the reflectivity r_0 in (6.35) (equation (7.5)). With r_0 = 0.06, that is typical for *L*-band (see Table 7.1), equation (7.5) yields β = 0.4-0.6 that is still greater than that in Fig. 7.1 but this is an effective value of β for a worse spatial resolution, at which an effective forest density was not big. One can see that a mosaic structure of forests can play a significant role in radiometric measurements at different spatial resolutions. This fact should be taken into account in global modeling of forest microwave brightness temperature at satellite scales.

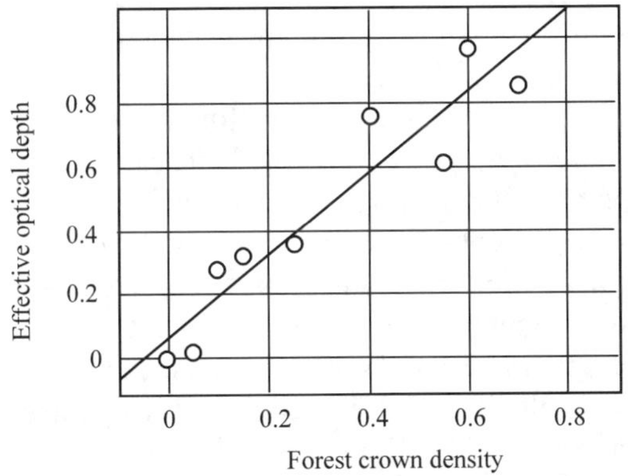

Fig. 7.5. Dependence of effective forest canopy optical depth on forest crown density at 1 GHz.

One can conclude from the aforesaid that, despite the relatively small amount of experimental data currently available, microwave radiometry has proved to be an efficient technique in monitoring vegetation and soil features in forests. Further studies of forest transmissivity in the entire microwave spectrum, together with the improvement of physical models for forest microwave emissivity, the development of appropriate retrieval algorithms, and experiments with airborne sensors directed to the validation of the models and algorithms developed, can significantly improve our knowledge in the field. Important work has already been accomplished, but much remains to be done before microwave radiometry of forests can be put to operational use.

7.3. STATISTICAL PROPERTIES OF MICROWAVE EMISSION FROM VEGETATION CANOPIES

The study of statistical properties of microwave emission from land surface is important for determining characteristic scales of brightness temperature spatial variability (these scales, in turn, put requirements on spatial resolution of radiometric sensors, on spatial frequency of ground truth measurements, and on sizes of sub-pixels in modeling microwave emission from land surface at low spatial resolution), for determining characteristic levels of brightness temperature fluctuations intrinsic to different landscapes (these levels are connected to the hydrological state of territory, for example, to forest fire risk), for revelation of statistical relationships between brightness temperatures at different frequencies and the geophysical quantities.

Shutko (1983) reported some data on the level of brightness temperature fluctuations for different landscapes in the wavelength range 0.8-3.4 cm. The data were obtained from airborne radiometers at a spatial resolution of 10-250 m for flat countries and 500-1000 m for mountain regions. It was shown that for steppe, semi-desert, and desert landscapes an effective value (rms = root mean square) of brightness temperature fluctuations is about 1-2 K. The biggest values of brightness temperature fluctuations were observed for partially vegetated swamp areas and moist soils (rms = 10-15 K). For steppes, the correlation length of brightness temperature spatial variations varied from some tens to some hundreds of meters. Later, Golovachev and Chukhlantsev (1986) studied spatial variations of brightness temperature over sprinkled agricultural fields. Data were obtained from airborne radiometers operating at 18 and 30 cm with a spatial resolution of ~50 m. It was found that the effective value of brightness temperature fluctuations can achieve 30 K in this case. The correlation length varied from ~200 m during irrigation period to ~500 m in spring. Shutko and Reutov (1983) built two-dimension (two-frequency) histograms of emissivity at 2 cm and 30 cm for regions with shallow subsoil waters. It was found that these histograms can be used for estimation of general hydrological situations at regional scales.

Several papers were devoted to estimations of forest fire risk from microwave radiometric data. According to Nesterov (1949), there are four levels of forest fire risk, which are determined by the moisture of flammable materials lying on soil surface (dead litter, moss, etc.). Kirdiashev (1983) and Valendik *et al.* (1980) found that the mean value of forest brightness temperature changes insignificantly during the transition from the low level of fire risk to the extreme level of fire risk. However, at that, the rms of brightness temperature variations changes up to 6-8 times. Kirdiashev and Savorsky (1986) have calculated and measured the rms of forest brightness temperature at 2.25 cm at different levels of fire risk. It was found that the ratio $\sigma / \Delta T_{b(f-w)}$, where σ is the dispersion of brightness temperature variations and $\Delta T_{b(f-w)}$ is the mean brightness temperature contrast between forest and smooth water surface, decreases drastically with an increase of fire risk. For a low level of fire danger, simulated and measured value of the ratio is >0.1, for an extreme level of fire danger, $\sigma / \Delta T_{b(f-w)} < 0.03$. A similar approach was further developed by Yakimov (1996).

At satellite scales, variations of brightness temperature are smoothed (see (6.51) and (6.52)). Nevertheless, Grankov *et al.* (1999) have shown that the level of these variations still can be an indicator of forest fire danger. Experimental data from SSM/I radiometer (85, 37, 22.235, and 19.35 GHz) were processed for April-September of 1988-1990, 1992, 1994, 1997, and 1998 for four forested areas ($1°\times1°$) in America (Alaska: 64.45°N-148°W and Oklahoma: 35°N-95°W) and Russia (Moscow region:55.7°N-39.5°E and Siberia: 57.3°N-87.2°E). To estimate annual and seasonal variability of brightness temperature, Grankov *et al.* (1999) used a variation coefficient that was defined as $\delta = \bar{\sigma} / \bar{T}_b$, where $\bar{\sigma}$ is the monthly averaged value of the brightness temperature rms at given forested area and $\bar{T}_b$ is averaged value of the brightness temperature of the area over the same period. The average brightness temperature itself was found to be strongly correlated with the air and canopy temperature but not with the level of forest fire danger (Milshin *et al.*, 1998). Monthly values of the variation coefficient were calculated from SMM/I data for all test areas for five years. The level of fire danger was estimated from meteorological data. For an indirect estimation of flammable materials moisture, the complex meteorological index is usually used, which is a function of daily data on the air temperature, precipitation, and dew point. The complex meteorological index was calculated for the test areas with the use of data obtained from NCDC (National Climate Data Center) Climate Service Branch for the years 1994, 1997, and 1998. It was found that average (climatic) values of variation coefficients at different polarizations are close to each other. Seasonal dependence of δ at different frequency SMM/I channels is rather similar. For a period from May to September δ variations are within 0.008-0.048 range. It was observed that maximum values of complex meteorological index for the given test area coincided with minimum values of variation coefficient and vice versa.

Chapter 8

VEGETATION EFFECT IN MICROWAVE REMOTE SENSING

8.1. ACCOUNTING FOR VEGETATION EFFECT IN MICROWAVE RADIOMETRY OF A SURFACE

8.1.1. Single Frequency Measurements

Accounting for vegetation screening effect is important in microwave radiometry of terrains and, especially, in microwave radiometry of soil moisture. Simple algorithms taking into account the vegetation screening effect were proposed in papers by Kirdiashev *et al.* (1979), Shutko and Chukhlantsev, 1982; Mo *et al.* (1982), Jackson *et al.* (1982), Ulaby *et al.* (1983a,b). These algorithms are based on the models for microwave emission from the Earth surface in the presence of vegetation canopies described in Chapter 6. An accuracy analysis of these algorithms was carried out in Chukhlantsev and Shutko (1988).

In single frequency radiometric measurements, an estimate of soil surface brightness temperature T_{bs}, when the soil is covered with vegetation, follows from expression (6.35):

$$T_{bs} = \frac{T_b - T'(1-\beta)}{\beta} \tag{8.1}$$

where $T' = (1-r_0)T$, T_b is the brightness temperature of the soil-vegetation system, $\beta = e^{-2\tau}$ is the vegetation layer transfer coefficient, r_0 is the reflectivity of the optically thick vegetation layer, T is the mean physical temperature of the system, and τ is the optical depth of the vegetation layer. It is seen from expression (8.1) that, to estimate the soil brightness temperature from radiometric measurements at a single frequency, a priori information on parameters T' and β is required. The derivatives of equation (8.1) with respect to T' and β characterize the influence of the uncertainty of these parameters on the accuracy of T_{bs} retrieval:

$$K^{T'} = \frac{\partial T_{bs}}{\partial T'} = -\frac{1-\beta}{\beta},$$ (8.2)

$$K^{\beta} = \frac{\partial T_{bs}}{\partial \beta} = -\frac{T'-T_{bs}}{\beta}.$$ (8.3)

The errors ΔT_{bs} of soil brightness temperature retrieval due to the uncertainty in T' and β, as well as due to errors in measurements of T_b, are calculated in Chukhlantsev and Shutko (1988) and are depicted in Fig. 8.1 for $T'-T_{bs} = 80\,\mathrm{K}$. As it is seen from Fig. 8.1, the main source of error in the T_{bs} retrieval is the uncertainty in the assignment of transfer coefficient β. When $\beta = 0,4...1$ and $\Delta\beta \leq 0,1$, the error of T_{bs} retrieval does not exceed 10-20 K. A drastic increase in this error at $\beta < 0,3...0,4$ makes it problematic to estimate the soil brightness temperature accurately at such small transfer coefficients. One can see that, to determine the soil surface brightness temperature with acceptable accuracy from microwave radiometric measurements at a single frequency, it is required to guarantee a low level of microwave attenuation in the vegetation layer ($\beta \geq 0,5$). Besides, the transfer coefficient should be assigned with accuracy not worse than 0.05-0.1. Results presented in previous chapters (see, particularly, Fig. 6.1 in Chapter 6) distinctly show that in the decimeter part of the microwave spectrum (wavelength $\lambda > 15\text{-}20$ cm) practically all types of vegetation canopies are semitransparent ($\beta \geq 0,5$). This circumstance makes it possible to retrieve soil surface parameters from microwave radiometric measurements in the L-band when the soil is covered with agricultural vegetation or with not so dense forest.

Equations (4.124) and (6.34) connect the transfer coefficient to a biometric characteristic of vegetation cover that is the vegetation water content W (or wet biomass) per unit area:

$$\beta = \exp[-2bW \sec \vartheta]$$ (8.4)

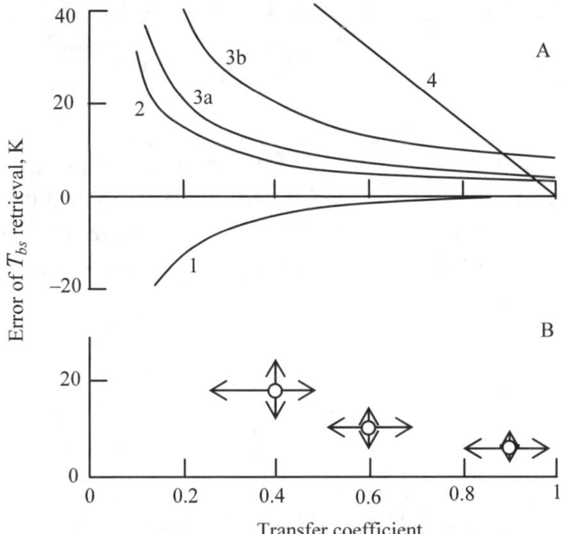

Fig. 8.1. Error of soil brightness temperature retrieval: A − due to the error in measured brightness temperature of vegetated soil (1), due to the uncertainty in physical temperature (2), due to the uncertainty of transfer coefficient (3), if vegetation effect is not accounted for at all (4). $\Delta T_b = \Delta T' = 3$ K; $T'-T_{bs} = 80$ K; $\Delta\beta = 0.05$ (3a) and $\Delta\beta = 0.1$ (3b), B − experimental data.

So, the transfer coefficient can be assigned if data on vegetation water content are available. Data, presented in Fig. 8.2, illustrate the change of transfer coefficient during the growth process for ear crops (wheat, rye, etc.) and fodder crops (alfalfa, clover, pea, etc.). These data were collected from radiometers mounted on a track (Chukhlantsev *et al.*, 1989). The dependence of the transfer coefficient uncertainty $\Delta\beta$, which appears if the vegetation biomass is assigned with relative accuracy of 30%, on the vegetation biomass is also depicted in Fig. 8.2. It is evident from Fig. 8.2 that at the *L*-band the presence of information about the type of vegetation and of approximate data on vegetation water content (or wet biomass) allows one to assign the vegetation transfer coefficient an absolute accuracy not worse than 0.1 for the values of the biomass up to 3 kg/m^2. Shutko (1982) derived an expression that connects the uncertainty $\Delta\beta$ with the error of soil moisture retrieval from radiometric measurements:

$$\frac{\Delta m_v}{m_v} \cong -1,5\frac{\Delta \beta}{\beta} \qquad (8.5)$$

From the aforesaid and equation (8.5) it follows that accounting for vegetation effect by assigning its transfer coefficient leads to an additional relative error in the soil moisture retrieval. The error is of the order of 10-30% when measurements are conducted at a single frequency in the *L*-band.

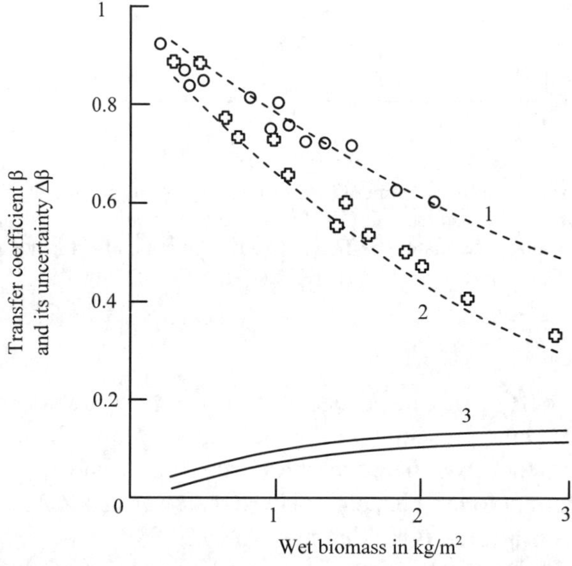

Fig. 8.2. Transfer coefficient of ear crops (1) and fodder crops (2) versus their wet biomass and uncertainty of these crops transfer coefficient (3) when the biomass value is assigned with accuracy 30%.

The efficiency of the described algorithm was experimentally tested by numerous researchers. Particularly, applicability of the algorithm was proved experimentally for different crops under natural conditions with the help of truck-mounted and airborne radiometers. Measurements with truck-mounted radiometers have shown that the use of the described algorithm, which is based on assigning the transfer coefficient or vegetation water content, allows one to retrieve volumetric soil moisture from radiometric measurements at a single frequency in the *L*-band with an error of 0.04-0.06 g/cm^3, while this error can reach 0.12 g/cm^3 if the vegetation effect is not accounted for. Airborne measurements conducted during several years (Chukhlantsev, 1981) allowed the author to collect extensive data on the transfer coefficient for different types of vegetation during growth periods. Some data on the transfer coefficient are presented in Table 8.1.

The algorithm that has been developed to take into account the reduction of brightness temperature sensitivity to soil moisture variations is widely used now in microwave radiometry of soil moisture and in interpretation of airborne and satellite radiometric data. However, the use of the algorithm requires knowledge of the vegetation transfer coefficient (or optical depth). This value can be estimated from equation (4.124) if data on vegetation water content are available. An estimate of vegetation water content or transfer coefficient can be obtained from some statistical data on the biomass accumulation in a crop during the growth period (Table 8.1). This approach provides, in principle, accepted accuracy of soil moisture retrieval because in the *L*-band the transfer coefficient is close to unity and its uncertainty does not affect significantly the accuracy of soil surface brightness temperature retrieval (Fig. 8.1). Nevertheless, for the retrieval of more accurate soil-moisture values, more exact data on the vegetation water content are required. These data can be obtained from ancillary remote sensors, particularly, from optical sensors. Several papers have been published that discuss the possibility of using different optical vegetation indices to improve soil moisture retrievals from microwave passive measurements.

Table 8.1. Transfer coefficient of different crops at 1.4 GHz.

Vegetation type	Wet biomass, kg/m^2	0-0.4	0.4-0.7	0.7-1	1-1.5	1.5-2	2-4
Ear crops	Time	May	June	June-July	July	July-August	August
Transfer coefficient		0.9	0.82	0.77	0.72		
Corn	Time	May	June	June-July	July	July-August	August
Transfer coefficient		0.95	0.88	0.82	0.74	0.66	0.5
Soybeans	Time	May	June	June-July	July	July-August	August
Transfer coefficient		0.93	0.8	0.7	0.6	0.5	
All crops	Time	May	June	June-July	July	July-August	August
Transfer coefficient		0.9	0.8	0.75	0.7	0.6	

Burke *et al.* (2002a, b) explored the feasibility of retrieving soil moisture from a single viewing angle, single frequency, and single polarization brightness temperature observations using a simple but physically based retrieval algorithm (with respect to accounting for vegetation as described above), with the optical depth estimated from a remotely sensed *NDVI* using an empirical relationship. The empirical relationship between optical depth and the *NDVI* used was derived from data collected at the El Reno site during the Southern Great Plains 1997 field study by calibrating the retrieval algorithm against field measurements of soil moisture measurements and airborne observations of microwave brightness temperature data. Radiometric data were collected using the Scanning Low Frequency Microwave Radiometer, a vertically polarized L-band push-broom microwave radiometer. Maps of brightness temperature for the site were obtained at different spatial resolutions varying from 200 to 600 m. The Normalized Difference Vegetation Index (*NDVI*) was taken from a Landsat Thematic Mapper and calculated using the equation:

$$NDVI = \frac{\rho_{NIR} - \rho_{RED}}{\rho_{NIR} + \rho_{RED}} \tag{8.6}$$

where ρ_{NIR} is the reflectance in the near-infrared wavelengths (0.78-0.90 μm), and ρ_{RED} is the reflectance in the visible wavelengths (0.63-0.69 μm). To obtain the relationship between the *NDVI* and optical depth it was necessary to estimate the latter. It was assumed that the difference between the retrieved and measured soil moisture was attributable solely to the optical depth of the canopy, the value of which was then adjusted in the retrieval algorithm to give a minimum difference between the measured and retrieved data. Then the independently derived value of optical depth was compared with the field-average *NDVI*. There was a significant linear relationship between the optical depth and $\log(1 - NDVI)$. It was noted, however, that, at high values of optical depth (greater than 0.6), the relation may become nonlinear. The relationship obtained was used then in the retrieval of soil moisture at some remaining sites. The agreement between measured and retrieved soil moisture was rather good. On the basis of the research, the authors noted that there seems to be some worthwhile potential in using remotely sensed vegetation indices in general, and *NDVI* in particular, as a surrogate measure of optical depth for soil moisture retrieval.

Jackson *et al.* (2004) evaluated the potential of using satellite spectral reflectance measurements to map and monitor vegetation water content for corn and soybean canopies. The goals were to provide high-quality vegetation water content estimates for the numerous microwave remote sensing soil moisture and evaporation studies conducted during the Soil Moisture Experiments 2002, and to contribute to the development of robust methods for providing near real-time information appropriate for global satellite applications. Two vegetation indices were explored to estimate vegetation water content: *NDVI* (8.6) and Normalized Difference Water Index (*NDWI*). The last index is of the form

$$NDWI = \frac{\rho_{NIR} - \rho_{SWIR}}{\rho_{NIR} + \rho_{SWIR}} \qquad (8.7)$$

where ρ_{SWIR} is the reflectance in a short wave infrared wavelength (*SWIR*) channel (1.2-2.5 μm). It was noted that the *NDVI* has limited capability for estimating vegetation water content because it is affected by other variables, particularly, by chlorophyll content, which is not always directly linked to vegetation water content. Additionally, *NDVI* saturates at intermediate values of the leaf area index, therefore it is not responsive to the full range of canopy vegetation water content. At the same time, the *NDWI* was found in previous research to be critical to estimating vegetation water content since the *SWIR* channel is sensitive to the vegetation water content and the near-infrared channel is needed to account for variation of leaf internal structure and dry matter content variations. Therefore, the *NDWI* is sensitive to the mass or volume of water and not to the fractional percentage of water. The relationships between both indices and vegetation water content of corn and soybeans canopies were obtained in the work on the basis of Landsat TM reflectance data and extensive ground-based measurements of vegetation water content in kg/m^2 (*VWC*). A functional form of these relationships was obtained as: for corn

$$VWC = 192.64 NDVI^5 - 417.46 NDVI^4 + 347.96 NDVI^3$$
$$- 139.93 NDVI^2 + 30.699 NDVI - 2.822 \qquad (8.8)$$

$$VWC = 9.82 NDWI + 0.05 \qquad (8.9)$$

for soybeans

$$VWC = 7.63 NDVI^4 - 11.41 NDVI^3$$
$$+ 6.87 NDVI^2 - 1.24 NDVI + 0.13 \qquad (8.10)$$

$$VWC = 1.44 NDWI^2 + 1.36 NDWI + 0.34 \qquad (8.11)$$

It could be noted that the relations obtained are quite different for investigated crops. It implies that the exploration of the *NDVI*- and *NDWI*- methods requires knowledge of vegetation type. Land cover classification was performed from Landsat TM data using supervised training procedures. It was found in the work that the *NDWI* is more accurate for *VWC* estimations. The root mean square error of *VWC* estimates was found to be 0.58 kg/m^2 for corn and 0.17 kg/m^2 for soybeans. The uncertainty of the transfer coefficient due to these errors can be found from equation (8.4):

$$\Delta\beta = -2b\Delta W\beta.$$

$$(8.12)$$

At the *L*-band, a characteristic value of b is ~ 0.15. It gives for the uncertainty $\Delta\beta = 0.05 - 0.15$ when $\beta \approx 0.8$ and $\Delta W = 0.2 - 0.6$. It follows from the previous analysis that this uncertainty is acceptable for soil moisture retrieval but it is comparable with that which occurs in simple estimation of vegetation water content from statistical data on the biomass accumulation in a crop during the growth period.

8.1.2. Two-Frequency Measurements

The necessity of a priori knowledge of the transfer coefficient (or vegetation water content) in single frequency measurements can be removed by conducting multi-configuration measurements, particularly, by measurements at two frequencies (wavelengths). An analysis of errors of soil surface brightness temperature retrieval that appear in these measurements was conducted by Chukhlantsev and Shutko (1988). The soil surface brightness temperature can be determined from equation (6.35) by measuring T_b at two wavelengths ($\lambda_1 < \lambda_2$, for clearness):

$$T_{bs} = T' - \frac{(T' - T_{b2})^{\frac{\theta}{\theta-1}}}{(T' - T_{b1})^{\frac{\theta}{\theta-1}}}$$

$$(8.13)$$

where $\theta = \dfrac{b_1}{b_2}$, b is the coefficient in (8.4), $T' = T_1' \approx T_2'$, and indices 1 and 2 are related to the measurements at λ_1 and λ_2, respectively. Equation (8.13) is obtained in an assumption that $T_{bs} = T_{bs2} \approx T_{bs1}$ that is realized in the case of not very rough surface. The error of T_{bs} retrieval is derived from (8.13):

$$\Delta T_{bs} = \frac{b_1}{b_1 - b_2} e^{2b_1 W} \Delta(T' - T_{b1}) - \frac{b_1}{b_1 - b_2} e^{2b_2 W} \Delta(T' - T_{b2}) \qquad (8.14)$$

and has a minimum value in measurements at two wavelengths that satisfy a condition

$$2(b_1 - b_2)W = 1, \quad b_2 = b_{\min} \qquad (8.15)$$

where $b_{\min}$ is a minimum value in the spectral dependence of b. When measurements are performed at these optimal wavelengths, the error of T_{bs} retrieval is

$$\Delta T_{bs} = -\frac{1 - \beta_2}{\beta_2} \Delta(T' - T_{b2}) - \frac{e \ln \beta_2}{\beta_2} \Delta(T' - T_{b1}). \qquad (8.16)$$

The dependence of ΔT_{bs} due to the measuring errors $\Delta(T' - T_{b1})$ and $\Delta(T' - T_{b2})$ on the transfer coefficient β_2 is depicted in Fig. 8.3.

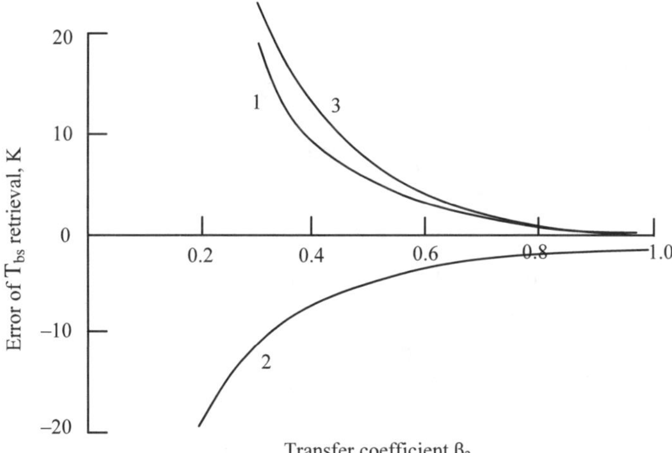

Fig. 8.3. Error of soil brightness temperature retrieval in two frequency measurements due to the error in measuring $T' - T_{b1}$ (1), due to the error in measuring $T' - T_{b2}$ (2), and due to the uncertainty of transfer coefficient ratio θ (3). $\Delta(T' - T_{b1}) = \Delta(T' - T_{b2}) = 3$ K; $T' - T_{bs} = 80$ K; $\Delta\theta = 0.04$.

Comparing data in Fig. 8.1 and Fig. 8.3, one can see that the errors of T_{bs} retrieval from single frequency and two-frequency measurements are close. In two-frequency measurements, it is essential to know how stable the ratio of optical depths at these two frequencies is. The error of T_{bs} retrieval due to this ratio uncertainty is

$$\Delta T_{bs} = (T' - T_{bs})(\ln \beta_2)^2 \Delta\theta . \qquad (8.17)$$

Calculations of ΔT_{bs}, according to equation (8.17), show (see Fig. 8.3) that the error ΔT_{bs} of the two-frequency algorithm does not exceed that of the single frequency algorithm when $\Delta\theta \le 0.5$. Available experimental data verify that variations of θ for a given type of canopy are within the required limit. For example, some data on optical depth of a pea canopy at wavelengths of 15 and 28 cm are presented in Fig. 8.4 (Chukhlantsev and Shutko, 1988). The root mean square of θ variation from the mean value is 0.1 with maximum deviation less than 0.6.

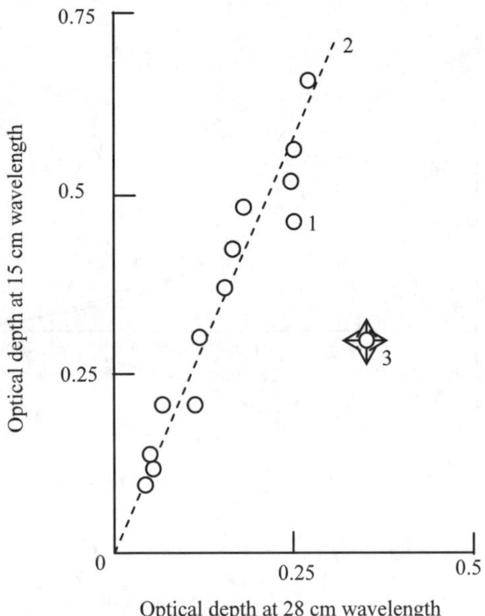

Fig. 8.4. Two-frequency diagram of pea canopy optical depths at wavelengths of 15 and 28 cm: experimental points (1), regression line with $\theta = 2.27$ (2), and measuring errors (3).

The analysis above points to a potential of multi-configuration radiometric measurements for soil moisture, as well as vegetation water content retrieval.

8.2. VEGETATION BIOMASS RETRIEVAL FROM MICROWAVE RADIOMETRIC MEASUREMENTS

8.2.1. Model Analysis

A feasibility of vegetation biomass determination from microwave radiometric measurements was displayed first in the work by Kirdiashev *et al.* (1979). Equation (6.35) was solved to find the coefficient β (and vegetation water content) from radiometric measurements at a single frequency for the same canopy with different T_{bs} (different soil moisture conditions). This approach was further used by many researchers for estimating vegetation optical depths (see Chapters 6 and 8). For two-frequency measurements, a model analysis of microwave radiometric algorithms as applied to vegetation biomass retrieval was performed by Chukhlantsev and Shutko (1987). These algorithms are based on models (equations (6.22) and (6.35)) for microwave emission from the Earth's surface in the presence of vegetation cover. From measurements of brightness temperature of soil-vegetation system at two different configurations (two wavelengths), one can get using equation (6.35):

$$\frac{\beta_1}{\beta_2} = \frac{T' - T_{b1}}{T' - T_{b2}}\left(1 + \frac{\Delta T_{bs}}{T' - T_{bs1}}\right) \tag{8.18}$$

where $\Delta T_{bs} = T_{bs1} - T_{bs2}$, $T' = (1 - r_0)T$, and indices 1 and 2 relates to the first and second measuring configuration, respectively. If conditions of measurements are chosen in such a way that

$$\Delta T_{bs} \approx 0, \tag{8.19}$$

an estimate of transfer coefficient ratio can be obtained from (8.18):

$$\frac{\beta_1}{\beta_2} \cong \frac{T' - T_{b1}}{T' - T_{b2}}. \tag{8.20}$$

On the other hand, according to (8.4), the ratio of transfer coefficients is

$$\frac{\beta_1}{\beta_2} \cong \exp[-2(b_1 - b_2)W] = \exp[-2(b_1 - b_2)mQ] \tag{8.21}$$

where m is the vegetation gravimetric moisture in wet weight basis and Q is the wet vegetation biomass. An estimate of vegetation biomass (the weight of vegetation per unit area) can be obtained, thus, from radiometric measurements at two configurations:

$$Q = -\frac{1}{2m(b_1 - b_2)} \ln\frac{T' - T_{b1}}{T' - T_{b2}}. \tag{8.22}$$

The temperature of vegetation can be equated with the air temperature or can be measured by an infrared sensor. The values of reflectivity r_0 for different types of vegetation canopies are known (see Fig. 6.2 in Chapter 6).

From equation (8.21), it is easy to obtain an equation that allows one to choose the measuring configurations 1 and 2, providing maximum sensitivity of estimated parameter β_1 / β_2 to biomass variations ($b_1 > b_2$, for clarity):

$$1 - 2(b_1 - b_2)mQ = 0. \tag{8.23}$$

An error of biomass retrieval ΔQ due to the error in β_1 / β_2 determination from measurements at configurations satisfying (8.23) has minimum value and is

$$\Delta Q = \frac{-\exp[2(b_1 - b_2)mQ]}{2m(b_1 - b_2)}\Delta\left(\frac{\beta_1}{\beta_2}\right) = -eQ\Delta\left(\frac{\beta_1}{\beta_2}\right), \tag{8.24}$$

and the relative error of biomass retrieval is

$$\frac{\Delta Q}{Q} = -e\,\Delta\left(\frac{\beta_1}{\beta_2}\right). \tag{8.25}$$

The equation for the measuring uncertainty $\Delta\left(\dfrac{\beta_1}{\beta_2}\right)$ is obtained from (8.20):

$$\Delta\!\left(\frac{\beta_1}{\beta_2}\right) = \frac{\Delta T'\left(1-\dfrac{\beta_1}{\beta_2}\right) - \Delta T_{b1} + \Delta T_{b2}\dfrac{\beta_1}{\beta_2}}{\beta_2(T'-T_{bs2})} + \frac{\beta_1}{\beta_2}\frac{\Delta T_{bs}}{T'-T_{bs1}} \qquad (8.26)$$

where $\Delta T'$, ΔT_{b1}, ΔT_{b2} are the errors in measuring "temperature" and brightness temperatures. At optimal measuring conditions (8.23), $\beta_1 / \beta_2 = e^{-1}$. The errors ΔT_{b1}, ΔT_{b2} are determined by the radiometer's sensitivity, the errors of absolute calibration of the radiometers, and the errors of radiation model (6.35). The uncertainty $\Delta\!\left(\dfrac{\beta_1}{\beta_2}\right)$ and the relative error of biomass retrieval $\dfrac{\Delta Q}{Q}$ can be evaluated under an adverse condition when $\Delta T_{b1} = - \Delta T_{b2} = \Delta T_b$ for different values of $\beta_2(T'-T_{bs2})$. Results of calculations are presented in Fig. 8.5. As it is seen from Fig. 8.5, when ΔT_b changes up to 3-4 K and when

$$\beta_2(T'-T_{bs2}) > 60...90 \text{ K}, \qquad (8.27)$$

the relative error of biomass retrieval does not exceed 20%. That implies that at least five gradations of vegetation biomass (water content) can be retrieved, in principle, from microwave radiometric measurements. One needs to consider the conditions (8.19), (8.23), and (8.27) that determine a practical realization of the retrieval. Condition (8.27) is satisfied for large values of soil moisture that are observed after rain fall or watering. Therefore, the biomass estimation algorithm considered can be used in the case when vegetation covers are above strongly moistened soils (Chukhlantsev and Shutko, 1985). The most favorable case is the radiometric observation of vegetation canopies when underlying surface is water (rice, rush) (Vorobeichik *et al.*, 1986). In this case, $(T'-T_{bs2}) \approx 200$ K, and the error of biomass retrieval is minimum 1 (Fig. 8.5). Besides, to satisfy the condition (8.27), vegetation canopy should be transparent at the second measuring configuration ($\beta_2 \approx 1$) that is observed in the decimeter wavelength band.

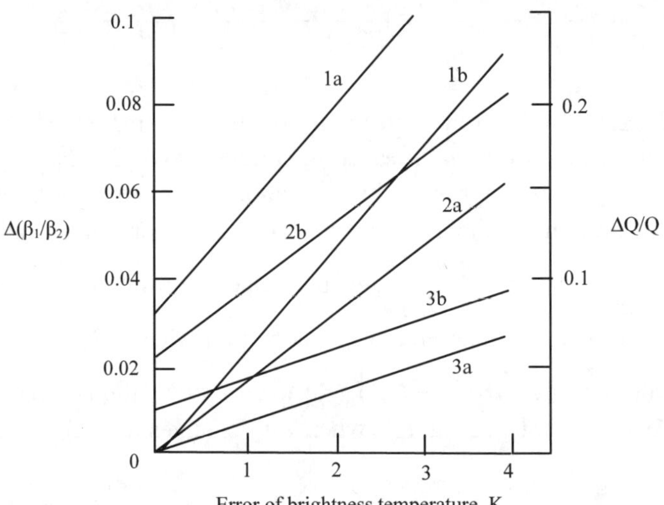

Fig. 8.5. Error of parameter $\Delta(\beta_1/\beta_2)$ retrieval and relative error of biomass retrieval versus error of brightness temperature measurements. $\beta_2(T' - T_{bs2}) = 60$ K (1), $\beta_2(T' - T_{bs2}) = 90$ K (2), and $\beta_2(T' - T_{bs2}) = 200$ K (1). $\Delta T' = 0$ K (a) and $\Delta T' = 2$ K (b).

Conditions (8.19) and (8.23) can be satisfied by conducting spectral radiometric measurements. Coefficient b has strong frequency dependence (see Chapters 4 and 5). In L-, S-, and C- bands, this dependence can be approximated by the expression (5.19):

$$b \approx \frac{b'}{\lambda} \tag{8.28}$$

where the factor b is known for different canopies (see Chapter 5). Parameter A can be introduced to describe the sensitivity of transfer coefficient ratio to biomass variations:

$$A = \frac{\Delta\left(\dfrac{\beta_1}{\beta_2}\right)}{\Delta Q}. \tag{8.29}$$

Dependence of this parameter on the biomass and wavelength λ_1 is presented in Fig. 8.6. From this dependence and from the previous analysis, the following conclusions can be made. To provide minimum error of biomass retrieval, the wavelength λ_2 should be chosen as big as possible (the maximum value of λ_2 is limited by the presence of galactic radiation and its influence on radiometric measurements that drastically increases at wavelengths bigger than 50 cm). With this choice of λ_2, $\beta_2 \approx 1$, and the best accuracy of vegetation biomass retrieval is achieved. The first wavelength λ_1 should be chosen from the condition (8.23) that gives a value $\lambda_1 \approx 2b'mQ$ increasing with the increase in the biomass. Condition (8.19) is satisfied in spectral measurements for not very rough soils. For rough soils, the algorithm discussed can be improved by using equation (8.18) instead of (8.20) and taking into account spectral behavior of rough soils emission (see Chapter 3).

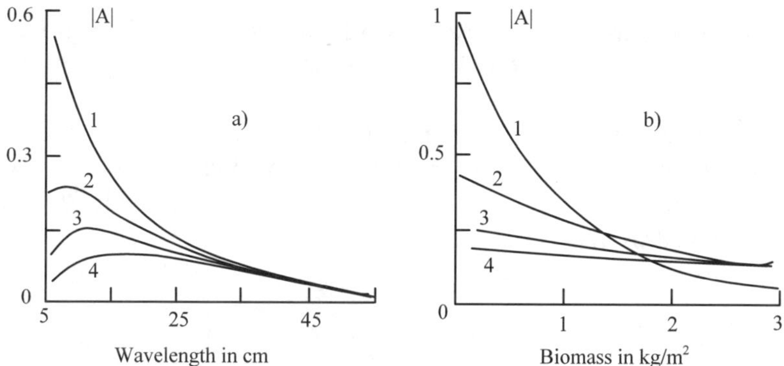

Fig. 8.6. Sensitivity of parameter $\Delta(\beta_1/\beta_2)$ to variations of biomass versus λ_1 (a) and Q (b). $\lambda_2 = 50$ cm. a): Q = 0.5 (1), 1.5 (2), 2.5 (3), and 3.5 kg/m² (4). b): $\lambda_1 = 5$ (1), 10 (2), 15 (3), and 20 cm (4).

It should be noted that algorithms considered in Sections 8.1.2 and 8.2.1 are, per se, a common algorithm that is based on a solution of the following system of equations:

$$\begin{cases} T_{b1} = T_{bs1}\beta_1 + (1 - r_{01})(1 - \beta_1)T \\ T_{b2} = T_{bs2}\beta_2 + (1 - r_{02})(1 - \beta_2)T \end{cases}. \tag{8.30}$$

In Section 8.1.2, this system is resolved with respect to T_{bs} (that is mainly determined by soil moisture), while in this section the system is resolved with respect to β that is a function of vegetation water content (wet biomass).

The system is very simple because the radiative model used is obtained on some simplifying assumptions (see Chapter 6). Nevertheless, the model is accurate enough and has an advantage that it can be resolved under some conditions analytically (equations (8.13) and (8.22)). It allows one to conduct easily a sensitivity analysis and to estimate the potential accuracy of parameters retrieval that is done above for the case of two-frequency measurements. Of course, the system can be also resolved using other two measuring configurations. Particularly, it can be realized by conducting polarization radiometric measurements (Owe *et al.*, 2001) or by conducting observations at different viewing angles (Wigneron *et al.*, 2004a).

8.2.2. Examples of Practical Realization

8.2.2.1. Vegetation Biomass Retrieval from Microwave Radiometric Measurements at a Single Frequency

Vegetation water content retrieval from single frequency measurements is possible, in principle, if the brightness temperature of the underlying soil surface is known. The transfer coefficient β (or transmissivity $t = \sqrt{\beta}$) can be determined from (6.35):

$$\beta = t^2 = \frac{T' - T_b}{T' - T_{bs}} \qquad (8.31)$$

where $T' = (1 - r_0)T$, T_b is the brightness temperature of the soil-vegetation system. The brightness temperature of the underlying soil surface is usually estimated from the brightness temperature of nearby bare soil surface having close soil moisture conditions. The retrieval technique described above was used by many authors (Kirdiashev *et al.*, 1979; Jackson *et al.*, 1982; Mo *et al.*, 1982; Wang *et al.*, 1982a; Brunfeldt and Ulaby, 1986; Pampaloni and Paloscia, 1986; Chukhlantsev *et al.*, 1989; Van de Griend *et al.*, 1996; Kruopis *et al.*, 1999; Milshin *et al.*, 1999).

As noted in the previous section, the most favorable situation is the radiometric observation of vegetation canopies when underlying surface is water (rice, rush). In this case, the error of vegetation biomass retrieval is minimum. Besides, well-studied emission properties of water surface allow us to obtain estimates of biomass from microwave radiometric measurements at a single frequency (Vorobeichik *et al.*, 1986). The biomass estimate is determined from measured brightness temperature by

$$Q = -\frac{1}{2mb_1} \ln \frac{T' - T_{b1}}{T' - T_{bw}}$$
(8.32)

where T_{bw} is the brightness temperature of open water surface. Biomass estimates can be also obtained using algorithmic dependences of emissivity on the under-water biomass (see Fig. 7.2 in Chapter 7). Exploration of the technique proposed was conducted for rice crops during 1980-83 in Krasnodar Region (Chukhlantsev and Shutko, 1987). A radiometer operating at a wavelength of 18 cm was installed onboard a light bi-plane Antonov-2. The flight height was 50-100 m providing a 35-70 m spatial resolution at the surface. To specify the dependence of coefficient b on the rice biomass and to validate the results obtained, several test sites were chosen where biomass sampling was performed. Biometric measurements were conducted by specialist of the Krasnodar Institute of Rice. Samples were taken by vegetation cut under-water from small square areas with a size of 0.5×0.5 m^2. To increase the accuracy of ground measurements, 12-21 samples were taken within a test site. The accuracy of biomass retrieval from radiometric data was estimated by comparing retrieved data with ground measurements at check sites with different rice species and crop heights varying within 60-90 cm. Results of the comparison are presented in Fig. 8.7.

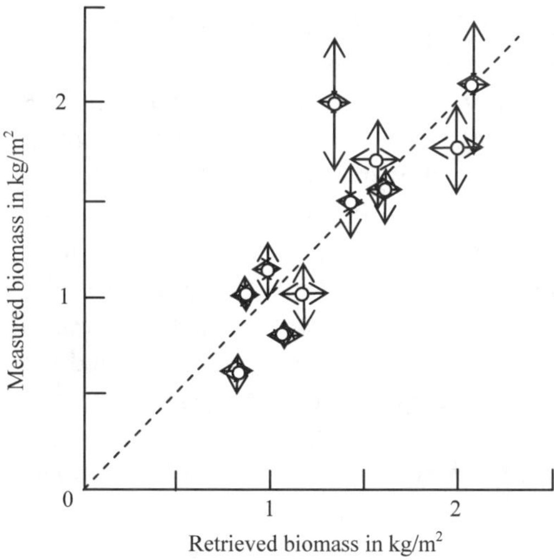

Fig. 8.7. Measured and retrieved biomass of rice crops.

Horizontal segments are due to spatial variations of the radiometric signal over the check site. Vertical segments represent the root mean square of biomass variations within the site. It is seen that the accuracy of biomass retrieval is 10-20% which is comparable with the accuracy of traditional ground measurements.

The described method of rice biomass retrieval was further developed by Yazerian (2000) for rice crop monitoring and rice yield prediction. The method can also be used for validation and calibration of satellite sensors with high spatial resolution (optical sensors, synthetic aperture radars) that are specialized for satellite monitoring of rice growth and global rice yield forecasting. Radiometric observations can also be useful for monitoring of lakes and rivers overgrown with rushes and canes.

8.2.2.2. Crop Water Content and Soil Moisture Retrieval from Spectral Measurements

Actually, there are few papers describing spectral retrieval algorithms. First of all, the work by Wigneron *et al.* (1995b) can be mentioned. The authors had available two extensive sets of radiometric data acquired over a soybean and a wheat crop. The measurements monitored the whole vegetation development during a 2.5 month period. Therefore, a large range of soil moisture and vegetation density conditions were observed. Both experiments were conducted on a plot located on the INRA (Institut National de Recherches Agronomiques) Avignon test site. The observations were acquired using the multifrequency passive microwave radiometer PORTOS (1.4, 5.05, 10.65, 23.8, 36.5, and 90 GHz). The radiometer was mounted on a 20-m crane boom and observations were carried out at different incidence angles (from 0° to 50°). The viewing direction of the radiometer was parallel to the row direction for both crops. During the experiments, absolute calibrations were made over calm water and over ecosorb slabs either at surrounding temperature or immersed into liquid nitrogen. During the measuring campaigns, soil and vegetation characteristics were sampled regularly. Twice a week, measurements of dry and wet biomass, water content, height, volume fraction, and geometry of the vegetation canopy were performed. Measurements of soil moisture content, soil and vegetation temperatures were performed concurrently with the radiometric measurements.

The retrieval algorithm was based on the $\tau - \omega$ model (6.22). The radiometric data used in the retrieval problem considered in the paper were acquired at 1.4 GHz (*L*-band) and 5.05 GHz (*C*-band). The range of incidence angles had been limited to $0° - 40°$, and four specific incidence angles were used in the study (8°, 18°, 28°, and 38°). Several configurations of the retrieval problem were tested based on the use of:

    A1. Two frequencies (1.4 GHz and 5.05 GHz) and four incidence angles,

    A2. Two frequencies (1.4 GHz and 5.05 GHz) and one incidence angle (38°),

    B1. One frequency (1.4 GHz) and four incidence angles,

    B2. One frequency (1.4 GHz) and one incidence angle (38°),

    C. One frequency (5.05 GHz) and four incidence angles.

For all above configurations, both vertical (v) and horizontal (h) polarizations were used. During the inversion process, two parameters were retrieved simultaneously: soil moisture and the optical depth for h polarization. These two parameters were retrieved for all the dates of radiometric observations. The inversion process was performed through a minimization procedure of the root mean square error between the measured and simulated microwave emissivities for every incidence angle and every frequency given by the retrieval configuration. It appeared that the soil roughness effects could be neglected for all the configurations. As no simple formulation was found to describe the dependence of the single scattering albedo on polarization and incidence angle, approximate values, $\omega = 0.04$ and $\omega = 0$, was used for wheat at 5 GHz and at 1.4 GHz, respectively. For the soybean, it was accepted that $\omega = 0.11$ and $\omega = 0$ at 5 GHz and at 1.4 GHz, respectively. The values of coefficient b (8.4) were found to be equal to $b_h = 0.12$ and $b_h = 0.18$, respectively, for wheat and soybean at 38° and at 1.4 GHz. At 5 GHz, $b_h = 0.4$ for both crops. These values are consistent with those reported earlier (see Chapter 5).

    It was found that the retrieval process fails for single frequency measurements (configurations B1, B2, and C). It was also shown that using simultaneous measurements at both L- and C-band for a range of incidence angles between 0° and 40°, good retrievals can be obtained (the precision error is about 15% for both soil moisture and vegetation water content). The results were found to be more sensitive to the accuracy of both the microwave data and the calibrated parameters (especially, to the accuracy of the ratio of optical depths at the frequencies explored). The retrieval accuracy obtained by Wigneron *et al.* (1995b) for a two-frequency measuring configuration is in general agreement with that predicted by model analysis conducted in previous sections.

    Later, the single frequency with a two-angles inversion procedure was tested and improved by Wigneron *et al.* (2004a). The approach was based on statistical relationships between the soil moisture and measured surface reflectivity at different observation angles and polarizations. It was found that only two observations at distinct incidence angles (ϑ_1 and ϑ_2) are required in this case. These relationships were calibrated from simulated datasets. It was found that best estimations of soil moisture could be made at horizontal polarization, for ϑ_1 varying between 15° and 30°, and with a difference ($\vartheta_2 - \vartheta_1$) larger than 30°. The method was tested against two experimental datasets acquired over crop fields (soybean and wheat). The average accuracy in the soil moisture retrievals during the whole crop cycle was found to be about 0.05 g/cm^3 for both crops.

In the paper by Calvet *et al.* (1995b), the use of higher frequencies for retrieval of surface parameters is discussed. The PORTOS measurements at 23.8, 36.5, and 90 GHz over a sparse sorghum and patchy wheat canopies were examined. It was found that the top surface-soil moisture can be retrieved with acceptable accuracy for sparse vegetation volume fraction below 0.5×10^{-3}, or dense patchy vegetation with fractional coverage below 0.5.

In the paper by Wigneron (1994), the PORTOS measurements at 1.4, 5.05, and 36.5 GHz over a soybean canopy were used to retrieve the soil moisture and the vegetation volume fraction on the basis of the continuous approach for vegetation (see Chapter 4). A good agreement between measured and inversed surface parameters was observed.

Wigneron *et al.* (1996) studied the feasibility of passive microwave observations for estimation of amount of water intercepted by crops during rain. The results presented in the paper were obtained from low-frequency (1.4 and 5 GHz) ground-based measurements over a wheat canopy (a part of PORTOS measurements). Observations were made twice during a several-hour irrigation phase. During this period, irrigation water was sprinkled over the canopy by regularly spaced sprinkler heads. The aim of irrigation was to simulate a heavy rainfall event. The microwave signature of the canopy was analyzed for different stages of the irrigation phase: before irrigation: just after irrigation: several hours after irrigation (after sun and wind dried the wheat field). The possibility of two-frequency radiometric measurements for retrieval soil moisture and vegetation water content was distinctly demonstrated. (It could be noted that artificial irrigation provides high values of soil moisture and, thus, the best conditions for vegetation water content retrievals as it follows from the foregoing discussion (Section 8.2.1)). It was found that the amount of water intercepted by the wheat canopy can achieve 0.74-1.32 kg/m^2 and represents about 70% of the amount of water contained within the vegetation material. These figures were consistent with available literature data on the magnitude of wheat rainfall storage capacity.

8.2.2.3. Surface Soil Moisture and Optical Depth Retrieval
from Polarization Measurements

Utilization of polarization measurements for soil moisture and optical depth (vegetation water content) retrieval at satellite scales was discussed, particularly, by Owe *et al.* (2001) and Meesters *et al.* (2005). The retrieving procedure is started from equation (7.22) for the brightness temperature of vegetated terrains that is written in the form of a system of equations for horizontal (h) and vertical (v) polarization:

$$\begin{cases} T_{bh} = T_s e_h t + T_v (1-t)(1-\omega) + T_v (1-t)(1-\omega)(1-e_h)t \\ T_{bv} = T_s e_v t + T_v (1-t)(1-\omega) + T_v (1-t)(1-\omega)(1-e_v)t \end{cases}. \tag{8.32}$$

The canopy transmissivity, t, is defined as

$$t = \exp\{-\tau \sec \vartheta\} \tag{8.33}$$

where τ is the vegetation optical depth and ϑ is the observation angle. The following assumptions are further made in the retrieval model. It is proposed that at satellite scales the optical depth τ and the single scattering albedo ω have minimal polarization dependence. Moreover, an average scattering albedo of 0.06 is used in the model. It is also assumed that surface roughness has a minimum effect at the satellite scale footprints considered. Therefore, the emissivities e_h and e_v are calculated by Fresnel formulas for a perfectly smooth surface and are considered as known functions of the observation angle and absolute value of the soil dielectric constant (ε) that is used instead of soil moisture in order to minimize the influence of soil physical properties. A further assumption, that

$$T_s \approx T_v \approx T, \tag{8.34}$$

is also made in the model. This is considered as a reasonable assumption for global locations with less dense or intermittent vegetation cover.

In the original retrieval model (Owe *et al.*, 2001), the surface soil moisture and optical depth were simultaneously restored using equations (8.32) and the h- and v-polarized brightness temperatures at 6.6 GHz from the SMMR (Scanning Multichannel Microwave Radiometer) instrument, which flew onboard the Nimbus-7 satellite. The 37 GHz v-polarized channel was used to derive surface temperature. With known values or estimates for the incidence angle, single scattering albedo, and surface temperature, the only remaining unknowns are ε and τ. With two equations (8.32) and two unknowns, it would appear logical to solve them as a system of simultaneous equations. However, the solution is far from straightforward, as both equations are a quadratic in t, and are non-monotonic within the range $0 \leq t \leq 1$. Solving for t yields two distinct expressions, therefore each one must be tested independently in order to determine which one yields the correct result. So, the Normalized Polarization Difference Index (3.9) is used in the model. It was called (Owe *et al.*, 2001) Microwave Polarization Difference Index (MPDI) and was defined as (3.9)

$$MPDI = \frac{T_{bv} - T_{bh}}{T_{bv} + T_{bh}}. \tag{8.35}$$

The MPDI is a monotonic function of τ and is assumed to be a better starting point in the retrieval procedure. The MPDI follows immediately from the satellite brightness temperature observations, and depends only on ε and τ (and known parameters). The dependence of MDPI on the optical depth was approximated by Owe *et al.* (2001) with a logarithmic function:

$$\tau = C_1 \ln(C_2 \cdot MDPI + C_3) \tag{8.36}$$

where C_1, C_2, and C_3 are the coefficients depending on the absolute value of the soil dielectric constant. The coefficients are expressed as a series of lengthy polynomials. Equation (8.36) was then substituted into the first equation of (8.32). The only remaining unknown was the dielectric constant of the soil. In this way, the canopy optical depth and the soil emissivity were defined in terms of the soil dielectric constant. Next, the model used a non-linear iterative procedure in a forward approach to solve the first equation of (8.32) by optimizing on the dielectric constant. The methodology was applied to the entire data set of nighttime SMMR brightness temperatures for Illinois. Six year time series of optical depth retrievals for this area have demonstrated a distinct annual course in the optical depth that coincided well with expected vegetation dynamics and the available NDVI data. A disadvantage of the model is a need for cumbersome recalculating of the curves (8.36) with a change in the observation angle and scattering albedo.

The original retrieval model (Owe *et al.*, 2001) was sufficiently improved by Meesters *et al.* (2005). This paper suggested an analytical solution, as an alternative to the numerical approximation. This approach is not only more elegant, but computationally more efficient as well. Furthermore, the curve-fitting procedure (Owe *et al.*, 2001) was valid for only one set of incidence angle and single scattering albedo values. Changing either value would require recalculation of the four polynomials. The new procedure is valid under any conditions. A new expression for t is derived from equations (8.32) and (8.35):

$$\frac{1}{MPDI} = \frac{e_v + e_h}{e_v - e_h} + \frac{2(1+t)(1-t)(1-\omega)}{\omega t + (1-\omega)t^2}\frac{1}{e_v - e_h}. \tag{8.37}$$

This equation was reworked to the more concise expression

$$2a = \frac{2(1-t^2)}{2dt + t^2} \tag{8.38}$$

where

$$a(\varepsilon,\vartheta) = \frac{1}{2}\left[\frac{e_v(\varepsilon,\vartheta) - e_h(\varepsilon,\vartheta)}{MDPI} - e_v(\varepsilon,\vartheta) - e_h(\varepsilon,\vartheta)\right] \tag{8.39}$$

and

$$d = \frac{1}{2}\frac{\omega}{1-\omega}. \tag{8.40}$$

A quadratic equation in t is obtained from (8.38):

$$(1+a)t^2 + 2adt - 1 = 0. \tag{8.41}$$

This equation is solved with respect to the optical depth that yields

$$\tau = \cos\vartheta \ln\left(ad + \sqrt{(ad)^2 + a + 1}\right). \tag{8.42}$$

This equation is used in the retrieval algorithm with the first equation in (8.32) in a forward modeling approach as in the original model (Owe *et al.*, 2001). The difference in computing time between the original and improved approaches varied considerably, with the new approach being consistently faster by an average factor of 7. That seems to be important in processing long term data sets and generating daily global maps of vegetation optical depth and soil moisture. The improved retrieval model was used to derive global maps of optical depth at ¼ degree resolution from 6.6 GHz Nimbus/SMMR nighttime observations. The global maps for January and July are presented in the paper and compared with NDVI global maps. Visual comparisons between the global optical depth maps and NDVI suggest similar spatial characteristics. The authors (Meesters *et al.*, 2005) have noted that certain land cover conditions interfere with or preclude some modeling applications. Particularly, excessive vegetation (optical thickness more than 0.8 for the *C*-band) begins to saturate the observed signal, such that soil signal fraction is reduced sufficiently, and the retrieval may become unreliable. Besides, a high density of artificial electronic signals over some parts of the world can interfere with the retrieval. Nevertheless, the retrieval model developed seems to be very prospective because it operates with a simple inversion algorithm and uses a simple measuring configuration (single frequency polarization measurements). The model can be used at lower frequencies (SMOS), where the radiometric signal saturates at higher values of the vegetation water content that can provide more accurate estimations of soil moisture.

8.3. SOIL MOISTURE AND VEGETATION BIOMASS RETRIEVAL FROM MULTI-CONFIGURATION MICROWAVE RADIOMETRIC MEASUREMENTS

A review of different approaches for soil moisture (and vegetation biomass) retrievals from microwave radiometric measurements was given, particularly, by Wigneron *et al.* (2003). The approaches used to account for effects caused by vegetation cover were broken down into three main categories:

- *Statistical techniques*: For each land cover type/biome or group of pixels (for space-borne observations), linear relationships between measured brightness temperature and surface soil moisture are established. (Potentials of this approach are described in Section 8.1.1.)
- *Forward model inversion*: In this approach, a model is used to simulate remotely sensed signatures (output) on the basis of land surface parameters (input). Inversion methods are developed to produce an "inverse model" in which outputs are the relevant land surface variables. (Potentials of this approach are described in Sections 8.1.2 and 8.2.1.)
- *Explicit inversion*: The explicit inverse of the physical process can be built by transferring input (remote sensing measurements) into output (land surface parameters). In most studies, neural networks are used to create this explicit inverse function.

In the first approach, a simple regression is used to relate measured brightness temperatures to geophysical quantities. For example, soil moisture can be retrieved from a linear regression:

$$m_V = a_0 + a_1 T_{b1} + a_2 T_{b2} + \ldots \tag{8.43}$$

where $T_{b,i}$ corresponds to measurements made for various configurations of the sensor, in terms of incidence angle, polarization, or frequency. In Pellarin *et al.* (2003a), volumetric soil moisture was proposed to be retrieved from the regression

$$m_V = a_0 + a_1 IND_1 + a_2 IND_2 + \ldots \tag{8.44}$$

where IND_i is the measured values of different polarization indices (see Chapter 3). The use of the first approach assumes that regression coefficients a_i are known for each pixel. The coefficient can be calculated on the basis of a forward model accounting for fractions of different surfaces within the mixed pixel; type and water content of vegetation inherent for the pixel. So, this approach is the simplest but it requires a lot of ancillary information. The interested reader can find a review of the statistical approach application in the paper by Wigneron *et al.* (2003).

The second approach allows one to retrieve, in principle, several parameters of vegetation-soil system. Methods of soil moisture and vegetation water content simultaneous retrieval are discussed in previous sections. As a rule, in these methods, some simplifying assumptions are made about vegetation and soil emission properties (neglecting vegetation reflectivity, neglecting polarization dependence of vegetation optical depth, introducing ancillary information on the system temperature, etc.). In the general case, the problem of forward model inversion to retrieve land surface parameters is formulated as follows: a radiative model is used to simulate the microwave radiometric measurements as a function of the land surface characteristics x_j:

$$T_{b,i} = T_{b,i}(x_1, x_2, \ldots x_p; s_i) + \varepsilon_i \tag{8.45}$$

where $T_{b,i}$ corresponds to measurements made for various configurations of the sensor, in terms of incidence angle, polarization, or frequency, s_i stands for configuration parameters, and ε_i is the residual error between the simulated and measured brightness temperature values. The retrieval methodology requires two main steps: selection of a forward model (8.45) providing minimum model errors; selection of a method for inversion (8.45) providing maximum accuracy of x_j estimation at known errors of the model and measurements and their statistical properties.

When the number of retrieved land surface characteristics is not big (for example, only two parameters should be retrieved), the number of measurements is limited, and the inverse problem is referred to the problem of estimating these parameters from the measurements. A very common algorithm for estimating the parameters is the minimization of the difference (discrepancy) between the measured brightness temperatures and the simulated brightness temperatures. In that algorithm, model parameters providing the minimum difference are accepted as the estimates of these parameters. The choice of minimization procedure depends on the statistical properties of the errors of measurements and the model. As a rule, these properties are not known, and the squared discrepancy between the measured and simulated brightness temperatures is minimized (Njoku and Li, 1999). But it is also known that the minimization of the absolute values of the discrepancy can provide better estimates in some cases.

When the number of retrieved land surface characteristics is big, the number of measurements is big too, and the inverse problem assumes features of the so-called ill-posed inverse problem (Tikhonov and Arsenin, 1979). There are some methods for solving these problems (regularization methods that make use of additional a priori information about the retrieved parameters) that are also used in the statistical inversion approach (Pulliainen *et al.*, 1993; Pardé *et al.*, 2004).

Application of the forward model inversion approach can be illustrated by the following examples. Retrieval of land surface parameters using passive microwave measurements at 6-18 GHz was considered by Njoku and Li (1999). The approach was evaluated for retrieval soil moisture, vegetation water content, and surface temperature and was proposed to be applicable to data from the Advanced Microwave Scanning Radiometer. The algorithm developed has been tested using data from the Nimbus-& Scanning Multichannel Microwave Radiometer (SMMR) for the years 1982-1985, over the African Sahel.

The retrieval algorithm was based on a radiative transfer model that relates parameters describing the surface and atmosphere to the observed brightness temperatures. The model represents these processes in a simplified form appropriate to the spatial scale of the satellite footprints. The parameterization of the model was designed with the retrieval algorithm in mind. Errors in the model approximations, and *a priory* uncertainties in the model parameters, were reflected in the resulting retrieval errors. These errors and uncertainties were estimated, and their influences on retrieval error were evaluated. The $\tau - \omega$ model (6.22) in the isothermal case was used to describe the emission from vegetated soils. The optical depth was calculated by (4.124). Soil emissivity was calculated by Fresnel formulas. Soil roughness was accounted for by introducing soil roughness parameters Q_s

and h_s (see Chapter 3). Model parameters Q_s, h_s, and ω were derived from measurements over the desert and forest calibration sites. For a heterogeneous scene, area-weighted averages over the scene components were accounted for using equation (6.53). The atmosphere element of the model was described in terms of weighted-mean temperature and atmosphere optical depth.

The radiative transfer model was written in the form of (8.45) as

$$T_{b,i} = \Phi_i(x_1, x_2, \ldots x_p).$$

(8.46)

The sensitivity of brightness temperature to moisture and vegetation was estimated and computed explicitly. The sensitivities were normalized such that the sensitivity S_{ij} of brightness temperature at channel i to geophysical parameter x_j was expressed as

$$S_{ij} = \left| X_j \left(\frac{\partial \Phi_i}{\partial x_j} \right)_{x=x_0} \right|$$

(8.47)

where X_j are within a typical parameter dynamic range and x_0 are baseline values of the parameter x, at which the sensitivities were evaluated. The sensitivity analysis has shown that the sensitivity to soil moisture decreases with increasing vegetation, and there was a little sensitivity at vegetation water contents greater than ~ 1.5 kg/m^2 even at 6.6 GHz, while the sensitivity to surface temperature remained high. Sensitivities to other variables (atmosphere variables) and model parameters (soil roughness and vegetation albedo) were less than to the free main variables, and, they were not dominant factors in the retrievals.

The retrieving procedure found values for the set of variables (soil moisture, vegetation water content, surface temperature, and atmosphere precipitable water) that minimize χ^2, i.e., the weighted-sum of squared differences between observed $T_{b,i}^{obs}$ and computed Φ_i brightness temperatures, where

$$\chi^2 = \sum_{i=1}^{6} \left(\frac{T_{b,i}^{obs} - \Phi_i}{\sigma_i} \right)^2.$$

(8.48)

The efficient Levenberg-Marquardt algorithm was used to search for the set of variables x_j that minimizes the χ^2 (Press *et al.*, 1989). At each retrieval point, the algorithm started with *a priori* values x_0 of the geophysical variables to be retrieved and adjusted these iteratively until convergence to the minimum χ^2 was achieved within specified criteria. The σ_i represented the measurement noise standard deviation in channel i. The simulation results have indicated that, even accounting for modeling errors, retrieval accuracies for soil moisture, vegetation water content, and surface temperature were better than 0.06 g/cm^3, 0.1 kg/m^2, and 2° C, respectively. The aforementioned procedure was then applied to SMMR data. The retrieval results have been compared to output from an operational numerical weather prediction model. The retrieval algorithm was shown to discriminate well between soil moisture, vegetation, and temperature variations and provided estimates consistent with the expected accuracies.

Another example of the forward model inversion approach is demonstrated in Pardé *et al.* (2004). The possibility to implement N-parameter retrievals from L-band microwave observations ($N = 2, 3, 4...$) was investigated in the paper based on experimental datasets acquired over a variety of crop fields. The datasets were collected during BARC (Wang *et al.*, 1982a), PORTOS-91 and -93 (Wigneron *et al.*, 1995, 1993; Calvet *et al.*, 1995b), and EMIRAD (Wigneron *et al.*, 2004b) experiments. The $\tau - \omega$ model was used in forward modeling emission properties of the crop fields. The retrievals were based on the inversion of this model. Seven model parameters were considered in the retrieval: soil moisture and the roughness parameter h_s, surface temperature, vegetation optical thickness, the single scattering albedo at both vertical and horizontal polarizations, and the C_{pol} parameter (see equation (5.26)). These parameters were retrieved by minimizing a cost function (C) using a generalized least square iterative algorithm (Marquardt, 1963) modified to account for *a priori* information available on the model input parameters. This *a priori* information consisted of: 1) the initial value of the parameters, which corresponds to an *a priori* estimate of the parameter, and 2) the uncertainty associated with these estimates. The minimization routine provided both the retrieved surface parameters and the theoretical estimates of the standard deviation associated to these retrievals, assuming that the brightness temperature measurements and the model-input parameters were affected by a Gaussian random error.

The cost function was the sum of the weighted squared differences between simulated and measured values (compare with (8.48))

$$C = \sum_i \left(\frac{T_{b,i}^{obs} - \Phi_i}{\sigma_i} \right)^2 + \sum_j \left(\frac{P_j^{ini} - P_j^{retr}}{\sigma_j} \right)^2 \qquad (8.49)$$

where P_j^{ini} and P_j^{retr} are the initial (*a priori* estimate) and retrieved value of the j-th parameter, respectively, and σ_j is the standard deviation of this parameter. (The second term in (8.49) represents a regularization of the inversion procedure.) Different sets of initial values for the parameters were used to determine the best retrieval configurations. It was shown that depending on the retrieval configuration, in terms of the number of retrieved parameters and constraints applied to the retrieved parameters, very different accuracies can be obtained, particularly, in the soil moisture retrievals. Nevertheless, the root mean squared error between the estimated and measured soil moisture was, as a rule, less than 0.06 g/cm^3.

Many different iterative algorithms are available to minimize the cost function. Recently, neural network algorithms, as another alternative approach to the forward model inversion, are being applied. A neural network can be trained by an appropriate set of input-output data generated with the use of the forward model. Thus, the neural network realizes rather complex and nonlinear relationships in the form of images that are retained in the neural network during the training process. The neural network algorithms can compete with those described above in productivity, because, if the neural network has been trained, parameter inversion can be accomplished very quickly. In the field of microwave radiometry of a vegetation-soil system, the application of neural networks was considered, particularly, in papers by Liou *et al.* (2001), Nazarov *et al.* (2003, 2004).

Another simple way to invert a forward model using a neural network is to train an inverse model by reversing the roles of the inputs and outputs: the input nodes of the neural network are the measured brightness temperature and the output nodes are the land surface parameters. This method, known as *explicit inversion*, was, particularly, used in Liu *et al.* (2002). Unfortunately, as noted in Wigneron *et al.* (2003), the forward model can be characterized by a "many-to-one mapping" (i.e., a set of measurements cannot be uniquely related to environmental variables). Several studies expressed concerns about the fact that the explicit inversion approach may lead to wrong results when the inverse image of the forward model is not convex. Besides, the explicit inversion approach (like statistical techniques) has some limitations as it can only be used for the regions and the time period during which it was calibrated.

8.4. VEGETATION EFFECT IN ACTIVE MICROWAVE REMOTE SENSING

Due to their high spatial resolution, Synthetic Aperture Radars (SARs) are the most attractive microwave instruments for remote sensing of land surface from space. The first experimental investigation demonstrating radar's strong sensitivity to soil moisture variations was published over three decades ago (Ulaby, 1974). In addition to its strong sensitivity to soil moisture content, radar backscatter is also dependent on the soil's surface roughness and vegetation cover that was distinctly demonstrated by SIR-B (the Shuttle Imagine Radar, 1984) exploitation (Dobson and Ulaby, 1986; Wang *et al.*, 1986). Several theoretical models are available now to describe radar backscatter from vegetated terrains. The most frequently used model represents the radar backscattering coefficient of a vegetation-covered soil surface as a sum of three types of contributions (Eom and Fung, 1984; Lang and Sidhu, 1983; Mo *et al.*, 1984; Ulaby *et al.*, 1981-1986; Dobson and Ulaby, 1998):

$$\sigma^0 = \sigma^0_{sur} + \sigma^0_{veg} + \sigma^0_{s-veg} \tag{8.50}$$

where $\sigma^0_{sur} = \sigma^0_s t^2$ represents the backscatter contribution of the bare soil surface σ^0_s attenuated by the vegetation layer (t is the vegetation transmissivity, σ^0_{veg} is the direct backscatter contribution of the vegetation layer, and σ^0_{s-veg} represents multiple scattering involving the vegetation elements and the ground surface. Sensitivities of backscattering coefficient to the model parameters were estimated, particularly, by Chukhlantsev and Vinokurova (1991). Theoretical models for the bare soil backscatter σ^0_s are based on small perturbation approach, physical optics (Kirchhoff) approach, geometric optics approach, or semi-empirical approach (Dobson and Ulaby, 1998). All the approaches express the soil backscatter in the form of the product of soil Fresnel reflectivity R and a function of root mean square of soil height variations σ_r, correlation length of soil surface l_r, and observation angle ϑ_i. Particularly, in the small perturbation approach the soil co-polarized backscatter at vertical or horizontal polarization is given by

$$\sigma^0_s = 4k_0\sigma_r^2 l_r^2 \cos^4\vartheta_i \exp\{-4k_0\sigma_r^2\cos^2\vartheta_i\}\exp\{-(k_0 l_r \sin\vartheta_i)^2\}R \tag{8.51}$$

where k_0 is the free space wave number. The backscattering coefficient is usually expressed in dB, and, since all models include the soil reflectivity as the coefficient, one can obtain

$$\sigma_s^0(R,\sigma_r,l_r,\vartheta_i)\,[dB] = 4.34\ln R + \sigma_s^0(\sigma_r,l_r,\vartheta_i)\,[dB] . \qquad (8.52)$$

The sensitivity of backscatter to the soil reflectivity (soil moisture) variations is, then, given by

$$\partial\sigma_s^0(R,\sigma_r,l_r,\vartheta_i)\,[dB]/\partial R = 4.34/R . \qquad (8.53)$$

The sensitivity does not depend on the soil roughness and weakly depends on the frequency and observation angle. Estimates performed (Chukhlantsev and Vinokurova, 1991) have shown that the dynamic range of σ_s^0 change due to variations of soil moisture from very dry to very moist condition does not exceed 8-9 dB. One can note that, at the same conditions, the range of brightness temperature variations is about 100 K. It is seen that the sensitivity to the soil moisture variations related to the possible measurement accuracy (~1 dB in radar measurements and ~1 K in radiometric measurements) in active microwave remote sensing is significantly worse than in microwave radiometry. Moreover, the range of σ_s^0 change makes up 18 dB, when the standard deviation of soil height varies in truly observed intervals from 0.5 cm to 4 cm. The above estimates are consistent with known experimental data. Some experimental regression dependencies of the backscattering coefficient on the volumetric moisture of soil top layer are presented in Fig. 8.8 (Chukhlantsev and Vinokurova, 1991). These dependencies were obtained from SIR-B experiment (Dobson and Ulaby, 1986: Wang *et al.*, 1986), and from ground based measurements (Mo *et al.*, 1984; Mo *et al.*, 1998). One can see from Fig. 8.8 that the effect of soil roughness is really the dominant one in the radar sensing of surface. To decouple this effect from that of soil moisture, dual-frequency measurements of the backscattering coefficient could be used (Mo *et al.*, 1998). Dobson and Ulaby (1998) described an algorithm for mapping soil moisture that is based on full-polarimetric SAR measurements. A two-dimensional diagram was proposed to be used for the simultaneous estimation of both soil moisture and soil roughness rms from radar measurements of the co-polarized and cross-polarized ratios.

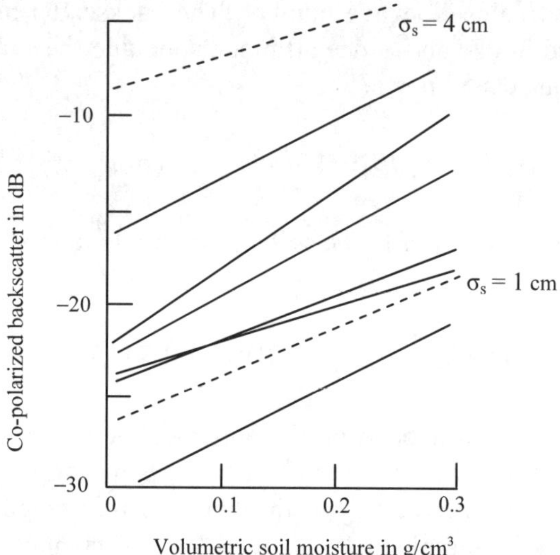

Fig. 8.8. Regression dependences of co-polarized backscatter at *C*- and *L*-band on soil moisture.

The influence of vegetation on the radar backscatter can be revealed on the basis of a simplified model for the backscattering coefficient. The model neglects the third term (accounting for interaction between the soil surface and canopy) in (8.50) and was proposed by Ulaby *et al.* (1983a). Moreover, the second term in (8.50) was proposed to express the form

$$\sigma^0_{veg} = \sigma^0_v (1 - t^2) \qquad (8.54)$$

where σ^0_v is the backscattering coefficient of an optically thick vegetation layer. The approach (8.54) directly follows from the "cloud" model developed by Attema and Ulaby (1978). Applicability of this approach is approved both theoretically (Vinokurova *et al.*, 1991) and experimentally (Wigneron *et al.*, 1999a). Some experimental data on σ^0_v for different crops are presented in Fig. 6.4 (Chapter 6). Substitution of (8.54) into (8.50) yields (Ulaby *et al.*, 1983a)

$$\sigma^0 = \sigma^0_v + (\sigma^0_v - \sigma^0_s)t^2 = \sigma^0_v + (\sigma^0_v - \sigma^0_s)\beta \qquad (8.55)$$

where β is the same slope reduction factor that is used in microwave radiometric remote sensing (see equation (7.35)). It is seen that the reduction in radar sensitivity to soil moisture caused by a canopy in frames of the simplified model considered is quite the same as the reduction in radiometric sensitivity. More accurate estimates of vegetation effect were obtained on the basis of a three-component model by Chukhlantsev and Vinokurova (1991). The third term in (8.50) was modeled in a single scattering approach as

$$\sigma^0_{s-veg} = 4\omega R \tau \cos \vartheta_i \, t^2 \tag{8.56}$$

where ω is the single scattering albedo and τ is the vegetation optical depth. In this case, the reduction in radar sensitivity to soil moisture can be expressed as

$$\partial \sigma^0(R)[dB]/\partial R = \frac{4.34}{R} S \tag{8.57}$$

where

$$S = \frac{1}{1 + \sigma^0_{veg} / (\sigma^0_{sur} + \sigma^0_{s-veg})}. \tag{8.58}$$

The quantity S has a sense of slope reduction factor in radar measurements of soil moisture in the presence of a vegetation canopy. This quantity, in contrast to the use of a β-factor, depends on the surface backscatter. Calculated values of S against the vegetation optical thickness (Chukhlantsev and Vinokurova, 1991) are presented in Fig. 8.9. Using the link (5.124) between the optical depth and vegetation water content ($\tau = bW / \cos \vartheta$; $b = {\sim}0.15$ at L-band), one can find from Fig. 8.9 that at L-band, radar measurements of soil moisture are possible ($S > 0.5$) at the values of vegetation water content less than 1-2 kg/m^2. Close estimates were obtained by Dobson and Ulaby (1998). They noted that for vegetation cover with biomass less than 0.5 kg/m^2, the effect of the vegetation on the radar backscatter may be ignored for the co-polarized responses at L-band. For higher biomass values and/or higher frequencies, it is not possible to separate the soil and vegetation backscatter contributions (the quantities σ^0_v and σ^0_s in (8.55) are rather close to each other), necessitating the use of multi-frequency multi-polarization observations in order to simultaneously estimate both the soil and vegetation-cover parameters governing the radar backscatter. Although some attempts have been made to develop such an algorithm, none exist at present.

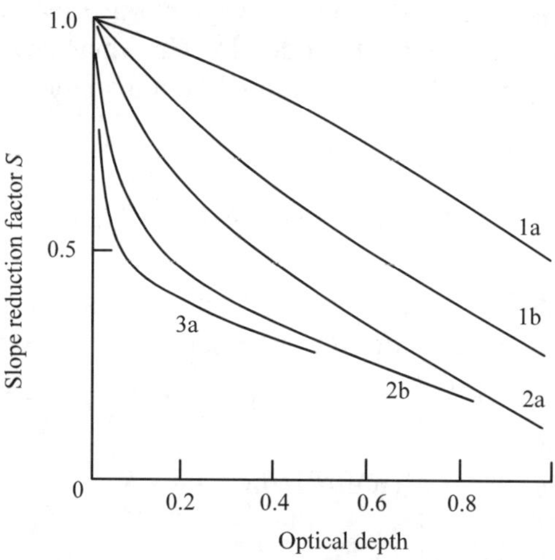

Fig. 8.9. The slope reduction factor S versus the vegetation optical depth. Soil backscatter is -5 dB (1); -15 dB (2); and -25 dB (3). Single scattering albedo is 0.1 (a) and 0.3 (b). Observation angle is 30°.

Combined use of active and passive microwave remote sensing for improvement of soil moisture estimation accuracy was studied by numerous researchers (Tsang *et al.*, 1982; Theis *et al.*, 1984; Ulaby *et al.*, 1983a; Fung and Eom, 1985; Eom, 1986; Wang *et al.*, 1987; Ferrazzoli *et al.*, 1988, 1989; Chauhan *et al.*, 1994; Chauhan and Lang, 1994; Chauhan, 1997; Du *et al.*, 2000; Wigneron *et al.*, 1999b; 2000; Njoku *et al.*, 2002; Bolten *et al.*, 2003). A combined active/passive technique could serve as a method to retrieve multiple surface variables: soil moisture, surface temperature, surface roughness, and vegetation water content. Unfortunately, at present, the active and passive components are still not merged into a single retrieval technique.

Chapter 9

MICROWAVE RADIOMETRY OF VEGETATION CANOPIES IN CONTEXT OF GLOBAL CHANGE RESEARCH

9.1. GLOBAL CLIMATE PROBLEMS AND THE CARBON CYCLE

During the last decades, the global carbon cycle has been under extensive study by numerous researches that connect the future climate change to an increase of carbon dioxide (CO_2) and other greenhouse gas concentrations in the atmosphere (Kondratyev *et al.*, 2002; Kondratyev *et al.*, 2003, 2004; Canadel *et al.*, 2003). Indeed, in 2001 the Intergovernmental Panel on Climate Change (IPCC, 2001a) concluded that most of the warming observed over the last half of the twentieth century can be attributed to human activities that have increased greenhouse gas concentrations in the atmosphere. They also warned that these changes will continue to drive rapid climate changes for several centuries to come. Chief among these greenhouse gases is CO_2, whose atmospheric concentrations have been dramatically altered by human perturbations to the global carbon cycle. The data of the Third IPCC Report (IPCC, 2001b) allow one to consider the dynamics of the global carbon cycle and climate during the last century (Kondratyev *et al.*, 2004):

- Atmospheric CO_2 concentration, in the pre-industrial epoch constituting 280 ± 10 ppmv (parts per million volume) during several thousand years, had been gradually increasing and by 1999 reached 367 ppmv.
- The principal cause of the CO_2 concentration growth is fossil fuel combustion. Global CO_2 emissions into the atmosphere between 1980 and 1989 constituted 5.4 ± 0.3 Gt C/yr, and during 1990-1999 they reached 6.3 ± 0.4 Gt C/yr.

- The annual growth of total atmospheric CO_2 content was 3.3 ± 0.1 Gt C/yr during 1980-1989 and decreased slightly to 3.2 ± 0.1 Gt C/yr during 1990-1999. The difference between the annual level of CO_2 emissions and its content in the atmosphere characterizes the scale of CO_2 assimilation by the World Ocean and land ecosystems.

Between 1991 and 1997 about 6.2 Gt C/yr of CO_2 were emitted to the atmosphere due to fossil fuel combustion. During this period the atmospheric CO_2 content increased only by 2.8 Gt C/yr. Using the observation data on atmospheric CO_2 concentration, Battle *et al.* (2000) have shown that during this period the biosphere of both the land and of the World Ocean assimilated, respectively, 1.4 ± 0.8 Gt C/yr and 2.0 ± 0.6 Gt C/yr. A rapid assimilation of carbon by the biosphere during this period contrasts with conditions in the 1980s, when it was practically neutral. The world Ocean had been a powerful sink for carbon until 1995, but in 1995-1996 the assimilation of carbon decreased. Apparently, the storage of carbon on land is more variable than in the oceans.

The term "greenhouse" effect implies a sum of results of simulating the effects caused in the climate system and is associated with a number of natural and anthropogenic processes (Kondratyev *et al.*, 2004). On the whole, this term refers to an explanation of changes in the atmosphere thermal regime caused by the impacts of some gases on the process of radiation transport. Many gases are characterized by a high stability and long residence time in the atmosphere, and carbon dioxide is one of them. To assess the level of the greenhouse effect due to CO_2, one should be able to predict its concentration with an account of all feedbacks in the global biogeochemical carbon cycle as well as a correlation of this cycle with the cycles of other greenhouse gases. According to available estimates, the contribution of carbon dioxide to the greenhouse effect is about 63.5%. It is known that the role of the World Ocean in the CO_2 cycle exceeds all the other reservoirs. Nevertheless, the CO_2 cycle in the atmosphere-plant-soil system is also important. This cycle has been described in detail by Krapivin *et al.* (1982), where the base carbon fluxes of land ecosystems were given. The photosynthesizing elements of these ecosystems assimilate CO_2 from the atmosphere and transform it into substratum and humus. Therefore, the intensity of carbon flux in a biocenose depends on its biomass and productivity. It is clear, thus, that accuracy of the estimates of land vegetation productivity as well as the reliability and details of its structural classification are very important parameters to specify the scheme and characteristics of the global CO_2 cycle. That is why, analyzing the CO_2 dynamics in the biosphere, it is important to take into account the maximum possible number of its reservoirs and fluxes as well as their spatial distributions (Kondratyev *et al.*, 2004). It is here that numerous global models of the carbon cycle differ. The present level of these studies does not allow one to answer the principle question about the extent of details of the database concerning the supplies and fluxes of carbon. Therefore many authors analyzing the dynamic characteristics of the global CO_2 cycle, rather arbitrarily utilize fragments of databases on the distribution of carbon sinks and sources.

The interaction between two carbon reservoirs – the atmosphere and the land – is expressed by carbon fluxes formed through ecological, geophysical, and geochemical processes, including photosynthesis, respiration of plants and animals, decomposition of dead organic matter, vegetation burning, etc. Which of these processes prevails depends on many factors. Particularly, relationships between the global CO_2 cycle and surface vegetation manifest themselves through the dependence of primary production and the rate of dead biomass decomposition on atmospheric temperature and CO_2 concentration. The temperature dependence is manifested most in northern latitudes, where mean annual temperature variations can reach high levels, and the vegetation period of plants varies from two to seven months. An impact of atmospheric CO_2 on the growth of plants depends on many factors. There are two basic types of plants which vary in their reaction to changes in the partial pressure of atmospheric CO_2. The first, most widely spread type of plants is characterized by photorespiration realized due to ferments, which can simultaneously assimilate and emit CO_2 and O_2. The efficiency of the initial use of light by these plants increases with increasing CO_2 concentration. The other type of plants, such as high tropical grass, reacts weakly to changes in the concentrations of carbon dioxide.

The most important aspect in studying the global carbon cycle is the contribution of the interaction between surface vegetation and the atmosphere to CO_2 exchange. A minimum requirement with ensured CO_2 assimilation is the availability of CO_2, H_2O, light, chlorophyll, and proper environmental conditions (temperature and humidity). The complex assimilation formula is written as (Kondratyev *et al.*, 2004)

$$\underbrace{CO_2}_{246\,g} + \underbrace{H_2O}_{216\,g} + \underbrace{675\,kcal}_{solar\ energy} \rightarrow \underbrace{C_6H_{12}O_6}_{180\,g} + \underbrace{O_2}_{192\,g} + \underbrace{H_2O}_{108\,g}. \qquad (9.1)$$

This formula can be used to calculate the balance between plants and the atmosphere for the CO_2 exchange only, but can not be used for water, since water is a limiting factor for photosynthesis, and plants (because transpiration has not been considered in this formula) use much more water. In the global models the process of carbon assimilation should be carefully detailed to avoid the violation of the balanced description of the processes. Usually this is realized through introducing corrections (Krapivin and Kondratyev,

2002). For instance, possible losses in the balanced relationship for photosynthesis are taken into account, which are assumed to be 20-30%.

It is also necessary to consider the spatial heterogeneity of the Earth' cover differing in biomass, density, and intensity of organic matter formation. On average, 90% of the total biomass ($\sim$830 Gt C) are forest ranges ($\sim$50 $\times$ 10^6 m^2), 50% of these constituting tropical forests ($\sim$24.5 $\times$ 10^6 m^2), with only 10% referring to bushes, savannahs, meadows, deserts, semi-deserts, marshes, and cultivated lands. The process of organic matter formation is highly inhomogeneous: forest ranges 33 Gt C/yr, the remaining vegetation 20 Gt C/yr. This heterogeneity leads to a mosaic pattern of bioproductivity and therefore should be taken into account in the synthesis of the model.

If $R_k(\varphi,\lambda,t)$ is the photosynthesis production for vegetation of the type k at a latitude φ and longitude λ at a time moment t, the CO_2 flux from the atmosphere into the living biomass is described with a simple model (Kondratyev *et al.*, 2004):

$$H_6^C(\varphi,\lambda,t) = C_{23}R_k(\varphi,\lambda,t)$$
(9.2)

where the coefficient C_{23} reflects the efficiency of the mechanism of the photosynthesis response and, on average, is estimated at $C_{23} \cong 0.549$. Bjorkstrom (1979) proposed to approximate an assimilation of CO_2 by vegetation with the formula

$$H_6^C(\varphi,\lambda,t) = k_b(1 + \beta_v \ln\{C_A/C_A^*\}C_k)$$
(9.3)

where the parameter $\beta_v \in [0, 1]$ is the measure of the ability of the vegetation system to react to an increase of the atmospheric CO_2 partial pressure; C_k is the carbon content in the biomass of the k-th type of vegetation; k_b is the coefficient of proportion depending on the temperature and type of vegetation; C_A is the concentration of carbon dioxide in the atmosphere; and C_A^* is the concentration of carbon dioxide in the atmosphere in the pre-industrial period.

Various authors estimate the total flux H_6^C as ranging from 16.7 to 35 Gt C/yr. This scatter of the estimates enables one to reliably estimate the coefficients on approximations for $H_6^C(\varphi,\lambda,t)$. A more detailed description of H_6^C requires construction of an additional unit of the model, taking into account the relationship between CO_2 concentration and functioning of the surface biomes on a given territory. These specifications have been made in publications by Krapivin and Vilkova (1990), Nefedova and Tarko (1993), and Krapivin *et al.* (1996). To specify the function $R_k(\varphi,\lambda,t)$ additional empirical dependencies were used, exemplified in Table 9.1. In the publication by Vilkova *et al.* (1998), regression formulas were given which enable one to calculate the vegetation productivity, humus supply, and phytomass supply as a function of mean annual atmospheric temperature T_A (in °C) and precipitation W_{pr} (in mm/yr). Particularly, the productivity F (in kg/m^2/yr) is given by

$$F = 4.25\times10^{-4}T_A^3 - 8.76W_{pr}^3 - 1.99T_A^2W_{pr} + 4.29T_AW_{pr}^2 + 2.29T_A^2$$
$$+19.05W_{pr}^2 - 8.79T_AW_{pr} + 4.56T_A - 14.16W_{pr} + 4.18 \tag{9.4}$$

Several rough models relating the annual vegetation productivity to the radiation balance, annual precipitation and temperature are also available (e.g., Bazilevich *et al.*, 1968; Leif, 1985; Riabchikov, 1972).

Table 9.1. Dependence of the annual production (kg/m^2/yr) on the mean annual temperature and total precipitation W_{pr} (in mm/yr).

W_{pr}	Atmospheric temperature (in °C)							
	2	6	10	14	18	22	26	30
3,125			3.4	3.5	3.7	3.8	3.9	4.0
2,875			3.2	3.3	3.5	3.6	3.7	3.8
2,625			3.0	3.2	3.3	3.4	3.5	3.6
2,375			2.8	2.9	3.0	3.1	3.2	3.3
2,125			2.5	2.6	2.7	2.9	2.9	3.0
1,875		1.6	2.3	2.3	2.4	2.5	2.6	2.7
1,625	0.6	1.3	2.0	2.1	2.1	2.2	2.3	2.4
1,375	0.7	1.1	1.7	1.9	1.9	2.1	2.1	2.0
1,125	0.6	1.0	1.6	1.8	1.9	1.8	1.8	1.7
875	0.8	0.9	1.5	1.4	1.3	1.3	1.2	1.2
625	0.9	0.9	0.9	0.8	0.8	0.7	0.7	0.7
375	0.6	0.6	0.6	0.5	0.5	0.5	0.4	0.4
125	0.2	0.2	0.2	0.2	0.2	0.1	0.1	0.1

From the above discussion it follows that the photosynthesis production and, in turn, the CO_2 flux from the atmosphere into the vegetation is strongly dependent on the air temperature and moisture regime of the vegetation (soil moisture can serve as a characteristic of this regime). But the air temperature and soil moisture are exactly those parameters that could be determined from microwave radiometric observations. Moreover, two main CO_2 fluxes from the vegetation to the atmosphere, i.e., the flux resulting from the process of respiration and the flux resulting from the humus decomposition, are also dependent on the surface temperature and soil humidity.

More accurate models of vegetation growth and productivity start from the energy balance equation that describes energy exchange on the border of vegetation and atmosphere. For example, in the model of the biosphere (simple biosphere model-2 – SiB2) derived by Sellers *et al.* (1996b) there are equations to determine temperature, humidity, and evaporation for a vegetation canopy and a 3-layer soil. Apart from this, models of radiative transfer (soil-leaf canopy-atmosphere) and photosynthesis were synthesized. The model of the leaf canopy photosynthesis considers a 1-layer vegetation species, surface and root layers, as well as a zone of recharging. Equations of the SiB2 are based on the concept of flux exchange in the soil-leaf canopy-atmosphere system. The model considered the fluxes of sensible and latent heat transport through evaporation of water vapor from a vegetation canopy and the surface soil layer, as well as the CO_2 flux. The AliBi model (Olioso *et al.*, 1999) is a two-layer model: energy balances are computed at the soil surface and for the vegetation canopy, allowing separate simulations of soil evaporation and plant transpiration. Reviews of different photosynthesis models are given in Kondratyev *et al.* (2004) and Olioso *et al.* (1999), and the interested reader is referred to these publications.

An example of the global model of the carbon cycle for the land surface is presented in Kondratyev *et al.* (2004). In the model, each i-th site of land of the area S_i is characterized by an amount of carbon per unit area in the biomass of living plants B_i and the dead organic matter of the soil D_i. The model is written as a system of balanced equations:

$$
\left.
\begin{aligned}
&\partial B_i / \partial t = F(T_i, P_i)\{1 + (\delta/10)(C_A/C_A^*)\} - K_i B_i \\
&\partial D_i / \partial t = K_i B_i - E(T_i, P_i) D_i \\
&\partial C_A / \partial t = \sum_{i=1}^{N} [F(T_i, P_i)\{1 + (\delta/10)(C_A/C_A^*)\} - E(T_i, P_i) D_i] S_i + V
\end{aligned}
\right\}
\qquad (9.5)
$$

where $F(T_i, P_i)$ is the annual production of plants per unit area; $E(T_i, P_i)$ is the specific rate of decomposition of dead organic matter of soil per unit area; T_i and P_i are the mean annual atmospheric temperature and the annual precipitation amount over the i-th site, respectively; δ is the indicator of the annual production increase (%) with a 10% increase of CO_2 content in the atmosphere; K_i is the specific intensity of the biomass dying-off of plants on the i-th site of land; $C_A^* = C_A(t_0)$, where t_0 is an initial time moment (in the global models it is usually assumed that t_0 represents the year 1850); and V is an anthropogenic input of carbon into the atmosphere. Kondratyev *et al.* (2004) supposed that in the absence of anthropogenic impacts ($V = 0$) the amount of carbon in the atmosphere-plant-system is constant, and in the early industrial period the system was in a state of equilibrium. Then at the moment t_0 the following relationships are valid:

$$K_i = B_i^0 / F(T_i^0, P_i^0), \quad E(T_i^0, P_i^0) = F(T_i^0, P_i^0) / D_i^0(T_i^0, P_i^0), \quad i = 1, N. \quad (9.6)$$

As a result, the model makes it possible to calculate the dynamics of carbon in three reservoirs with their spatial distribution taken into account. Depending on the details of the types of vegetation covers, calculations of the concentration of carbon in the vegetation cover, soil and atmosphere from equation (9.5) requires knowledge of the functions F and E, as well as the δ coefficient. One can see that the model parameters are strongly dependent on the surface temperature and soil moisture. These variables can be retrieved from microwave radiometric measurements. Moreover, the vegetation biomass (or its increase) can be retrieved from the radiometric measurements and, in principle, these data can be used as a calibration parameter in the first equation in (9.5).

As it was mentioned above, climate change is directly connected to the CO_2 concentration in the atmosphere. Particularly, Mintzer (1987) proposed the following relation between the change of atmospheric temperature and the CO_2 concentration in the atmosphere:

$$\Delta T_{CO_2} = -0.677 + 3.019 \ln[C_A(t)/338.5]. \quad (9.7)$$

It should be noted that different global carbon cycle models yield different estimates of the atmosphere temperature increase. Numerous researchers reported an estimate $\Delta T_{CO_2} \leq 4.2°C$ that is accepted in the Kyoto Protocol. However, recent estimates accounting for the spatial distribution of different plant formations and specifying their seasonal productivity have shown (Kondratyev *et al.*, 2004) that the increase in the atmospheric temperature could be less ($\Delta T_{CO_2} \leq 2.4°C$).

9.2. ASSIMILATION OF REMOTE SENSING DATA INTO GLOBAL CARBON CYCLE MODELS

Study of energy and mass transfer of soils and vegetation canopies is very important for understanding of CO_2 assimilation and emission by a vegetation-soil system. Several semi-empirical models have been developed that connect remote sensing data obtained in visible and infrared domains to vegetation evapotranspiration and photosynthesis (see review in Olioso *et al.* (1999)). Estimation of evapotranspiration and photosynthesis from remote sensing data may also be performed with the use of soil-vegetation- atmosphere transfer models (SVAT models). In these models, the energy balance equation for a vegetated soil is usually written in the following form (Chudnovsky, 1976):

$$R_n = H + LE + G \tag{9.8}$$

where R_n is the net radiation, H is the sensible heat flux, LE is the latent heat flux, and G is the ground heat flux. The part of the net radiation that goes directly to the photosynthesis is usually omitted in the right side of (9.8), since it constitutes only a small percent of the R_n. The vegetation-soil system can be separated into some sub-layers, and the model (9.8) is written for each layer in this case. The advantage of SVAT models is that they give access to a detailed description of soil and vegetation canopy processes and simulate intermediary variables linked to hydrological and physiological processes. Thus, they are often proposed to estimate soil moisture from remote sensing data and then used as an interface with other models, such as atmospheric and hydrological models. The other advantage, compared to more empirical approaches, consists of the fact that they may be operated without a systematic use of remote sensing data; the model intrinsically provides the mean for interpolating fluxes between remote sensing data acquisitions. A description of different SVAT models and of assimilation of remote sensing data into these models is given by Olioso *et al.* (1999). The interested reader is referred to this paper.

Few works consider the coupling between SVAT models and microwave radiometric measurements. In Russia, it was Professor Chudnovsky (1979a, 1979b) who first noted prospects of combining SVAT models with microwave and infrared remote sensing data.

Reutov (1991) considered the interconnection between microwave and infrared brightness temperatures and the condition of agricultural fields. He assumed that the optimum condition for crop development is $R_n = LE$, which implies that all energy is consumed in evapotranspiration (certainly, it is not always true but can be used as an approach). In this case, the difference between the current condition and the "optimum" one may be estimated as

$$\Delta J = R_n / LE - 1 = (H + G) / LE .$$ (9.9)

It was shown in the work (Reutov, 1991) that ΔJ value may be estimated from microwave and infrared remote sensing data and some ancillary data. That, in principle, allows one to estimate the total biomass productivity of crops.

It is necessary to mention the works by Burke *et al.* (1997, 1998), and Burke and Simmonds (2001), where a soil water and energy budget model was coupled with a microwave emission model producing the MICRO-SWEAT model. The model couples the simulation of heat, water and microwave transfers in the soil-vegetation-atmosphere system and can be used to predict the relationship between near-surface water content and microwave brightness temperature.

Despite a limited amount of research, assimilation of microwave radiometric data into SVAT models has shown to be an effective instrument for the estimation of evapotranspiration and photosynthesis in vegetation canopies, and, hence, for the estimation of carbon fluxes between the vegetation and atmosphere. Knowledge of the state of the *soil-plant formation* (SPF) allows one to have a real picture of the spatial distribution of the carbon sinks and sources on the Earth's surface.

One prospective approach to the solution of the problems arising here is GIMS-technology (GIMS = GIS + Model) (Kondratyev *et al.*, 2004; Krapivin and Chukhlantsev, 2004; Krapivin *et al.*, 2005). A combination of an environmental acquisition system, a model of the functioning of the typical geo-ecosystem, a computer cartography system, and a means of artificial intelligence will result in the creation of a geo-information monitoring system for the typical natural element that is capable of solving many tasks arising in the microwave radiometry of the global vegetation cover. The GIMS-based approach allows the synthesis of a knowledge base that establishes the relationships between the experiments, algorithms, and models. The links between these areas have an adaptive character giving an optimal strategy for experimental design and model structure. The application of the GIMS method to the tasks of reconstructing the spatial and temporal distribution of the SPF microwave radiative characteristics is considered in Krapivin *et al.* (2005).

Methods of local environmental diagnostics cannot give complex estimations for the state of natural objects or processes, especially in the case when the environmental element occupies an enormous area. Any technical means of environmental data collection provides information that is spotty in time and fragmentary in space. In particular, microwave radiometric systems of remote sensing, widely used as the equipment of airborne laboratories and satellites, give data sets that are connected geographically along the traces. Reconstruction of the information within the inter-trace space is possible only by applying spatial-temporal interpolation algorithms (the development of which is a specific problem). One effective technique is a combination of monitoring data and a model describing the functioning of the environmental system within the studied area. Such an approach to environmental monitoring problems is developed in the framework of ecoinformatics (Kondratyev *et al.*, 2004; Krapivin and Kondratyev, 2002). Ecoinformatics suggests the development of a set of models for various processes in the biosphere, taking into account their spatial inhomogeneity, and the combination of existing databases with already functioning systems for environmental observations. This allows one to answer the following questions:

- what kind of instruments are to be used for conducting the so-called ground-truth and remote-sensing measurements?
- what is the cost to be paid for on-site and remote-sensing information?
- what kind of balance is to be taken into consideration between the information contents and the costs of on-site and remote-sensing data?
- what kind of mathematical models can be used for both data interpolation and their extrapolation in terms of time and space in order to reduce the frequency (and thus the cost) of the observations and to increase the reliability of forecasting the environmental behavior of the observed items?

These and other problems are solved by using a monitoring system based on combining the functions of environmental data acquisition, control of the data archives, data analysis, and forecasting characteristics of the most important processes in the environment. This unification forms the new information technology called GIMS-technology. The term *Geo-Information Monitoring System* (GIMS) is used to describe the formula mentioned above. There are two views of the GIMS. In the first view, the term "GIMS" is synonymous with "GIS". In the second view, the definition of GIMS expands on the GIS. In keeping with the second view, the main units of the GIMS are considered in Armand *et al.* (1987), Kondratyev *et al.* (2004), Krapivin and Kondratyev (2002), and Krapivin and Phillips (2001). The basic component of the GIMS is considered as a natural subsystem interacting through biospheric, climatic, and socio-economic connections with the global nature-society system. A model has been created describing this interaction and the functioning of different levels of the spatial and temporal hierarchy of the whole combination of processes in the subsystem. The model encompasses characteristic features for typical elements of the natural and anthropogenic processes and the model development is based on the existing information base. The model structure is oriented to the adaptive regime of its use (Fig. 9.1).

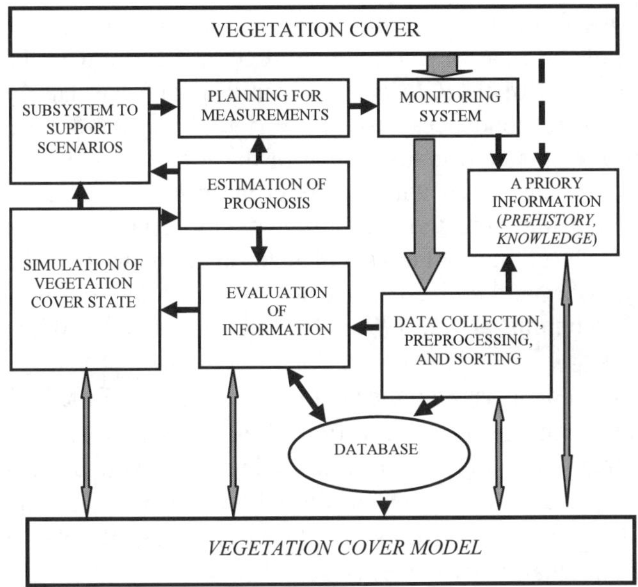

Fig. 9.1. Sketch of the adaptive regime of geoinformation monitoring with a combined use of the vegetation cover model and experimental measurements.

The GIMS includes a set of items the functions of which are determined by a collection of standard tasks. These items form the informational architecture of the monitoring system that includes the base of models describing the environmental subsystems in order to provide the possibility of environmental diagnostics. The GIMS-oriented system shell is formed to formalize the input information. An additional database level is synthesized to have multiple assignments of semantic structures to the real environmental subsystems with variable dimensions. These structures are designated by the matrix symbol $A_i = \|a_{i1,...,is}\|$, where the element $a_{i1,...,is}$ matches the object, process, phenomenon, event or other environmental bifurcation. Actually, matrix A_i is the parametrical image of the real environmental subsystem with its specific features. Parameter s reflects the dimension of the informational type for the subsystem section. The structures $\{A_i\}$ identify both the spatial distribution of the subsystem components and their types and parameters. The basic structures of $\{A_i\}$ have four dimensions: i_1 by latitude, i_2 by longitude, i_3 by height, and i_4 by time. Other structures of $\{A_i\}$ determine model coefficients, types of SPFs, precipitation, temperature, radiation, etc.

The semantic structures $\{A_i\}$, called identifiers, are used by the basic models for the formation of initial fields, validation of model output, and for preparation of the final or intermediate reports. Land cover classification is the main function of the GIMS. The identifier of basic land cover classes provides a correspondence between different types of SPFs and their parameters and spatial structure. Each of the identifier elements can have a vector structure connected with the description of various classifications of land cover and allowing the formation of global land cover classification maps by means of interpolation and extrapolation through the analysis of satellite data (Armand *et al.*, 1987; Chen *et al.*, 2003; Krapivin and Kondratyev, 2002; Running *et al.*, 1995; Zarco-Tejada *et al.*, 2003; Zhan *et al.*, 2002).

The GIMS database together with the structures of the $\{A_i\}$ consists of information about the model coefficients and a set of scenario fragments. The structures $\{A_i\}$ link up the knowledge base with the database. Each symbol of A_i is decoded in conformity with the depth hierarchy and reflects the reliability of the description of the environmental subsystem both on qualitative and quantitative levels.

The estimation and forecast of the state of the environmental system can be given by a simulation experiment, the conceptual and functional scheme of which is shown in Fig. 9.1. This scheme assumes various approaches to the synthesis of a model set to describe all aspects of the interactions of environmental bodies with their physical, biological, and chemical properties. The main idea of the GIMS-based simulation procedure is in the estimation of the deviation between the measured and predicted trends of the environmental subsystem, for use in making a decision on planning a monitoring regime or for correcting the model characteristics. Using this process for organization of dynamical monitoring of a specific natural object requires the adaptation and possible addition of new data and model items. That is why the basic model structure must be oriented to the adaptive regime of its use. A specialized identifier controls this regime making it possible to select between various regimes both manually and automatically relocatable. Moreover one of the GIMS subsystems has an intelligent support function. Software-mathematical algorithms are realized for providing the user with intelligent support in performing the complex analysis of objective information formed in the framework of the simulation experiment. The necessary information for an objective dialogue with the basic model is provided in a convenient form for the user. The introduction of data processing corrections is also provided.

The orientation of the GIMS, for example, on the problems of microwave emission from vegetation canopies or on the classification of terrestrial carbon sinks, requires the formation of a specific knowledge base and the synthesis of a subsystem with functions for modeling and evaluation of the soil-plant cover within the area studied.

In the common case, the GIMS input data are formed in conformity with spatial discretization of the Earth's surface. The basic type is a geographic grid with latitude and longitude steps of $\Delta\varphi$ and $\Delta\lambda$ respectively. The realization of an actual application of the existing database within the framework of the GIMS leads to non-uniform structures for different units. Depending on the specific features of the task under consideration, the real grid structure may be linked to regional features. The interactive mode of the GIMS provides for adaptation of the regime to the real natural system through a set of identifiers. The GIMS database has identifier sets for the spatial distribution of SPFs. The user can form such identifiers according to the concrete discretization of the space ($\Delta\varphi, \Delta\lambda$) using his own data sets. Thus, the GIMS problem shell formalizes the image of the environmental subsystem based on the standards set and allows the use of informational channels for the simulation experimental investigations.

One of the principal aspects of the anthropogenic impact on the environment is an evaluation of the consequences of CO_2 emissions into the atmosphere. The published results estimating the greenhouse effect and excess CO_2 distribution in the biosphere, which bear on this problem, are widely varying and sometimes contradictory or else they are too flatly stated. This is a natural consequence of all kinds of simplifications adopted in modeling the global CO_2 cycle. The GIMS makes it possible to create an effective monitoring system allowing an estimation of the spatial distribution of the carbon sinks and sources in real time.

Before this is attempted some problems should be solved to assess the role of the anthropogenic use of the Earth's surface. In particular, among these problems is that of the formalized description of the processes of change of the structure of the Earth land covers, such as afforestation, forest reconstruction, deforestation and the associated carbon supplies. Understanding of the meteorological processes as functions of greenhouse gases is one of the key problems of humankind in the first decade of the third Millennium. Only an adequate knowledge of the meteorological phenomena on various spatial-temporal scales under conditions of varying supplies of CO_2 and other greenhouse gases will enable one to make correct and constructive decisions in the field of global environmental protection.

The dynamics of surface ecosystems depends on interactions between the biogeochemical cycles, which during the last decade of the 20th century suffered significant anthropogenic modification, especially to the cycles of carbon, nitrogen, and water. The surface ecosystems, in which carbon remains in the living biomass, decomposing organic matter, and the soil, play an important role in the global CO_2 cycle. Carbon exchanges between these reservoirs and the atmosphere take place through photosynthesis, respiration, decomposition, and burning. Human interference into this process takes place through changing the structure of the vegetation covers, pollution of the water basin surfaces and of the soil areas, as well as through direct emissions of CO_2 into the atmosphere.

The role of various ecosystems in the formation of carbon supplies to the biospheric reservoirs determines the rate and direction of changes to the regional meteorological situations and to the global climate. The accuracy of assessment of the level of these changes depends on reliability of the data on the inventory of surface ecosystems.

Existing environmental data show that knowledge of the rates and trends of carbon accumulation in surface ecosystems is rather uncertain. However, it is clear that surface ecosystems are important assimilators of excess CO_2. Understanding the details of such assimilation is only possible through modeling of the process of plant growth, i.e., considering the effect of the nutrient elements of the soil and other biophysical factors on plant photosynthesis. Therefore, the forest ecosystems and associated processes of natural afforestation, forest reconstruction, and deforestation should be studied in detail. In a forested area, the volume of the reservoir of CO_2 from the atmosphere is a function of the density of the forest canopy, and in a given period of time a change of this volume is determined by the level and character of the dynamic processes of transition of one type of forest into another. The causes of this transition can be natural, anthropogenic, or mixed. Biocenology tries to create a universal theory of such transitions, but so far there is only a qualitative description of the observed transitions.

The GIMS comprises 30 models for SPF (Krapivin *et al.*, 2005). The list of SPFs given in Kondratyev *et al.* (2004) and Krapivin *et al.* (2005) was adapted to a spatial resolution of $4° \times 5°$ according to the classification by Bazilevich and Rodin (1967). The specific biomass Q_i of the i-th type of vegetation at time t can be parameterized by means of the following equation:

$$\partial Q_i / \partial t = R_i - M_i - E_i \qquad (9.10)$$

where R_i is the biomass productivity and M_i and E_i are the biomass losses at the expense of withdrawal and transpiration, respectively.

The function $M_i(\varphi,\lambda,t)$ reflects the set of natural M_{Ni} and anthropogenic M_{Ai} processes leading to vegetation biomass losses ($M_i = M_{Ni} + M_{Ai}$):

$$M_i(\varphi,\lambda,t) = \mu_i(t)Q_i(\varphi,\lambda,t) \qquad (9.11)$$

where φ and λ are the latitude and longitude, respectively.

The flux E_i is calculated by means of the formula (Sellers *et al.*, 1996a):

$$E_i(\varphi,\psi,t) = \frac{\rho c_p\left[e^*(T_c)-e_a\right]}{\gamma_p(r_c+r_b)} \qquad (9.12)$$

where $e^*(T_c)$ is the saturated vapor pressure inside the canopy foliage (in units of Pa), e_a is the vapor pressure in the canopy air space (Pa), r_c is canopy resistance (sm^{-1}), r_b is the bulk leaf boundary layer resistance of the canopy (sm^{-1}), ρ is air density (kg·m^{-3}), c_p is the air specific heat (J·kg^{-1}·K^{-1}) and γ_p is the psychrometric constant (Pa·K^{-1}).

The actual plant productivity is approximated as follows:

$$R_i = \delta_c^i\left(1+\alpha_T^i\cdot\Delta T/100\right)\exp\left(-\beta_i/Q_i\right)\min\left\{\delta_e^i,\delta_Z^i,\delta_W^i,\delta_B^i\right\} \qquad (9.13)$$

where α_T^i and β_i are indices of dependence of production on the temperature variation ΔT and biomass Q_i, respectively; δ_ζ^i is the index of limitation of production by the factor ζ: e = illumination, Z = pollution, W = soil moisture, B = nutrient salts of the soil and c = atmosphere CO_2 concentration. The δ_ζ^i functions actually used in the framework of real situations are calculated based on existing or preliminary receiving data. Thus, the role played by the atmospheric CO_2 concentration C_A in photosynthesis is described by the relation $\delta_c^i = b_iC_A/\left(C_A+C_{0.5}^i\right)$, where $C_{0.5}^i$ is the CO_2 concentration for which $\delta_c^i = b_i/2$. The influence of the solar radiation intensity $e(\varphi,\lambda,t)$ on photosynthesis is parameterized by the relation $\delta_e^i = \delta_i^*\exp(1-\delta_i^*)$, where $\delta_i^* = e/e_i^*$ and e_i^* is the optimal illuminance for i-th type of plant. A more detailed description of the correlations between the biocenotic processes is given in Kondratyev *et al.* (2004) and Krapivin and Kondratyev (2002).

The GIMS-based approach was used for the reconstruction of microwave attenuation by vegetation of different types based on fragmentary measurements (Krapivin *et al.*, 2005). Following the vegetation classification admitted and taking into account the quantitative characteristics of the forest ecosystems, the attenuation was estimated and tabulated. The values of attenuation obtained had a dispersion of about 5.5%. Certainly, this example showed just a possibility of having operative global information about the spatial distribution of the electromagnetic properties of the vegetation cover. More detailed calculations taking into consideration various microwave ranges with H and V polarizations may be useful in view of many potential applications.

As has been shown by many authors (e.g., Armand *et al.*, 1987; Kondratyev *et al.*, 2004; Krapivin and Kondratyev, 2002; Krapivin and Phillips, 2001), balanced criteria exist for selection of information covering the hierarchy of causal-investigative constraints in the biosphere. These include the coordination of measurement tolerances, the depth of spatial quantization in the course of describing the land covers, the degree of detailing of the biome, etc. At an empirical level, as expressed in evaluations of the results of computing experiments by experts, these criteria allow the selection of the informational structure of the geo-information monitoring system representing the hierarchical subordination of the models at various levels.

It is becoming clear that the GIMS-based approach opens a new perspective in the framework of microwave remote-sensing monitoring giving the possibility of combining theoretical and field investigations of the role of vegetation in the microwave emission from the Earth's surface. To achieve practical results it is necessary to fill the GIMS-model base with different items. The scope of the GIMS will be determined by a set of natural phenomena such as the energy balance in the biosphere that is directly related to vegetation covers, especially to the forests. That is why the nearest working regimes of the author and his colleagues in this area foresee the formulation and solution of the following series of problems:

- development of models to describe the dynamics of SPFs;
- analysis of the possibilities of different monitoring systems with respect to their efficiency for estimating the vegetation cover parameters;
- theoretical and experimental studies of microwave radiation and propagation in the atmosphere-vegetation-soil systems;
- synthesis of the GIMS's database including the theoretical and experimental estimations of the model coefficients, vegetation radiometric characteristics and the spatial distributions of the biomes.

REFERENCES

Abdulla S.A., Mohammed A.-K.A., and Al-Rizzo H.M. (1988). The complex dielectric constant of Iraqi soils as a function of water content and texture. *IEEE Trans. Geosci. Remote Sensing*, **26**:882-885.

Allen C.T. and Ulaby F.T. (1989). Modelling the polarization dependence of the attenuation in vegetation canopies. *Proc. of IGARSS'84, August 27-30, 1984, Strasbourg*, pp. 119-124.

Ambarnikov V.N. and Elagin A.H. (1989). Use of microwave radiometry for research of forest canopy opacity at laboratory conditions. *Proc. 2nd Symp. "Microwave Remote Sensing of Land Covers"*: Leningrad, 1989 (in Russian).

Armand N.A., Krapivin V.F., and Mkrtchyan F.A. (1987). *Methods for Data Processing in the Radio-physical Investigations of the Environment*. Moscow: Nauka Press, 270 pp. (in Russian).

Armand N.A. and Polyakov, V.M. (2005). *Radio Propagation and Remote Sensing of Environment*. Boca Raton, Florida: CRC Press.

Attema E.P.T. and Ulaby F.T. (1978). Vegetation modeled as a water cloud. *Radio Science*, **13**:357-364.

Barabanenkov Yu.N. (1975). Radiative transfer theory and multiple scattering by an ensemble of particles. *Sov. Phys. Usp.*, **117**:49-92.

Barton I.J. (1978). A case study comparison of microwave radiometer measurements over bare and vegetated surfaces. *J. Geophys. Res.*, **83**:3513-3517.

Basharinov A.E., Tuchkov L.T., Polyakov V.M., and Ananov N.I. (1968). *Measurement of Radiothermal and Plasma Radiation in the Microwave Band*. Moscow: Sovetskoe Radio Publ. House, 320 pp. (in Russian).

Basharinov A.E. and Shutko A.M. (1971). Measurements of land surface moisture by means of microwave radiometry. *Meteorologia i Gidrologia*, **9**:17-23 (in Russian).

Basharinov A.E., Gurvich A.S., and Egorov S.T. (1974). *Radioemission of the Earth as a Planet*. Moscow: Nauka Publ. 188 pp. (in Russian).

Basharinov A.E. and Shutko A.M. (1975). *Simulation Studies of the SHF Radiation Characteristics of Soils under Moist Conditions*. Greenbelt, USA: NASA TT, 1975.

Basharinov A.E., Zotova E.N., Naumov M.I., and Chuklantsev A.A. (1979). Radiation characteristics of vegetation in the microwave band. *Radiotekhnika*, **34**:16-24 (in Russian).

Battle M., Bender M.L., Tans P.P., White J.W.C., Ellis J.T., Conway T., and Francey R.J. (2000). Global carbon sinks and their variability inferred from atmospheric O_2 and $\delta^{13}C$. *Science*, **287**(5662):2467-2470.

Baturin V.E., Grankov A.G., Milshin A.A. et al. (1985). Experimental study of vegetation effect on microwave radiometry of soil moisture. *Proc. of the 1st Symposium "Microwave Remote Sensing in Agriculture"*, Irkutsk, 1985 (in Russian).

Bazilevich N.I. and Rodin L.E. (1967). Maps of biological productivity for pivotal types of land vegetation. *Izvestia Vsesoyuznogo Geograficheskogo Obshchestva*, **99(3)**:190-194 (in Russian).

Bazilevich N.I., Drozdov A.V., and Rodin L.E. (1968). Productivity of the Earth's vegetation: general features and connection to climate. *J. of General Biology*, **29**:261-271 (in Russian).

Birchak J.R., Gardner C.G., Hipp J.E., and Victor J.M. (1974). High dielectric constant microwave probes for sensing soil moisture. *Proc. IEEE*, **62**:92-98.

Bjorkstrom A.A. (1979). "A model of CO2 interaction between atmosphere, ocean and land biota." In *Global Carbon Cycle* (Scope-13, pp. 403-458), New York: John Wiley and Sons.

Bobrov P.P., Maslennikov N.M., Sologubova T.A., and Etkin V.S. (1989). Microwave dielectric properties of soils in the range of bound-to-free water transition. Reports of USSR Academy of Sciences, **304**:1116-1119.

Bogorodskii V.V., Kozlov A.I., and Tuchkov L.T. (1977). *Radioemission of the Earth's Covers*. Leningrad: Gidrometeoizdat, 223 pp. (in Russian).

Bogorodskii V.V., Kanareikin D.B., and Kozlov A.I. (1981). *Polarization of Self- and Scattered Emission from the Earth's covers*. Leningrad: Gidrometeoizdat, 280 pp. (in Russian).

Bolten J.D., Lakshmi V., and Njoku E.G. (2003). Soil moisture retrieval using the passive/active L- and S-band radar/radiometer. *IEEE Trans. Geosci. Remote Sensing*, **41**:2792-2801.

Borodin L.F. (2001). Microwave radiometry of peat and forest fires and forest fire risk. *Uspekhi sovremennoi radioelectroniki*, **No. 11**:59-64 (in Russian).

Borodin L.F., Kirdiashev K.P., Stakankin Yu.P., and Chukhlantsev A.A. (1976). On the application of microwave radiometry to forest fire surveys. *Radiotekhnika i Elektronika*, **21**:1945-1950 (in Russian).

Boyarskii D.A. and Tikhonov V.V. (1995). Effective permittivity microwave model for wet and frozen soils. *Journal of Communications Technology and Electronics*, **40**:51-54.

Boyarskii D.A., Tikhonov V.V., and Komarova N.Yu. (2002). Model of dielectric constant of bound water in soil for applications of microwave remote sensing. *PIER*, **35**:251-269.

Brown G.S. and Curry W.J. (1982). A theory and model for wave propagation through foliage. *Radio Science*, **17**:1027-1036.

Brunfeldt D.V. and Ulaby F.T. (1984). Measured microwave emission and scattering in vegetation canopy. *IEEE Trans. Geosci. Remote Sensing*, **22**:520-524.

Brunfeldt D.V. and Ulaby F.T. (1986) Microwave emission from row crops. *IEEE Trans. Geosci. Remote Sensing*, 24:353-359.

Burke E.J., Gurney R.J., Simmonds L.P., and Jackson T.J. (1997). Calibrating a soil water and energy budget model with remotely sensed data to obtain quantitative information about the soil. *Water Resour. Res.*, **33**:1689-1697.

Burke E.J., Gurney R.J., Simmonds L.P., and O'Neill P.E. (1998). Using a modeling approach to predict soil hydraulic properties from passive microwave measurements. *IEEE Trans. Geosci. Remote Sensing*, **36**:454-462.

Burke E.J. and Simmonds L.P. (2001). A simple physically based soil moisture retrieval algorithm and its sensitivity to measurement error. *Hydrol. Earth Syst. Sci.*, **5**:39-48.

Burke E.J., Bastidas L.A., and Shuttleworth W.J. (2002a). Exploring the potentials for multipatch soil moisture retrievals using multiparameter optimization techniques. *IEEE Trans. Geosci. Remote Sensing*, **40**:1114-1120.

Burke E.J., Shuttleworth W.J., and French A.N. (2002b).Using vegetation indices for soil-moisture retrieval from passive microwave radiometry. *Hydrology and Earth System Sciences*, **5(4)**:671-677.

Burke H-H.K. and Schmugge T.J. (1982). Effects of varying soil moisture contents and vegetation canopies on microwave emission. *IEEE Trans. Geosci. Remote Sensing*, **20**:268-274.

Burke W.J., Schmugge T.J., and Paris J.E. (1979). Comparison of 2.8- and 21-cm microwave radiometer observations over soils with emission model calculations. *J. Geophys. Res.*, **84**:197-207.

Calvet J.-C., Wigneron J.-P., Chanzy A., Raju S., and Laguerre L. (1995a) Microwave dielectric properties of a silt-loam at high frequencies. *IEEE Trans. Geosci. Remote Sensing,* **33**:634-642.

Calvet J.-C., Wigneron J.-P., Chanzy A., and Haboudane D. (1995b). Retrieval of surface parameters from microwave radiometry over open canopies at high frequencies. *Remote Sens. Environ.,* **53**:46-60.

Camps A. J., Corbella I., Torres F., Bará J., and Capdevila J. (2000). RF interference analysis in aperture synthesis interferometric radiometers: Application to L-band MIRAS instrument. *IEEE Trans. Geosci. Remote Sensing,* **38**:942-949.

Canadel I.G., Dickinson R., Hibbard K. et al. (eds.). (2003). *Global Carbon Project. The Science Framework and Implementation. Report No.1. Earth System Science Partnership*. Canberra.

Castel T., Beaudoin A., Floury N., Le Toan T., Caraglio Y., and Barczi J.-F. (2001) Deriving forest canopy parameters for backscatter models using the AMAP architectural plant model. *IEEE Trans. Geosci. Remote Sensing*, **39**:571-583.

Chanzy A. and Wigneron J.-P. (2000). "Microwave Emission from Soil and Vegetation." In *Radiative Transfer Models for Microwave Radiometry*, Ch. Maetzler, ed. Bern, Switzerland: COST, 2000, pp. 89-102.

Chanzy A., Raju S., and Wigneron J.-P. (1997). Estimation of soil microwave effective temperature at L and C bands. *IEEE Trans. Geosci. Remote Sensing*, **35**:570-580.

Chanzy A., Wigneron J.-P., Calvet J.-C., Laguerre L., and Raju S. (2000). "Surface emissivity data from PORTOS-Avignon experiment." In *Radiative Transfer Models for Microwave Radiometry*, Ch. Maetzler, ed. Bern, Switzerland: COST, 2000, pp. 171-172.

Chauhan N.S. (1997). Soil moisture estimation under a vegetation cover: Combined active and passive microwave remote sensing approach. *Int. J. Remote Sens.*, **18**:1079-1097.

Chauhan N.S. and Lang R.H. (1989). Polarization utilization in the microwave inversion of leaf angle distribution. *IEEE Trans. Geosci. Remote Sensing*, **27**:395-402.

Chauhan N.S. and Lang R.H. (1994). Use of discrete scatter model to predict active and passive microwave sensor response to corn: Comparison of theory and data. *IEEE Trans. Geosci. Remote Sensing*, **32**:416-426.

Chauhan N.S., Lang R.H., and Ranson K.J. (1991). Radar modeling of a boreal forest. *IEEE Trans. Geosci. Remote Sensing*, **29**:627-638.

Chauhan N., Lang R., Ranson J., and Kilic O. (1993). Multi-stand radar modeling from a boreal forest: results from the Boreas intensive field campaign – 1993. *Proc. of IGARSS'95, Firenze, Italy,* Vol. 2, pp. 981-983.

Chauhan N.S., Le Vine D.M., and Lang R.H. (1994). Discrete scatter model for microwave radar and radiometric response to corn: Comparison of theory and data. *IEEE Trans. Geosci. Remote Sensing*, **32**:416-426.

Chen J.M., Lin J., Leblanc S.G., Lacaze R., and Roujean J.-L. (2003). Multi-angular optical remote sensing for assessing vegetation structure and carbon absorption. *Remote Sensing Environ.,* **84(4)**:516-525.

Choudhury B.J., Schmugge T.J., Chang A., and Newton R.W. (1979). Effect of surface roughness on the microwave emission from soils. *J. Geophys. Res.*, **84**:5699-5706.

Choudhury B.J., Schmugge T.J., Mo T. (1982). A parameterization of effective soil temperature for microwave emission. *J. Geophys. Res.*, **87**:1301-1304.

Choudhury B.J., Owe M., Goward S.N., Golus R.E., Ormsby J.P., Chang A.T.C., and Wang J.R. (1987). Quantifying spatial and temporal variabilities of microwave brightness temperature over the U.S. southern great plains. *Int. J. Remote Sensing*, **8**:177-191.

Chuah H.T., Tjuatja S., Fung A.K., and Bredow J.W. (1997). Radar backscatter from a dense discrete random medium. *IEEE Trans. Geosci. Remote Sensing*, **35**:892-900.

Chuang S.L., Kong J.A., and Tsang L. (1982). Radiative transfer theory for passive microwave remote sensing of a two layer random medium with cylindrical structure. *J. Appl. Phys.*, **51**:5588-5593.

Chudnobsky A.F. (1976). *Thermal Physics of Soils*. Moscow: Nauka Press (in Russian).

Chudnobsky A.F. (1979a). A quantitative estimate of soil thermal regime from airborne measurements of soil moisture and temperature. In Science and Technology Bulletin No.37 *"Airborne Remote Sensing of Agricultural fields"*, pp. 3-7. Leningrad: Agro-Physical Institute, 1979 (in Russian).

Chudnobsky A.F. (1979b). Estimation of evapotranspiration from airborne measurements. In Science and Technology Bulletin No.37 *"Airborne Remote Sensing of Agricultural fields"*, pp. 30-33. Leningrad: Agro-Physical Institute, 1979 (in Russian).

Chukhlantsev A.A. (1981). *Microwave Emission from Vegetation Canopies. Ph.D. Dissertation*. Moscow: The Moscow Institute of Physics and Technology (State University), 1981 (in Russian).

Chukhlantsev A.A. (1985). On microwave radiometry of vegetation canopies. *Proceedings of 1985 International Symposium on Antennas and EM Theory, 1985 August 26-28, Beijing,* China: Academic Publishers, 1985, pp. 532-536.

Chukhlantsev A.A. (1986). Scattering and absorption of microwave radiation by elements of plants. *Radiotekhnika i Elektronika*, **31**:1095-1104 (in Russian).

Chukhlantsev A.A. (1988). On effective dielectric permittivity of vegetation in the microwave band. *Radiotekhnika i Elektronika*, **33**:2310-2319 (in Russian).

Chukhlantsev A.A. (1989b). On modeling vegetation as a collection of scatterers. *Radiotekhnika i Elektronika*, **34**:240-244 (in Russian).

Chukhlantsev A.A. (1989a). A possible model approach to vegetation cover classification from radar sounding data. *Issledovanie Zemli iz Kosmosa (Research of the Earth from Space)*, **No. 4**:84-90 (in Russian).

Chukhlantsev A.A. (1992). Microwave emission and scattering from vegetation canopies. *Journal of Electromagnetic Waves and Applications*, **6**:1043–1068.

Chukhlantsev A.A. (2002). Microwave attenuation by coniferous branches. *Lesnoi vestnik (Forest news)*, **No. 1(21)**:110-112 (in Russian).

Chukhlantsev A.A. and Golovachev S.P. (1989) Microwave attenuation in a vegetation canopy. *Radiotekhnika i Elektronika*, **34**:2269-2278 (in Russian).

Chukhlantsev A.A. and Golovachev S.P. (2002). Estimates of 3…300 cm radio wave attenuation by vegetation canopies. *Lesnoi vestnik (Forest news)*, **No. 1(21)**:112-117 (in Russian).

Chukhlantsev A.A. and Shutko A.M. (1982). Vegetation screening effect in passive microwave remote sensing. *Izvestia Vuzov, Geodezia i Aerofotosiomka*, **2**:80-82 (in Russian).

Chukhlantsev A.A. and Shutko A.M. (1985). The method of vegetation biomass determination. *USSR invention No. 1255905*.

Chukhlantsev A.A. and Shutko A.M. (1987). Application of microwave radiometry for determining biometric parameters of vegetation canopies. *Issledovanie Zemli iz Kosmosa (Research of the Earth from Space)*, **No. 5**:42-48 (in Russian).

Chukhlantsev A.A. and Shutko A.M. (1988). Microwave radiometry of the Earth's surface: effect of vegetation. *Issledovanie Zemli iz Kosmosa (Research of the Earth from Space)*, **No. 2**:67-72 (in Russian).

Chukhlantsev A.A. and Vinokurova S.I. (1991). On application of radar equipment for remote sensing of vegetation and soil. *Issledovanie Zemli iz Kosmosa (Research of the Earth from Space)*, **No. 4**:21-26 (in Russian).

Chukhlantsev A.A., Shutko A.M., Golovachev S. P., et al. (2003a). "Conductivity of leaves and branches and its relation to the spectral dependence of attenuation by forests in meter and decimeter band," *Proc. of IGARSS 2003, 21-25 July 2003, Toulouse, France.*

Chukhlantsev A.A., Yu. G., Tishchenko, S.V., Marechek et al. (2003b). "Laboratory complex for measuring of EM waves attenuation by vegetation fragments," *Proc. of IGARSS 2003, 21-25 July 2003, Toulouse, France.*

Chukhlantsev A.A., Shutko A.M., and Golovachev S.P. (2003c). Attenuation of electromagnetic waves by vegetation canopies. *Radiotekhnika i Elektronika*, **48**:1283-1311 (in Russian).

Chukhlantsev A.A., Golovachev S.P., Shutko A.M. (1989). Experimental study of vegetable canopy microwave emission. *Adv. Space Res.*, **9**:(1)317-(1)321.

Chukhlantsev A.A., Marechek S.V., Golovachev S.P., Novichikhin E.P., Tishchenko Yu.G., and Shutko A.M. (2004a). Continuous microwave attenuation spectra of trees fragments. *Proceedings of MicroRad 2004 Conference, February 24-27, 2004, Rome, Italy.*

Chukhlantsev A.A., Shutko A.M., and Chukhlantsev A.A., (Jr.). (2004b). Modeling of polarization properties of microwave emission from moist soils. *Problems of Environment and Natural Resources*, **12**:3-8 (in Rusian).

Chukhlantsev A.A., Krapivin V.F., Chukhlantsev A.A., Jr. (2005). Microwave emission from rough moist soils. *Problems of Environment and Natural Resources*, **9**:48-59.

Coppo P., Luzi G., Paloscia S., and Pampaloni. P. (1991). Effect of soil roughness on microwave emission: Comparison between experimental data and models. *Proc. IGARSS'91, Helsinki, 1991 June 3-6*, pp. 1167-1170.

Crosson W.L., Laymon C.A., Inguva R., and Bowman C. (2002). Comparison of two microwave radiobrightness models and validation with field measurements. *IEEE Trans. Geosci. Remote Sensing*, **40**:143-152.

Crow W.T., Drusch M., and Wood E.F., (2001). An observation system simulation experiment for the impact of land surface heterogeneity on AMRS-E soil moisture retrieval. *IEEE Trans. Geosci. Remote Sensing*, **39**:1622-1631.

Crow W.T., Chan S.T.K., Entekhabi D., Houser P.R., Hsu A.Y., Jackson T.J., Njoku E.G., O'Neill P.E., Shi J., and Zhan X. (2005). An observing system simulation experiment for Hydros radiometer-only soil moisture product. *IEEE Trans. Geosci. Remote Sensing*, **43**:1289-1303.

Curtis J.O. Moisture effect on the dielectric properties of soils. (2001). *IEEE Trans. Geosci. Remote Sensing*, **39**:125-128.

Davenport I.J., Fernández-Gálvez J., and Gurney R.J. (2005). A sensitivity analysis of soil moisture retrieval from the tau-omega microwave emission model. *IEEE Trans. Geosci. Remote Sensing*, **43**:1304-1316.

De Loor G.P. (1960). Dielectric properties of heterogeneous mixtures containing water. *J. Microwave Power*, **3**:67-73.

De Roo R., Kuga Y., Dobson M.C., and Ulaby F.T. (1991). Bistatic radar scattering from organic debris of a forest floor. *Proc. of IGARSS'91 Symposium, Helsinki, 1991 June 3-6*, Vol. 1, pp. 15-18.

De Roo R.D., Du Y., Ulaby F.T., and Dobson M.C. (2001). A semi-empirical backscattering model at L-band and C-band for a soybean canopy with soil moisture inversion. *IEEE Trans. Geosci. Remote Sensing*, **39**:864-872.

Dobson M.C. (1988). Diurnal and seasonal variations in the microwave dielectric constant of selected trees. *Proceedings of the IGARSS'88, Edinburgh, UK*, Vol. 3, pp. 1754.

Dobson M.C. and Ulaby F.T. (1986). Preliminary evaluation of the SIR-B response to soil moisture, surface roughness, and crop canopy cover. *IEEE Trans. Geosci. Remote Sensing*, **24**:517-526.

Dobson M.C. and Ulaby F.T. (1998). "Mapping soil moisture distribution with imaging radar." in *Principles and Applications of Imaging Radar*, Edited by F.M. Henderson and A.J. Lewis, Wiley, 1998, pp. 407-433.

Dobson M.C., Ulaby F.T., Hallikainen M.T., and El-Rayes M.A. (1985). Microwave dielectric behavior of wet soil – Part II: Dielectric mixing models. *IEEE Trans. Geosci. Remote Sensing*, **23**:35-46.

Dobson M.C., de la Sierra R., and Christensen N. (1991). Spatial and temporal variations of the microwave dielectric properties of loblolly pine trunks. *Proceedings of IGARSS'91, Espoo, Finland.*

Dobson M.C., Ulaby F.T., Le Toan T., Beaudoin A., Kasischke E.S., and Christensen Jr. N.L. (1992). Dependence of radar backscatter on coniferous forest biomass. *IEEE Trans. Geosci. Remote Sensing,* **30**:412-415.

Du L.-J. and Peake W.H. (1969). Rayleigh scattering from leaves. *Proc. of IEEE,* **57**: 1227-1233.

Du Y., Ulaby F.T., and Dobson M.C. (2000). Sensitivity to soil moisture by active and passive microwave sensors. *IEEE Trans. Geosci. Remote Sensing,* **38**:105-114.

Durden S.L., Klein J.D., and Zebker H.A. (1991). Polarimetric radar measurements of a forested area near Mt. Shasta. *IEEE Trans. Geosci. Remote Sensing,* **29**:441-450.

Eagman J. and Lin W. (1976). Remote sensing of soil moisture by a 21 cm passive radiometer. *J. Geophys. Res.,* **81**:3660-3666.

El-Rayes M.A. and Ulaby F.T. (1987). Microwave dielectric spectrum of vegetation – part I: experimental observations. *IEEE Trans. Geosci. Remote Sensing,* **25**:541-549.

England A.W., Galantowicz J.F., and Schretter M.S. (1992). The radiobrightness thermal inertia measure of soil moisture. *IEEE Trans. Geosci. Remote Sensing,* **30**:132-139.

Engman E.T. and Chauhan N. (1995). Status of microwave soil moisture measurements with remote sensing. *Remote Sens. Environ.,* **51**:189-198.

Entekhabi D., Nakamura H., and Njoku E.G. (1994). Solving the inverse problem for soil moisture and temperature profiles by sequential assimilation of multifrequency remotely sensed observations. *IEEE Trans. Geosci. Remote Sensing,* **32**:438-448.

Entekhabi D., Njoku E., Houser P., Spencer M., Doiron T., Smith J., Girard R., Belair S., Crow W., Jackson T., Kerr Y., Kimball J., Koster R., McDonald K., O'Neill P., Pulz T., Running S., Shi J., Wood E., and van Zyl J. (2004). The Hydrosphere State (HYDROS) mission concept: An earth system pathfinder for global mapping of soil moisture and land freeze/thaw. *IEEE Trans. Geosci. Remote Sensing,* **42**:2184-2195.

Eom H.J. (1986). Regression models for vegetation radar-backscattering and radiometric emission. *Remote Sensing Environ.,* **19**:151-157.

Eom H.J. and Fung A.K. (1984). A scatter model for vegetation up to Ku-band. *Remote Sens. Environ.,* **15**:185-200.

Eom H.J. and Fung A.K. (1986). Scattering from a random layer embedded with dielectric needles. *Remote Sens. Environ.,* **19**:139-149.

Ferrazzoli P. and Guerriero L. (1994). "Modeling microwave emission from vegetation-covered surfaces: a parametric analysis". In *ESA/NASA International Workshop.* Choudhury B.J., Kerr J.H., Njoku E.G., Pampaloni P., eds. Utrecht, The Netherlands: VSP, 1994, pp. 389-402.

Ferrazzoli P. and Guerriero L. (1996). Passive microwave remote sensing of forest: a model investigation. *IEEE Trans. Geosci. Remote Sensing,* **34**:433-443.

Ferrazzoli P., Luzi G., Paloscia S., Pampaloni P., Shiavon G., and Solimini D. (1989). Comparison between the microwave emissivity and backscattering coefficient of crops. *IEEE Trans. Geosci. Remote Sensing,* **27**:772-777.

Ferrazzoli P., Shiavon G., Solimini D., Luzi G., Pampaloni P., and Paloscia S. (1988). Comparison between microwave emissivity and backscattering coefficient of agricultural fields. *Proc. IGARSS'88, Edinburgh, Scotland, 13-16 Sept.*, pp. 667-772.

Ferrazzoli P., Guerriero L., Paloscia S., Pampaloni P., and Solimini D. (1992a). Modelling polarization properties of emission from soil covered with vegetation. *IEEE Trans. Geosci. Remote Sensing*, **30**:157-165.

Ferrazzoli P., Paloscia S., Pampaloni P., Schiavon G., Solimini D., and Coppo P. (1992b). Sensitivity of microwave measurements to vegetation biomass and soil moisture content: A case study. *IEEE Trans. Geosci. Remote Sensing*, **30**:750-756.

Ferrazzoli P., Wigneron J.P., Guerriero L., and Chanzy A. (2000). Multifrequency emission of wheat: Modeling and applications. *IEEE Trans. Geosci. Remote Sensing*, **38**: 2598-2607.

Ferrazzoli P., Guerriero L., and Wigneron J.-P. (2002). Simulating L-band emission of forests in view of future satellite applications. *IEEE Trans. Geosci. Remote Sensing*, **40**: 2700-2708.

Finkelberg V.M. (1968). Wave propagation in a random medium. The correlation group method. *Soviet Physics JETP*, **26**:268-277.

Franchois A., Pineiro Y., and Lang R.H. (1998). Microwave permittivity measurements of two conifers. *IEEE Trans. Geosci. Remote Sensing*, **36**:1384-1395.

Fung A.K. (1979). Scattering from a vegetation layer. *IEEE Trans. Geosci. Remote Sensing*, **17**:1-6.

Fung A.K. (1994). *Microwave Scattering and Emission Models and Their Application.* Norwood, MA: Artech House, 1994.

Fung A.K. and Eom H.J. (1985). A comparison between active and passive sensing of soil moisture from vegetated terrain. *IEEE Trans. Geosci. Remote Sensing*, **23**:768-775.

Fung A.K. and Fung H.S. (1977). Application of first order renormalization method to scattering from a vegetation-like half-space. *IEEE Trans. Geosci. Remote Sensing*, **15**:189-195.

Fung A.K. and Ulaby F.T. (1978). A scatter model for leafy vegetation. *IEEE Trans. Geosci. Remote Sensing*, **16**:281-285.

Galantowicz J.F., Entekhabi D., and Njoku E.G. (1999). Test of sequential data assimilation for retrieving profile soil moisture and temperature from observed L-band radiobrightness. *IEEE Trans. Geosci. Remote Sensing*, **37**:1860-1870.

Golovachev S.P., Reutov E.A., Chukhlantsev A.A., and Shutko A.A. (1989). Experimental investigation of microwave emission of vegetable crops. *Izvestia VUZ'ov, Radiofizika*, **32**:551-556 (in Russian).

Golovachev S.P. and Chukhlantsev A.A. (1986). Statistical properties of microwave emission from irrigated fields. *Proc. All-Union Conference on Statistical Processing of Remote Sensing Data*, Riga: RIIGA.

Goita K.,Walker A.E., Goodison B.E., and Chang A.T.C. (1997). Estimation of snow water equivalent in the boreal forests using passive microwave data. *Proc. Int. Symp. Geomatics in the Era of Radarsat, Ottawa, Canada, 26-30 May 1997.*

Grankov A.G., Liberman B.M., Reutov E.A., Chukhlantsev A.A., and Shutko A.M. (1979). On combined use of microwave and infrared passive remote sensing. In Science and Technology Bulletin No.37 *"Airborne Remote Sensing of Agricultural fields"*, pp. 48-54. Leningrad: Agro-Physical Institute, 1979 (in Russian).

Grankov A.G., Dragun V.S., Zakrevsky V.I. et al. (1992). Use of microwave radiometric methods for ecological monitoring of Belarus. *RNTC "Ecomir" Preprint No. 10*, 30 pp.

Grankov A.G., Kuznetsob O.O., Milshin A.A., and Shelobanova N.K. (1999). Seasonal and annual dynamics of microwave emission from forests at different fire danger state by SSM/I data. *Problemy Okruzhaushchei Sredy i Prirodnykh Resursov (Environmental Problems and Natural Resources)*, **No. 10**:2-14.

Grankov A.G., Milshin A.A., and Chukhlantsev A.A. (2004). Modeling microwave emission from forests in satellite observations. *Proc. of the Sixth International Symposium "Ecoinformatics Problems", Moscow, 1-3 December 2004*, Moscow: IRE RAS, 2004, pp. 38-42.

Grankov A.G., Milshin A.A., and Chukhlantsev A.A. (2005). Experimental estimations of radio wave attenuation in a tree crown at millimeter, centimeter, and decimeter wavelengths. *Issledovanie Zemli iz Kosmosa (Research of the Earth from Space)*, **No. 2**:20-26 (in Russian).

Guha A., J.M. Jackobs, T.J. Jackson et al. (2003). Soil moisture mapping using ESTAR under dry conditions from the Southern Great Plains experiment (SGH99). *IEEE Trans. Geosci. Remote Sensing*, **41**:2392-2397.

Hallikainen M.T., Ulaby F.T., Dobson M.C., El-Rayes M.A., and Wu L.-K. (1985). Microwave dielectric behavior of wet soil – Part I: Empirical models and experimental observations. *IEEE Trans. Geosci. Remote Sensing*, **23**: 25-34.

Hallikainen M.T., Tares T., Hyyppa J., Somersalo E., Ahola P., Toikka M., and Pullianen J.T. (1990). Helicopter-borne measurements of radar backscatter from forests. *Int. J. Remote Sensing*, **11**:1179-1191.

Hallikainen M., Jääskeläinen V.S., Pulliainen J., and Koskinen J. (2000). Transmissivity of boreal forest canopies for microwave radiometry of snow. *Proc. IGARSS-2000, Honolulu, Hawaii, July 2000*, pp. 1564-1566.

Herbstreit J.W. and Crichlow W.Q. (1964). Measurement of the attenuation of radio signals by jungles. *Radio Science*, **68D**:903-906.

Hilhorst M.A., Dirksen C., Kampers F.W.H., and Feddes R.A. (2001). Dielectric relaxation of bound water versus soil matric pressure. *Soil Sci. Soc. Am. J.*, **65**: 311-314.

Hipp J.E. (1974). Soil electromagnetic parameters as a function of frequency, soil density, and soil moisture. *Proc. IEEE*, **62**: 98-103.

Hirosava H., Matsuzaka Y., and Kobayashi O. (1989). Measurement of microwave backscatter from a cypress with and without leaves. *IEEE Trans. Geosci. Remote Sensing*, **27**:698-701.

Hoekman D.H. (1987). Measurements of the backscatter and attenuation properties of forest stands at X-, C-, and L-band. *Remote Sensing Environ.*, **23**:397-416.

Hoekman D.H. and Quinones M.J. (2002). Biophysical forest type characterization in the Colombian Amazon by airborne polarimetric SAR. *IEEE Trans. Geosci. Remote Sensing*, **40**:1288-1300.

Hoekstra P. and Delaney A. (1974). Dielectric properties of soils at UHF and microwave frequencies. *J. Geophys. Res.*, **79**:1699-1708.

Hornbuckle B.K., England A.W., De Roo R.D., Fishman M.A., and Boprie D.L. (2003). Vegetation canopy anisotropy at 1.4 GHz. *IEEE Trans. Geosci. Remote Sensing*, **41**:2211-2223.

Ilyin V.A. and Sosnovskiy Yu.M. (1994). Laboratory research of salinity effect upon the microwave dielectric properties of sand. *Radiotekhika i Elektronika*, **40**: 48-54 (in Russian).

Imhoff M.L. (1995). A theoretical analysis of the effect of forest structure on Synthetic Aperture Radar backscatter and the remote sensing of biomass. *IEEE Trans. Geosci. Remote Sensing*, **33**:341-352.

Inguva R. and Bowman C. (2002). Comparison of two microwave radiobrightness models and validation with field measurements. *IEEE Trans. Geosci. Remote Sensing*, **40**:143-152.

IPCC (2001a). *Climate Change 2001: The Scientific Basis.* Contribution of Working Group I to the Third Assessment report of the IPCC. Cambridge: Cambridge University Press.

IPCC (2001b). *Climate Change 2001: Mitigation.* Contribution of Working Group III to the Third Assessment report of the IPCC. Cambridge: Cambridge University Press.

Ishimaru A. (1978). *Wave Propagation and Scattering in Random Media. Vol. 2.* New York: Academic, 1978.

Jackson T.J. (1990). Laboratory evaluation of a field-portable dielectric/soil-moisture probe. *IEEE Trans. Geosci. Remote Sensing*, **28**: 241-245.

Jackson T.J. (1993). Measuring surface soil moisture using passive microwave remote sensing. *Hydrol. Process.*, 7:139-152.

Jackson T.J. (1997). Soil moisture estimation using special satellite microwave/imager satellite data over a grassland region. *Water Resources Res.*, **33**:1475-1484.

Jackson T.J. and Le-Vine D.M. (1996). Mapping surface soil moisture using an aircraft-based passive microwave instrument: Algorithm and example. *J. Hydrol.*, **184**:84-99.

Jackson T.J. and O'Neill P.E. (1986). Microwave dielectric model for aggregated soils. *IEEE Trans. Geosci. Remote Sensing*, **24**:920-929.

Jackson T.J. and O'Neill P.E. (1987a). Salinity effects on the microwave emission from soils. *IEEE Trans. Geosci. Remote Sensing*, **25**:214-220.

Jackson T.J. and O'Neill P.E. (1987b). Temporal observations of surface soil moisture using a passive microwave sensor. *Remote Sensing Environ.*, **21**:281-296.

Jackson T.J. and Schmugge T.J. (1989). Passive microwave remote sensing system for soil moisture: Some supporting research. *IEEE Trans. Geosci. Remote Sensing*, **27**:225-235.

Jackson T.J. and Schmugge T.J. (1991). Vegetation effect on the microwave emission of soils. *Rem Sens. Environ.*, **36**:203-212.

Jackson T.J., Schmugge T.J., and Wang J.R. (1982). Passive microwave sensing of soil moisture under vegetation canopies. *Water Resources Research*, **18**:1137-1142.

Jackson T.J., Schmugge T.J., and O'Neill P.E. (1984). Passive microwave remote sensing of soil moisture from an aircraft platform. *Remote Sensing Environ.*, **14**:135-151.

Jackson T.J., O'Neill P.E., and Wang J.R. (1986). Passive microwave soil moisture research. *IEEE Trans. Geosci. Remote Sensing*, **24**:12-22.

Jackson T.J., Schmugge T.J., Parry R., Kustas W.P., Ritchie J.C., Shutko A.M., Haldin A., Reutov E., Novichikhin E., Liberman B.,Shiue J.C., Davis M.R., Goodrich D.C., Amer S.B., and Bach L.B. (1992). Multifrequency passive microwave observations of soil moisture in an arid rangeland environment. *Int. J. Remote Sensing*, **13**:573-580.

Jackson T.J., LeVine D.M., Griffis A.J., Goodrich D.C., Schmugge T.J., Swift C.T., and O'Neill P.E. (1993). Soil moisture and rainfall estimation over a semiarid environment with the ESTAR microwave radiometer. *IEEE Trans. Geosci. Remote Sensing*, **31**: 836-841.

Jackson T.J., LeVine D.M., Swift C.T., Schmugge T.J., and Schiebe F.R. (1995). Large area mapping of soil moisture using the ESTAR passive microwave radiometer in Washita'92. *Remote Sens. Environ.*, **53**:27-37.

Jackson T.J., O'Neill P.E., and Swift C.T. (1997) Passive microwave observation of diurnal surface soil moisture. *IEEE Trans. Geosci. Remote Sensing*, **35**:1210-1222.

Jackson T.J., Le Vine D.M., Hsu A.Y., Oldak A., Starks P.J., Swift C.T., Isham J.D., and Haken M. (1999). Soil moisture mapping at regional scales using microwave radiometry: the Southern Great Plains Hydrology Experiment. *IEEE Trans. Geosci. Remote Sensing*, **37**:2136-2151.

Jackson T.J. and Hsu A.Y. (2001). Soil moisture and TRMM microwave imager relationships in the Southern Great Plains 1999 (SGP99) experiment. *IEEE Trans. Geosci. Remote Sensing*, **39**:1632-1642.

Jackson T.J., Hsu A.Y., Shutko A.M. et al. (2002a). Priroda microwave radiometer observation in the Southern Great Plains 1997 hydrology experiment. *Int. J. Remote Sensing*, **23**:231-248.

Jackson T.J., Gasiewski A.J., Oldak A., Klein M., Njoku E.G., Yevfrafov A., Christiani S., and Bindlish R. (2002b). Soil moisture retrieval using C-band polarimetric scanning radiometer during the Southern Great Plains 1999 experiment. *IEEE Trans. Geosci. Remote Sensing*, **40**:2151-2161.

Jackson T.J., Chen D., Cosh M., Li F., Anderson M., Walthall C., Doriaswamy P., and Hunt E.R. (2004). Vegetation water content mapping using Lansat data derived normalized difference water index for corn and soybeans. *Remote Sens. Environ.*, **92**:475-482.

Judge J., Galantowicz J.F., and England A.W. (2001). A comparison of ground-based and satellite-borne microwave radiometric observations in the Great Plains. *IEEE Trans. Geosci. Remote Sensing*, **39**:1686-1696.

Karam M.A. and Fung A.K. (1983). Scattering from randomly oriented circular discs with application to vegetation. *Radio Science*, **18**:557-565.

Karam M.A. and Fung A.K. (1988). Electromagnetic scattering from a layer of finite length, randomly oriented, dielectric, circular cylinders over a rough interface with application to vegetation. *Int. J. Remote Sensing*, **9**:1109-1134.

Karam M.A. and Fung A.K. (1989). Leaf-shape effects in electromagnetic wave scattering from vegetation. *IEEE Trans. Geosci. Remote Sensing*, **27**:687-697.

Karam M.A., Fung A.K., and M.Antar Y.M. (1988). Electromagnetic wave scattering from some vegetation samples. *IEEE Trans. Geosci. Remote Sensing*, **26**:799-808.

Karam M.A., Fung A.K., Lang R.H., and Chauhan N.S. (1992). A microwave scattering model for layered vegetation. *IEEE Trans. Geosci. Remote Sensing*, **30**:767-784.

Kasischke E.S., Christensen N.L., and Haney. (1994). Modeling of geometric properties of loblolly pine tree stand characteristics for use in radar backscattering studies. *IEEE Trans. Geosci. Remote Sensing*, **32**:800-822.

Kerr Y.H. and Njoku E.G. (1990). A semi-empirical model for interpreting microwave emission from semiarid land surfaces as seen from space. *IEEE Trans. Geosci. Remote Sensing*, **28**:384-393.

Kerr Y.H. and Wigneron J.P. (1994). "Vegetation Model and Observations. A Review," In *ESA/NASA International Workshop*. Choudhury B.J., Kerr J.H., Njoku E.G., Pampaloni P., eds. VSP, 1994, pp.317-344.

Kerr Y.H., Waldteufel P., Wigneron J.P., Martinuzzi J.-M., Font J., and Berger M. (2001). Soil moisture retrieval from space: The soil moisture and ocean salinity (SMOS) mission. *IEEE Trans. Geosci. Remote Sensing*, **39**:1729-1735.

Kira T., Ogava H. (1971)."Assessment of Primary Production in Tropical and Equatorial Forests," In *Productivity of Forest Ecosystems. Proceedings of the Brussels Symposium 27-31 October 1969*. Duvigneaud P, ed. Paris: UNESCO, 1971, pp. 309-321.

Kirdiashev K.P. (1983). On stability of natural objects brightness temperature fields. . *Proc. of 1st Interdepartmental Meeting "Statistical methods of remote sensing data processing", Sept. 1980, Minsk*. Moscow: IRE, 1983, pp. 13-16.

Kirdiashev K.P. and Savorsky V.P. (1986). Statistical estimate of forest fire risk from microwave radiometric measurements. *Radiotekhnika I Elektronika*, **31**:1239-1241.

Kirdiashev K.P., Chukhlantsev A.A., and Shutko A.M. (1979). Microwave radiation of the Earth's surface in the presence of vegetation cover. *Radio Eng. Electron. Phys.*, **2**: 37-56.

Kleshchenko V.N. (2002). *Dielectric Properties of Wet and Salted Soils at Positive and Negative Temperatures. Ph.D. Dissertation*, Barnaul: The Altai State University, 2002 (in Russian).

Kondratyev K. Ya., Kpapivin V.F., and Phillips G.W. (2002). *Global Environmental Change: Modeling and Monitoring*. Berlin: Springer-Verlag.

Kondratyev K. Ya., Kpapivin V.F., and Varotsos C.A. (2003). *Global Carbon Cycle and Climate Change*. Chichester: Springer/PRAXIS.

Kondratyev K. Ya., Melentiev V.V., and Nazarkin V.A. (1992). *Remote Sensing of Aqua Tories from Space*, S.-Petersburg: Gidrometeoizdat, 1992 (in Russian).

Kondratyev K. Ya., Kpapivin V.F., Savinykh V.P., and Varotsos C.A. (2004). *Global Ecodynamics: A Multidimensional Analysis*. Chichester: Springer/PRAXIS.

Koskinen J.T., Pulliainen J.T., Maekynen M.P., and Hallikainen M.T. (1999). Seasonal comparison of HUTSCAT ranging scatterometer and ERS-1 SAR microwave signatures of boreal forest zone. *IEEE Trans. Geosci. Remote Sensing*, **37**:2068-2079.

Kosolapov V.V. and Kozoderov V.V. (1994). New applications of airborne and satellite remote sensing to forest ecosystems research. *Issledovanie Zemli iz kosmosa*, **No.5**: 68-80 (in Russian).

Krapivin V.F. and Phillips G.W. (2001) A remote sensing-based expert system to study the Aral-Caspian aquageosystem water regime. *Remote Sensing Environ.*, **75**:201-215.

Krapivin V.F. and Vilkova L.P. (1990). Model estimation of excess CO_2 distribution in the biosphere structure. *Ecological Modelling*, **50**:57-78.

Krapivin V.F., and Kondratyev K. Ya. (2002). *Global Changes of the Environment: Ecoinformatics*. St. Petersburg: St. Petersburg University Publishing.

Krapivin V.F. and Chukhlantsev A.A. (2004). Microwave radiometry of vegetation and soil in a context of global change of environment. *Ecological Systems and Instruments*, **9**: 37-45 (in Russian).

Krapivin V.F., Shutko A.M., Chukhlantsev A.A., Golovachev S.P., and Phillips G.W. GIMS-based method for vegetation microwave monitoring. *Environmental Modelling and Software*, available at hppt://www.sciencedirect.com/science

Krapivin V.F., Svirezhev Yu.M., and Tarko A.M. (1982). *Mathematical Modeling of the Global Biospheric Processes*. Moscow: Science (in Russian).

Krapivin V.F., Vilkova L.P., Rochon G.L., and Hicks D.R. (1996). Model estimation of the role of urban areas in global CO_2 dynamics. *Proc. of Eco-Informa'96, 4-7 November 1996, Florida*. Cambridge, Massachusetts: MIT Press.

Krotikov V.D. (1962). Dielectric properties of dry soils. *Izv. Vyss. Ucheb. Zaved. Radiofiz.*, **5**:1057-1061 (in Russian).

Kruopis N., Praks J., Arslan A.N., Alasalmi H., Koskinen J.T., and Hallikainen M.T. (1999). Passive microwave measurements of snow-covered forest areas in EMAC'95. *IEEE Trans. Geosci. Remote Sensing*, **37**:2699-2705.

Kurvonen L. and Hallikainen M. (1997). Influence of land-cover category on brightness temperature of snow. *IEEE Trans. Geosci. Remote Sensing*, **35**:367-377.

Kurvonen L., Pulliainen J., and Hallikainen M. (1998). Monitoring of boreal forests with multitemporal special sensor microwave imager. *Radio Science*, **33**: 731-744.

Lane J.A. and Saxton J.A. (1952). Dielectric dispersion in pure polar liquids at very high radar frequencies, III, The effect of electrolytes in solution. *Proc. Roy. Soc. London*, **214**:531-545.

Lang R.H. (1981). Electromagnetic backscattering from a sparse distribution of lossy dielectric scatterers. *Radio Science*, **16**:15-30.

Lang R.H. (2004). Scattering from a layer of discrete random medium over a random interface: application to microwave backscattering from forests. *Waves in Random Media*, **14**:s359-s391.

Lang R.H. and Sidhu J.S. (1983). Electromagnetic backscattering from a layer of vegetation: a discrete approach. *IEEE Trans. Geosci. Remote Sensing*, **21**:62-71.

Lang R.H., Seker S.S., and Le Vine D.M. (1986). Vector solution for the mean electromagnetic fields in a layer of random particles. *Radio Science*, **21**:771-786.

Lang R.H., Utku C., De Matthaeis P., Chauhan N., and Le Vine D.M. (2001). L-band radiometry measurements of conifer forests. *Proc. IGARSS 2000, Honolulu, Hawaii, July 2000*.

Laursen B. and Skou N. (1998). Synthetic aperture radiometry evaluated by a two-channel demonstration model. *IEEE Trans. Geosci. Remote Sensing*, **36**:822-832.

Laymon C.A., Crosson W.L., Jackson T.J., Manu A., and Tsegaye T.D. (2001). Ground-based passive microwave remote sensing observations of soil moisture at S-band and L-band with insight into measurement accuracy. *IEEE Trans. Geosci. Remote Sensing*, **39**:1844-1858.

Le Toan T., Beaudoin A., Riom J., and Guyon D. (1992). Relating forest biomass to SAR data. *IEEE Trans. Geosci. Remote Sensing,* **30**:403-411.

Levin, M.L. and Rytov, S.M. (1967). *Theory of Equilibrium Thermal Fluctuations in Electrodynamics*. Moscow: Nauka Press, 308 pp. (in Russian).

Le Vine, D.M. (1990). The sensitivity of synthetic aperture radiometers for remote sensing applications from space. *Radio Science*, **25**(4):441-453.

Le Vine D.M., Schneider A., Lang R.H., and Carter H.G. (1985). Scattering from thin dielectric disks. *IEEE Trans. Antennas Propagat.*, **33**:1410-1413.

Le Vine D.M. and Karam M.A. (1996). Dependence of attenuation in a vegetation canopy on frequency and plant water content. *IEEE Trans. Geosci. Remote Sensing*, **34**:1090-1096.

Leckie D.G. and Ranson K.J. (1998). "Forestry Applications Using Imaging Radar," in *Principles and Applications of Imaging Radar*, F.M. Henderson and A.J. Lewis, ed. Wiley, 1998, pp. 435-509.

Leith H. (1985). A dynamic model of the global carbon flux through the biosphere and its relations to climatic and soil parameters. *Int. J. Biometeorol.*, **29**:17-31.

Leschanskiy Y.I., Lebedeva G.N., and Schumilin V.D. (1971). Electrical parameters of sandy and loamy soils in the range of centimeter, decimeter, and meter wavelength. *Izv. Vyss. Ucheb. Zaved. Radiofiz.*, **14**: 562-569 (in Russian).

Li L.-W., Koh J.-H., Yeo T.-S., Leong M.-S., and Kooi P.-S. (1999). Analysis of radiowave propagation in a four-layered anisotropic forest environment. *IEEE Trans. Geosci. Remote Sensing*, **37**: 1967-1979.

Li Q., Tsang L., Shi J., and Chan C.H. (2000). Application of physic-based two-grid method and sparse matrix canonical grid method for numerical simulations of emissivities of soils with rough surfaces at microwave frequencies. *IEEE Trans. Geosci. Remote Sensing*, **38**:1635-1643.

Liberman B.M., Grankov A.G., Milshin A.A., Maximov A.E., and Braun S.V. (1995). Application of microwave remote sensing methods for estimation of grounds and water states in Southern Bavaria region. *Proc. 7th British Electromagnetic Measurements Conference, The winter Gardens Malvern, UK, 1995*.

Liberman B., Grankov A., Milshin A., Golovachev S., and Vishniakov V. (1996). Experimental study of fire risk by means of passive microwave and infrared remote sensing methods. *Proc. IGARSS 1996, 27-31 May 1996, Lincoln, Nebraska*.

Liou Y.-A. and England A.W. (1996). Annual temperature and radiobrightness signatures for bare soils. *IEEE Trans. Geosci. Remote Sensing*, **34**:981-990.

Liou Y.-A., Chen, K.-S., and Wu T.-D. (1999). Reanalysis of L-band brightness predicted by LSP/R model for prairie grassland: Incorporation of rough surface scattering. *IEEE Trans. Geosci. Remote Sensing*, **37**:1848-1859.

Liou Y.-A., Galantowicz J.F., and England A.W. (2001). A land surface process/radiobrightness model with coupled heat and moisture transport for prairie grassland. *IEEE Trans. Geosci. Remote Sensing*, **39**:129-135.

Liou Y.-A., Liu S.-F., and Wang W.-J. (2001). Retrieving soil moisture from simulated brightness temperatures by a neural network. *IEEE Trans. Geosci. Remote Sensing*, **39**:1662-1672.

Liu S.-F., Liou Y.-A., Wang W.-J., Wigneron J.-P., and Lee J.-B. (2002). Retrieval of crop biomass and soil moisture from measured 1.4 and 10.65 GHz brightness temperature, *IEEE Trans. Geosci. Remote Sensing*, **40**:1260-1268.

Lopes A. and Mougin E. (1990). Microwave propagation in cylindrical-shaped forest components: interpretation of attenuation observations. *IEEE Trans. Geosci. Remote Sensing*, **28**:315-324.

Macelloni G., Paloscia S., Pampaloni P., and Ruisi R. (1998). Microwave emission features of crops with vertical stems. *IEEE Trans. Geosci. Remote Sensing*, **36**:332-337.

Macelloni G., Paloscia S., Pampaloni P., and Ruisi R. (2001). Airborne multifriquency L- to Ka- band radiometric measurements over forests. *IEEE Trans. Geosci. Remote Sensing*, **39**:2507-2513.

Macelloni G., Paloscia S., Pampaloni P., Ruisi R., and Santi E. (2003). Microwave radiometric features of Mediterranean forests: seasonal variations. *Proc. IGARSS-2003, Toulouse, July 2003.*

Magagi R., Bernier M., and Ung C.-H. (2002). Quantitative analysis of RADARSAT SAR data over a sparse forest canopy," *IEEE Trans. Geosci. Remote Sensing*, **40**:1301-1313.

Manual for Portable Dielectric Probe, Lawrence, KS: Applied Microwave Corp., 1989.

Marliani F., Paloscia S., Pampaloni P., and Kong J.A. (2002). Simulating coherent backscattering from crops during the growing cycle. *IEEE Trans. Geosci. Remote Sensing*, **40**:162-177.

Marquardt D.W. (1963). An algorithm for least-squares estimation of non-linear parameters. *J. Soc. Indust. Appl. Math.*, **11**:431-444.

Masson V., Champeaux J.L., Chauvin F., Meriguet C., and Lacaze R. (2003). A global database of land surface parameters at 1 km resolution in meteorological and climate models. *J.Clim.*, **16**:1261-1282.

Mätzler C. (1990). Seasonal evolution of microwave radiation from an oat field. *Remote Sens. Environ.*, **31**:161-173.

Mätzler C. (1994). Microwave (1-100 GHz) dielectric model of leaves," *IEEE Trans. Geosci. Remote Sensing*, **33**:947-949.

Mätzler C. (1995). "Microwave dielectric model of leaves," in *Microwave Radiometry and Remote Sensing of Environment*, Ed. D.Solimini, Utrecht, The Netherlands: VSP, 1995, pp. 389-390.

Mätzler C. (1998). Microwave permittivity of dry sand. *IEEE Trans. Geosci. Remote Sensing*, **36**:317-319.

Mätzler C. (2000). "Microwave emission from covered surfaces: zero-order versus multiple scattering," In *Radiative Transfer Models for Microwave Radiometry*. C. Mätzler, ed. Bern, Switzerland:VSP, 2000, pp. 73-81.

Mätzler C. and Sume A. (1989). Microwave radiometry of leaves. in *Microwave Radiometry and Remote Sensing Applications*, Ed., P.Pampaloni, Utrecht, The Netherlands: VSP, 1989.

Mätzler C., Wegmüller U. (1994). "Progress in multi-frequency radiometry of natural objects." *ESA/NASA International Workshop*. Choudhury B.J., Kerr J.H., Njoku E.G., Pampaloni P., eds. Utrecht: VSP, 1994, pp. 345-355.

McDonald K.C., Dobson M.C., and Ulaby F.T. (1990). Using MIMICS to model L-band multiangle and multitemporal backscatter from a walnut orchard. *IEEE Trans. Geosci. Remote Sensing*, **28**:477-491.

McDonald K.C., Dobson M.C., and Ulaby F.T. (1991). Modeling multi-frequency diurnal backscatter from a walnut orchard. *IEEE Trans. Geosci. Remote Sensing*, **29**:852-863.

McDonald K.C., Zimmermann R., Way J., and Chun W. (1999). Automated instrumentation for continuous monitoring of the dielectric properties of woody vegetation: system design, implementation, and selected *in situ* measurements. *IEEE Trans. Geosci. Remote Sensing*, **37**:1880-1894.

McDonald K.C., Zimmermann R., and Kimball J.K. (2002). Diurnal and spatial variations of xylem dielectric constant in Norway spruce as related to microclimate, xylem sap flow, and xylem chemistry. *IEEE Trans. Geosci. Remote Sensing*, **40**:2063-2082.

Meesters A.G.C.A., de Jeu R.A.M., and Owe M. (2005). Analytical derivation of the vegetation optical depth from the microwave polarization difference index. *IEEE Trans. Geosci. Remote Sensing Letters*, **2**:121-123.

Melon P., Martinez J.-M., Toan T. Le, Ulander L.M.H., and Beaudoin A. (2001). On the retrieving of forest stem volume from VHF SAR data: observation and modeling, *IEEE Trans. Geosci. Remote Sensing*, **39**:2364-2372.

Milshin A.A. and Grankov A.G. (2000). Some experimental results of microwave emission of forests in L-band. *Issledovanie Zemli iz Kosmosa (Research of the Earth from Space)*, **No. 3**:50-57 (in Russian).

Milshin A.A., Grankov A.G., and Shelobanova N.K. (1998). A study of seasonal dynamics of microwave emission from land surface-forest-atmosphere system. Preprint No.3 (624). Moscow: IRE RAS, 1998, 61 pp.

Milshin A.A., Grankov A.G., and Mishanin V.G. (1999). Mapping of the temperature and moisture regime of forests. *Issledovanie Zemli iz Kosmosa (Research of the Earth from Space)*, **No. 5**:85-93 (in Russian).

Mintzer I.M. (1987). *A Matter of Degree: The Potential for Controlling the Greenhouse Effect*. (Res.Rep. No. 15, 70 pp.) New York: World Resources Institute.

Mironov V.L., Komarov S.A., Rychkova N.V., and Kleshchenko V.N. (1995). Study of the dielectric properties of wet grounds at microwave frequencies. *Earth Obs. Remote Sens.*, **12**:495-504.

Mironov V.L., Komarov S.A., and Kleshchenko V.N. (1997). Effect of bound water on the dielectric properties of wet frozen soils. *Earth Obs. Remote Sens.*, **13**:347-356.

Mironov V.L., Dobson M.C., Kaupp V.H., Komarov S.A., and Kleshchenko V.N. (2004). Generalized refractive mixing dielectric model for moist soils. *IEEE Trans. Geosci. Remote Sensing*, **42**:773-785.

Mo T. and Schmugge T.J. (1987). A parameterization of the effect of surface roughness on microwave emission. *IEEE Trans. Geosci. Remote Sensing*, **25**:47-54.

Mo T., Choudhury B.J., Schmugge T.J., Wang J.R., and Jackson T.J. (1982). A model for microwave emission from vegetation-covered fields. *J. Geophys. Res.*, **87**:11229-11237.

Mo T., Schmugge T.J., and Jackson T.J. (1984). Calculation of radar backscattering coefficient of vegetation-covered soil. *Remote Sens. Environ.*, **15**:119-133.

Mo T., Wang J.R., and Schmugge T.J. (1998). Estimation of surface roughness parameters from dual-frequency measurements of radar backscattering coefficient. . *IEEE Trans. Geosci. Remote Sensing*, **26**:574-579.

Moghaddam M. and Saatchi S.S. (1999). Monitoring tree moisture using an estimation algorithm applied to SAR data from BOREAS. *IEEE Trans. Geosci. Remote Sensing*, **37**:901-916.

Mougin E., Lopes A., and Le Toan T. (1990). Microwave propagation at X-band in cylindrical-shaped forest components: attenuation observations. *IEEE Trans. Geosci. Remote Sensing*, **28**:60-69.

Murata M., Aiba H., Tonoike K., Komai J., Hirosava H., and Nakada K. (1987). Experimental results of L-band microwave penetration properties of trees. *Proceedings of the IGARSS'87*, pp. 815-820.

Nashashibi A.Y., Ulaby F.T., Frantzis P., and De Roo R.D. (2002). Measurements of the propagation parameters of tree canopies at MMW frequencies. *IEEE Trans. Geosci. Remote Sensing*, 40:298-304.

Nazarov L.E., Chukhlantsev A.A., Shutko A.M., and Golovachev S.P. (2003). Neural network algorithms of soil moisture retrieval from microwave radiometric measurements. *Neurocomputers: Development and Applications*, **No. 12**: 20-29 (in Russian).

Nazarov L.E., Chukhlantsev A.A., and Shutko A.M. (2004). Neural networks in soil moisture retrieval from microwave radiometric measurements. *Issledovanie Zemli iz Kosmosa*, **No. 6**:21-29.

Nefedova E.I. and Tarko A.M. (1993). Study of the global carbon cycle in the atmosphere-ocean system. *Annals of the Russian Academy of Sciences*, **333**:645-647 (in Russian).

Nesterov V.G. (1949). *Forest flammability and methods of its determination.* Moscow-Leningrad: Goslesbumizdat, 1949.

Newton R.W. and Rouse J.W., Jr. (1980). Microwave radiometer measurements of soil moisture content. *IEEE Trans. Antennas Propagat.*, **28**:680-686.

Njoku E.G. and Kong J.-A. (1977). Theory for passive microwave remote sensing of near-surface soil moisture. *J. Geophys. Res.*, **82**:3108-3118.

Njoku E.G. and Li Li. (1999). Retrieval of land surface parameters using passive microwave measurements at 6-18 GHz. *IEEE Trans. Geosci. Remote Sensing*, **37**:79-93.

Njoku E.G. and O'Neill P.E. (1982). Multifrequency microwave radiometer measurements of soil moisture. *IEEE Trans. Geosci. Remote Sensing*, **20**:468-475.

Njoku E.G., Wilson W., Yueh S., Li F., Dinardo S., Jackson T., Lakshmi V., and Bolten J. (2002). Observations of soil moisture using a passive and active low-frequency microwave airborne sensor during SGP99. *IEEE Trans. Geosci. Remote Sensing*, **40**:2659-2673.

Njoku E.G., Jackson T.J., Lakshmi V., Chan T.K., and Son V.N. (2003). Soil moisture retrieval from AMSR-E. *IEEE Trans. Geosci. Remote Sensing*, **41**:215-229.

O'Neil P.E., Jackson T.J., Blanchard B.J., Wang J.R., and Gould W.I. (1984). Effect of corn stalk orientation and water content on passive microwave remote sensing of soil moisture. *Remote Sens. Environ.*, **16**:55-67.

O'Neill P.E., Joseph A., De Lannoy G., Lang R., Utku C., Kim E., Houser P., and Gish T. (2003). Soil moisture retrieval through changing corn using active/passive microwave remote sensing. *Proc. IGARSS 2003, Toulouse, France.*

Olioso A., Chauki H., Courault D, and Wigneron J.P. (1999). Estimation of evapotranspiration and photosynthesis by assimilation of remote sensing data into SVAT models. *Remote Sens. Environ.*, **68**:341-356.

Orlov R.A. and Torgashin B.D. (1978). *Modeling radar returns from the Earth's surface.* Leningrad: Leningrad State University Publishing, 1978 (in Russian).

Owe M., Van de Griend A.A., and Chang A.T.C. (1992). Surface moisture and satellite microwave observation in semiarid southern Africa. *Water Resources Res.*, **28**:829-839.

Owe M., de Jeu R., and Walker J. (2001). A methodology for surface soil moisture and vegetation optical depth retrieval using the microwave polarization difference index. *IEEE Trans. Geosci. Remote Sensing*, **39**:1643-1654.

Paloscia S. and Pampaloni P. (1984). Microwave remote sensing of plant water stress. *Remote Sens. Environ.*, **16**:249-255.

Paloscia S. and Pampaloni P. (1987). X-band features of canopy cover: an up to date summary of active and passive measurements. *Adv. Space. Res.*, **7**:(11)305-(11)308.

Paloscia S. and Pampaloni P. (1988). Microwave polarization index for monitoring vegetation growth. *IEEE Trans. Geosci. Remote Sensing*, **26**:617-621.

Paloscia S. and Pampaloni P. (1992). Microwave vegetation indexes for detecting biomass and water conditions of agricultural crops. *Remote Sens. Environ.*, **49**:15-26.

Pampaloni P. (2004). Microwave radiometry of forests. *Waves in Random Media*, **14**: S275-S298.

Pampaloni P. and Paloscia S. (1985a). Experimental relationships between microwave emission and vegetation features. *Int. J. Remote Sensing*, **6**:315-323.

Pampaloni P. and Paloscia S. (1985b).Microwave emission from vegetation: General aspects and experimental results. *Proccedings of EARSeL Workshop "Microwave remote sensing applied to vegetation" Amsterdam, 10-12 December 1984*: ESA SP-227, 1985, pp. 53-60.

Pampaloni P. and Paloscia S. (1986). Microwave emission and plant water content: a comparison between field measurements and theory. *IEEE Trans. Geosci. Remote Sensing*, **24**:900-905.

Pardé M., Wigneron J.P., Waldteufel P., Kerr Y.H., Chanzy A., Søbjærg S.S., and Skou N. (2004). N-parameter retrieval from L-band microwave observations acquired over a variety of crop fields. *IEEE Trans. Geosci. Remote Sensing*, **42**:1168-1178.

Pardé M., Goïta K., Royer A., and Vachon F. (2005). Boreal forest transmissivity in the microwave domain using ground-based measurements. *IEEE Trans. Geosci. Remote Sensing Letters*, **2**:169-171.

Paris J.F. (1986). Probing thick vegetation canopies with a field microwave scatterometer. *IEEE Trans. Geosci. Remote Sensing*, **24**:886-893.

Peake W.H. (1959). Interaction of electromagnetic waves with some natural surfaces. *IRE Trans. Antennas. Propagat.*, **AP-7**:s325-s329.

Pellarin T., Calvet J.-C., and Wigneron J.-P. (2003a). Surface soil moisture retrieval from L-band radiometry: A global regression study. *IEEE Trans. Geosci. Remote Sensing*, **41**:2037-2051.

Pellarin T., Wigneron J.-P., Calvet J.-C., Berger M., Douville H., Ferrazzoli P., Kerr Y.H., Lopez-Baesa E., Pulliainen J., Simmonds L.P., and Waldteufel P. (2003b) Two-year global simulation of L-band brightness temperatures over land. *IEEE Trans. Geosci. Remote Sensing*, **41**:2135-2139.

Peplinski N.R., Ulaby F.T., and Dobson M.C. (1995). Dielectric properties of soils in the 0.3-1.3 GHz range. *IEEE Trans. Geosci. Remote Sensing*, **33**: 803-807. (Corrections: *IEEE Trans. Geosci. Remote Sensing*, 1995, **33**:1340).

Pitts D.E., Badhwar G.D., and Reyna E. (1988). The use of a helicopter mounted ranging scatterometer for estimation of extinction and scattering properties of forest canopies, part II: experimental results for high-density aspen. *IEEE Trans. Geosci. Remote Sensing*, **26**:144-152.

Polyakov V.M., Tishchenko Yu.G., and Chukhlantsev A.A. (1994). On microwave radiometry of frozen soils temperature and moisture. *Radiotekhika i Elektronika*, **39**: 405-409 (in Russian).

Press W.H., Flannery B.P., Teukolsky S.A., and VetterlingW.T. (1989). *Numerical Recipes.* New York: Cambridge Univ. Press, 1989.

Promes P.M., Jackson T.J., and O'Neill P.E. (1988). Significance of agricultural row structure on the microwave emissivity of soils. *IEEE Trans. Geosci. Remote Sensing*, **26**: 580-589.

Pulliainen J., Kärnä J., and Hallikainen M. (1993). Development of geophysical retrieval algorithms for the MIMR. *IEEE Trans. Geosci. Remote Sensing*, **31**:268-277.

Pulliainen J.T., Kurvonen L., and Hallikainen M.T. (1999). Multitemporal behavior of L- and C-band SAR observations of boreal forests. *IEEE Trans. Geosci. Remote Sensing*, **37**:927-937.

Raju S., Chanzy A., Wigneron. J.-P., Calvet J.-C., Kerr Y., and Laguerre L. (1995). Soil moisture and temperature profile effects on microwave emission at low frequencies. *Remote Sens. Environ.*, **54**:85-97.

Redkin B.A. and Klochko V.V. (1977). Calculation of averaged dielectric permittivity tensor for vegetation media. *Radiotekhnika i Elektronika*, **22**:1596-1599 (in Russian).

Redkin B.A., Klochko V.V., and Ocheret G.G. (1973). Theoretical and experimental study of reflection coefficient of vegetation canopies at small sliding angles. *Izvestia VUZ'ov, Radiofizika (Engl. Transl. Radiophys, Quantum Electron.)*, **16**:1178-1181 (in Russian).

Reutov E.A. (1991). On interconnection between microwave and infrared radiation fields and the condition of natural objects. *IEEE Trans. Geosci. Remote Sensing*, **29**:191-193.

Reutov E.A. and Shutko A.M. (1986a). Determining soil moisture of soils with transient layer from multi-frequency radiometric measurements. *Issledovanie Zemli iz Kosmosa (Research of the Earth from Space)*, **1**:71-78 (in Russian)

Reutov E.A. and Shutko A.M. (1986b). Prior-knowledge-based soil-moisture determination by microwave radiometry. *Sov.J. Remote Sensing*, **5(1)**:100-125.

Reutov E.A. and Shutko A.M. (1989). Soil moisture determination by radiometric measurements with use of a priory information. *Issledovanie Zemli iz Kosmosa (Research of the Earth from Space)*, **4**:84-90 (in Russian).

Riabchikov A.M. (1972). *Structure and Dynamics of Geosphere: Natural Development and Anthropogenic Change*. Moscow: Mysl Ed. (in Russian).

Richards P.U. *Tropical rain forest*. Moscow: Mir, 1961 (in Russian).

Ruf C.S., Swift C.T., Tanner A.B., and Le Vine D.M. (1988). Interferometric synthetic aperture microwave radiometry for the remote sensing of the Earth. *IEEE Trans. Geosci. Remote Sensing*, **26**:597-611.

Running S.W., Loveland T.R., Pierce L.L., Nemani R.R., and Hunt E.R. (1995). A remote sensing based vegetation classification logic for global land cover analysis. *Remote Sensing Environ.*, **51**(1):39-48.

Rytov S.M. (1953). *The Theory of Electrical Fluctuations and Thermal Emission*. Moscow: USSR Academy of Sciences Publ. House, 230 pp. (in Russian).

Rytov S.M., Kravtsov Yu. A., and Tatarskii V.I. (1978). *Introduction to Statistical Radiophysics*. Part II. *Random Fields*. Moscow: Nauka Press, 462 pp. (in Russian).

Ryzhov Yu.A., Tamoikin V.V., and Tatarskii V.I. (1965). Electromagnetic wave propagation in a random medium with strong fluctuations of dielectric permittivity. *JETP*, **48**: 656-673.

Ryzhov Yu.A. and Tamoikin V.V. (1970). Radiation and propagation of electromagnetic waves in randomly inhomogeneous media. *Radiophys. Quantum. Electron.*, **13**:273-300.

Saatchi S.S. and McDonald K.C. (1997). Coherent effects in microwave backscattering models for forest canopies. *IEEE Trans. Geosci. Remote Sensing*, **35**:1032-1044.

Sabburg J., Ball A.R., and Hancock N.H. (1997). Dielectric behavior of moist swelling clay soils at microwave frequencies. *IEEE Trans. Geosci. Remote Sensing*, **35**: 784-787.

Salas W.A., Ranson K.J., Rock B.N., and Smith K.T. (1994). Temporal and spatial variations in dielectric constant and water status of dominant forest species from New England. *Remote Sens. Environ.*, **47**:109-119.

Satoo T. (1971). "Primary Production Relations of Coniferous Forests in Japan," In *Productivity of Forest Ecosystems. Proceedings of the Brussels Symposium 27-31 October 1969*. Duvigneaud P, ed. Paris: UNESCO, 1971, pp. 191-205.

Savage N., Ndzi D., Seville A., Vilar E., and Austin J. (2003). Radio wave propagation through vegetation: Factors influencing signal attenuation. *Radio Science*, **38**:9-1-9-14.

Schiffer R. and Thielheim K.O. (1979). Light scattering by dielectric needles and discs. *J. Appl. Phys.*, **50**:2476-2483.

Schmugge T.J. (1983). Remote sensing of soil moisture: Recent advances. *IEEE Trans. Geosci. Remote Sensing*, **21**:336-344.

Schmugge T.J. (1998). Applications of passive microwave observations of surface soil moisture. *J. Hydrol.*, **212-213**:188-197.

Schmugge T.J. and Choudhury B.J. (1981). A comparison of radiative transfer models for predicting the microwave emission from soils. *Radio Science*, **16**:927-938.

Schmugge T.J. and Jackson T.J. (1992). A dielectric model of the vegetation effects on the microwave emission from soils. *IEEE Trans. Geosci. Remote Sensing*, **30**:757-760.

Schmugge T.J., Gloersen P., Wilheit T., and Geiger F. (1974). Remote sensing of soil moisture with microwave radiometer. *J. Geophys. Res.*, **79**:317-323.

Schmugge T.J., O'Neil P.E., and Wang J.R. (1986). Passive microwave soil moisture research. *IEEE Trans. Geosci. Remote Sensing*, **24**:12-22.

Seker S.S. and Schneider A. (1988). Electromagnetic scattering from a dielectric cylinder of finite length. *IEEE Trans. Antennas Propagat.*, **36**:303-307.

Sellers P., Hall F., Margolis H., et al. (1995). The boreal ecosystem-atmosphere study (BOREAS): an overview of early results from the 1994 field year. *Bulletin of the American Meteorological Society*, **76**:1549-1577.

Sellers P.J., Bounoua L., Collatz G.L. , Randall D.A., Dazlich D A., Los S.O., Berry J.A., Fung I., Tucker C.J., Field C.B., and Jensen . T.G. (1996a). Comparison of radiative and physiological effects of doubled atmospheric CO_2 on climate. *Science*, **271**: 1402-1406.

Sellers P.J., Los S.O., Tucker C.J., Justice C.O., Dazlich D.F., Collatz G.J., and Randall D.A. (1996b). A revised land surface parameterization (SiB2) for atmospheric GCMs. Part II: The generation of global fields of terrestrial biophysical parameters from satellite data. *Journal of Climate*, **9**:708-737.

Senior T.B.A., Sarabandy K., and Ulaby F.T. (1987). Measuring and modeling the backscattering cross section of a leaf. *Radio Science*, **22**:1109-1116.

Shadrina I.A. (2001). Microwave dielectric properties of woody materials. *Proc. of All-Russian Conference "Remote Sensing of Terrains and Atmosphere.* pp. 237-241 (in Russian).

Sharkov E.A. (2003). *Passive Microwave Remote Sensing of the Earth. Physical Foundations.* Chichester, UK: Springer/Praxis.

Shinohara H., Homma T., Nohmi H., Hirosava H., and Tagawa T. (1992). Relation between L-band microwave penetration/backscattering characteristics and state of trees. *Proceedings of the IGARSS'92*, pp. 539-541.

Shutko A.M. (1982). Microwave radiometry of lands under natural and artificial moistening. *IEEE Trans. Geosci. Remote Sensing*, **20**:18-26.

Shutko A.M. (1983). Statistical properties of microwave emission from land surface at centimetric waves. *Proc. of 1st Interdepartmental Meeting "Statistical methods of remote sensing data processing", Sept. 1980, Minsk.* Moscow: IRE, 1983, pp. 39-45.

Shutko A.M. (1986). *Microwave Radiometry of Sea Surface and Soil.* Moscow: Nauka, 1986 (in Russian).

Shutko A.M. and Chukhlantsev A.A. (1982). Microwave radiation peculiarities of vegetative covers. *IEEE Trans. Geosci. Remote Sensing*, **20**:27-29.

Shutko A.M. and Reutov E.A. (1982). Mixture formulas applied in estimation of dielectric and radiative characteristics of soil and grounds at microwave frequencies. *IEEE Trans. Geosci. Remote Sensing*, **20**:29-32.

Shutko A.M. and Reutov E.A. (1983). Statistical processing of microwave radiometric observations over regions with shallow subsoil waters. . *Proc. of 1st Interdepartmental Meeting "Statistical methods of remote sensing data processing", Sept. 1980, Minsk.* Moscow: IRE, 1983, pp. 34-38.

Sieber A.J. (1985). Forest signatures in imaging and non-imaging microwave scatterometer data. *ESA Journal*, **9**:431-448.

Sihvola A., Sharma R., and Avelin J. (2000). Modeling of the electromagnetic response of geophysical media with complicated geometries: analytical and numerical approaches. In *Microwave Radiometry and Remote Sensing of the Earth's Surface and Atmosphere.* Editors: P. Pampaloni and S. Paloscia, VSP, 2000, pp. 81-88.

Silvestrin P., Berger M., Kerr Y.H., and Font J. (2001). ESA's second Earth explorer opportunity mission: The soil moisture and ocean salinity mission – SMOS. *IEEE Geosci. Remote Sensing Society Newsletter, March* 2001:11-14.

Smirnov V.V. (1971). *Organic mass in some forest phytocenosis of Europe part of USSR.* Moscow: Nauka, 1971 (inRussian).

Stakankin Yu.P. (1986). *Microwave emission of forest fires. PhD Dissertation.* Leningrad State University.

Stiles J. and Sarabandi K. (1996). A scattering model for thin dielectric cylinder of arbitrary cross section and electrical length. *IEEE Trans. Antennas Propagat.*, **44**:260-266.

Stogrin A. (1971). Equation for calculating the dielectric constant of saline water. *IEEE Trans. Microwave Theory Tech.*, **19**:733-736.

Sun G., Simonett D.S., and Strahler A.H. (1991). A radar backscatter model for discontinuous coniferous forests. *IEEE Trans. Geosci. Remote Sensing,* **29**:639-648.

Tamasanis D.T. (1992). Application of volumetric multiple scattering approximations to foliage media. *Radio Science,* **27**:797-812.

Tamir T. (1967). On radio-wave propagation in forest environments. *IEEE Trans. Antennas Propagat.*, **15**:806-817.

Tan H.S. (1981). Microwave measurements and modeling of the permittivity of tropical vegetation samples. *Appl. Phys.*, **25**:351-355.

Tewary R.K., Swarup S., and Roy M.N. (1990). Radio wave propagation through rain forest of India. *IEEE Trans. Antennas Propagat.*, **38**:433-449.

Theis S.W., Blanchard B.J., and Blanchard A.J. (1984). Utilization of active microwave roughness measurements to improve passive microwave moisture estimates over bare soils. *Proceedings of IGARSS'84 Symposium, Strasburg 27-30 August 1984,* pp. 257-262.

Tikhonov A.N. and Arsenin V.O. (1979). *Metods of ill-posed problems solution.* Sec. ed., Moscow: Nauka, 1979.

Tikhonov V.V., (1994). "Model of complex dielectric constant of wet and frozen soil in the 1-40 GHz frequency range," *Proc. IGARSS' 94*, pp. 1576-1578.

Tikhonov V.V. (1996). *Electrodynamic Models of Natural Dispersed Media in the Microwave Range, PhD Thesis.* Moscow: The Moscow State Pedagogical University (in Russian).

Tsang L. and Ding K.-H. (1991). Polarimetric signatures of a layer of random nonspherical discrete scatterers overlying a homogeneous half-space based on first- and second-order vector radiative transfer theory. *IEEE Trans. Geosci. Remote Sensing,* **29**:242-253.

Tsang L. and Kong J.A. (1981). Application of strong fluctuation random medium theory to scattering from vegetation-like half space. *IEEE Trans. Geosci. Remote Sensing,* **19**: 62-69.

Tsang L. and Newton R.W. (1982). Microwave emission from soils with rough surfaces. *J. Geophys. Res.*, **87**:9017-9024.

Tsang L., Blanchard A.J., Newton R.W., and Kong J.A. (1982). A simple relation between active and passive microwave remote sensing measurements of earth terrain. *IEEE Trans. Geosci. Remote Sensing,* **20**:482-485.

Tsang L., Kong J.A., and Shin R.T. (1985). *Theory of Microwave Remote Sensing.* New York: Wiley-Interscience, 1985.

Tsargorodtsev Yu.P., Chukhlantsev A.A., Kobylyansky V.I. and Chukhlantsev A.A. (2004). Dielectric permittivity of forest branches in the 30-300 MHz frequency band. *Proc. of Second All-Russian Conference "Remote Sensing of Terrains and Atmosphere".* S-Petersburg, *16-18 June* 2004, (in Russian).

Ulaby F.T. (1974). Radar measurements of soil moisture content. *IEEE Trans. Antennas Propagat.*, **22**:257-265.

Ulaby F.T. and Bush T.F. (1976). Monitoring wheat growth with radar. *Photogrammetric Engineering and Remote Sensing,* **42**:557-568.

Ulaby F.T. and El-Rayes M.A. (1987). Microwave dielectric spectrum of vegetation – part II: dual dispersion model. *IEEE Trans. Geosci. Remote Sensing,* **25**:550-557.

Ulaby F.T. and Jedlicka R.P. (1984). Microwave dielectric properties of plant material. *IEEE Trans. Geosci. Remote Sensing,* **22**:406-415.

Ulaby F.T. and Wilson E.A. (1985). Microwave attenuation properties of vegetation canopies. *IEEE Trans. Geosci. Remote Sensing,* **23**:746-753.

Ulaby F.T., Dobson M.C., and Brunfeldt D.R. (1983a). Improvement of moisture estimation accuracy of vegetation-covered soil by combined active/passive microwave remote sensing. *IEEE Trans. Geosci. Remote Sensing*, 21:300-307.

Ulaby F.T., Moore R.K., and Fung A.K. (1981, 1982, 1986). *Microwave Remote Sensing, Active and Passive*. Vol. 1-3, Dedham, MA: Artech House, 1982-1986.

Ulaby F.T., Razani M., and Dobson M.C. (1983b). Effect of vegetation cover on the microwave radiometric sensitivity to soil moisture. *IEEE Trans. Geosci. Remote Sensing*, 21:51-61.

Ulaby F.T., Sarabandy K., McDonald K., Whitt M., and Dobson M.C. (1990). Michigan microwave canopy scattering model. *Int. J. Remote Sensing*, 11:1223-1253.

Ulaby F.T., Tavakoli A., and Senior T.B.A. (1987). Microwave propagation constant for a vegetation canopy with vertical stalks. *IEEE Trans. Geosci. Remote Sensing*, 25: 714-725.

Ulaby F.T., Witt M.W., and Dobson M.C. (1990). Measuring the propagation properties of a forest canopy using a polarimetric scatterometer. *IEEE Trans. Antennas Propagat.*, 38:251-258.

Valendik E.N., Kisliakhov E.K., and Sukhinin A.I. (1980). Estimate of forest fire risk from microwave radiometric data. *Issledovanie Zemli iz Kosmosa (Research of the Earth from Space)*, **No. 3**:14-19 (in Russian).

Van de Griend A.A., Owe M., de Ruiter J., and Gouweleeuw B.T. (1996). Measurement and behavior of dual-polarization vegetation optical depth and single scattering albedo at 1.4- and 5-GHz microwave frequencies. *IEEE Trans. Geosci. Remote Sensing*, 34: 957-965.

Van de Griend A.A. and Wigneron J.P. (2004a). The b-factor as a function of frequency and canopy type at H-polarization. *IEEE Trans. Geosci. Remote Sensing*, 42:786-794.

Van de Griend A.A. and Wigneron J.P. (2004b). On the measurement of microwave vegetation properties. *IEEE Trans. Geosci. Remote Sensing*, 42:2277-2289.

Van de Hulst H.C. (1957). *Light Scattering by Small Particles*. New York: Wiley, 1957.

Varekamp C. and Hoekman D.H. (2002). High-resolution InSAR image simulation for forest canopies. *IEEE Trans. Geosci. Remote Sensing*, 40:1648-1655.

Vichev B.I., Krasteva E.N., and Kostov K.G. (1995). Study of seasonal evolution of tree emission using zenith-looking microwave radiometers. *Proceedings of the IGARSS'95*, pp. 981-983.

Vilkova L.P., Novichikhin E.P., Santalov N.P., and Yakovleva G.D. (1998). Estimation of the parameters of the biogeochemical cycle of carbon on global and regional scales. *Problems of Environmental and Natural Resources*, 7:24-37 (in Russian).

Vinokurova S.I., Smirnov M.T., and Chukhlantsev A.A. (1991). Microwave radiative model for scattering layer above a rough surface. *Izvestia VUZ'ov, Radiofizika (Engl. Transl. Radiophys, Quantum Electron.)*. 36:472-476 (in Russian).

Vorobeichik E.A., Kibalnikov S.V., Chukhlantsev A.A., Shutko A.M., and Yazerian G.G. (1986). The method of rice biomass determination. *USSR invention No. 1408986*.

Vorobeichik E.A., Petibskaia V.S., Chukhlantsev A.A., and Yazerian G.G. (1988). Microwave radiation of the rice fields. *Radiotekhnika i Elektronika*, 33:2420-2421 (in Russian).

Wagner W., Lemoine G., Borgeaud M., and Rott H. (1999). A study of vegetation cover effects on ERS scatterometer data. *IEEE Trans. Geosci. Remote Sensing*, 37:938-948.

Wait J.R. (1955). Scattering of a plane wave from a circular dielectric cylinder at oblique incidence. *Can. J. Phys.*, 33:189-196.

Wait J.R. (1965). The long wavelength limit in scattering from a dielectric cylinder at oblique incidence. *Can. J. Phys.*, 43:2212-2214.

Wang J.R. (1980). The dielectric properties of soil-water mixtures at microwave frequencies. *Radio Science*, 15:977-985.

Wang J.R. (1983). Passive microwave sensing of soil moisture content: The effect of soil bulk density and surface roughness. *Remote Sens. Environ.*, **13**:329-344.

Wang J.R. (1985). Effect of vegetation on soil moisture sensing observed from orbiting microwave radiometers. *Remote Sens. Environ.*, **17**:141-151.

Wang J.R. and Choudhury B.J. (1981). Remote sensing of soil moisture content over bare field at 1.4 GHz frequency. *J. Geophys. Res.*, **86**:5277-5282.

Wang J.R. and Schmugge T.J. (1980). An empirical model for the complex dielectric permittivity of soils as a function of water content. *IEEE Trans. Geosci. Remote Sensing*, **18**:288-295.

Wang J.R., Newton R.W., and Rouse Jr., J.W. (1980). Passive microwave remote sensing of soil moisture: The effect of tilled row structure. *IEEE Trans. Geosci. Remote Sensing*, **18**:296-302.

Wang J.R., McMurtrey J.E.,III, Engman E.T., Jackson T.J., Schmugge T.J., Gould W.I., Fuchs J.E., and Glazar W.S. (1982a). Radiometric measurements over bare and vegetated fields at 1.4-GHz and 5-GHz frequencies. *Remote Sens. of Environ.*, **12**: 295-311.

Wang J., Schmugge T., Gould W., Fuchs J., Glazer W., and McMurtrey J.E. (1982b). A multi-frequency radiometric measurements of soil moisture content over bare and vegetated fields. *Gephys. Res. Lett.*, **9**:416-419.

Wang J.R., O'Neill P.E., Jackson T.J., and Engman E.T. (1983). Multifriquency measurements of the effects of soil moisture, soil texture, and soil roughness. *IEEE Trans. Geosci. Remote Sensing*, **21**:44-51.

Wang J.R., Shiue J.C., Chuang S.L., Shin R.T., and Dombrowski M. (1984). Thermal microwave emission from vegetated fields: A comparison between theory and experiment. *IEEE Trans. Geosci. Remote Sensing*, **22**:143-149.

Wang J.R., Engman E.T., Shiue J.C. et al. (1986). The SIR-B observations of microwave backscatter dependence on soil moisture, surface roughness, and vegetation covers. *IEEE Trans. Geosci. Remote Sensing*, **24**:510-516.

Wang J.R., Engman E.T., Mo T., Schmugge T.J., and Shiue J.C. (1987). The effects of soil moisture, surface roughness, and vegetation on *L*-band emission and backscattering. *IEEE Trans. Geosci. Remote Sensing*, **25**:825-833.

Wang J.R., Shiue J.C., Schmugge T.J., and Engman E.T. (1989). Mapping surface soil moisture with L-band radiometric measurements. *Remote Sensing Environ.*, **27**: 305-312.

Wang J.R., Shiue J.C., Schmugge T.J., and Engman E.T. (1990). The L-band PBMR measurements of surface soil moisture in FIFE. *IEEE Trans. Geosci. Remote Sensing*, **28**:906-913.

Wang Y., Day J., and Sun G. (1993). Santa Barbara microwave backscattering model for woodlands. *Int. J. Remote Sensing*, **14**:1477-1493.

Way J., Paris J., Dobson M.C., McDonald K.C., et al. (1991). Diurnal change in trees as observed by optical and microwave sensors: the EOS synergism study. *IEEE Trans. Geosci. Remote Sensing*, **29**:807-821.

Wegmüller U., and Mätzler C. (1999). Rough bare soil reflectivity model. *IEEE Trans. Geosci. Remote Sensing*, **37**:1391-1395.

Wegmüller U., Mätzler C., and Schanda E. (1989). Microwave signatures of bare soil. *Adv. Space Res.*, **9**:307-316.

Wegmüller U., Mätzler C., and Njoku E. (1994). *"Canopy opacity model"* In *ESA/NASA International Workshop*. B.J. Choudhury, J.H. Kerr, E.G. Njoku, P. Pampaloni, eds. Utrecht, The Nitherland: VSP, 1994, pp. 375-387.

Wen B., Tsang L., Winnerbrener D.P., and Ishimaru A. (1990). Dense medium radiative transfer theory: comparison with experiment and application to microwave remote sensing and polarimetry. *IEEE Trans. Geosci. Remote Sensing*, **28**:46-59.

Whale H.A. (1968). Radio propagation through New Guinea rain forest. *Radio Science*, **3**:1038-1043.

Wigneron J.P. (1994). "Retrieval of geophysical parameters from multifrequency passive microwave measurements over a soybean canopy." In *ESA/NASA International Workshop,p.* B.J. Choudhury, J.H. Kerr, E.G. Njoku, P. Pampaloni, eds. Utrecht, Netherland: VSP, 1994, pp. 403-420.

Wigneron J.P., Kerr Y.H., Chanzy A., and Jin Y.Q. (1993a). Inversion of surface parameters from passive microwave measurements over a soybean field. *Remote Sens. Environ.*, **46**:61-72.

Wigneron J.P., Calvet J.C., Kerr Y., Chanzy A., and Lopes A. (1993b). Microwave emission of vegetation: Sensitivity to leaf characteristics. *IEEE Trans. Geosci. Remote Sensing*, **31**:716-726.

Wigneron J.P., Calvet J.C., Chanzy A., Grosjean O., and Laguerre L. (1995a). A composite discrete-continuous approach to model the microwave emission of vegetation. *IEEE Trans. Geosci. Remote Sensing*, **33**:201-211.

Wigneron J.P., Chanzy A., Calvet J.C., and Brugier N. (1995b). A simple algorithm to retrieve soil moisture and vegetation biomass using passive microwave measurements over crop fields. *Remote Sens. Environ.*, **51**:331-341.

Wigneron J.P., Calvet J.C., and Kerr Y. (1996). Monitoring water interception by crop fields from passive microwave observations. *Agricultural and Forest Meteorology*, **80**: 177-194.

Wigneron J.P., Guyon D., Calvet J.C., Courrier G., and Bruguier N. (1997). Monitoring coniferous forest characteristics using a multifrequency (5-90 GHz) microwave radiometer. *Remote Sen. Environ*, **60**:2728-2733.

Wigneron J.P., Ferrazzoli P., Olioso A., Bertuzzi P., and Chanzy A. (1999a). A simple approach to monitor crop biomass from C-band radar data. *Remote Sen. Environ*, **69**:179-188.

Wigneron J.P., Ferrazzoli P., Calvet J.C., Kerr Y., and Bertuzzi R. (1999b). A parametric study on passive and active microwave observations over a soybean crop. *IEEE Trans. Geosci. Remote Sensing*, **37**:1697-1707.

Wigneron J.P., Waldteufel P., Chanzy A., Calvet J.C., and Kerr Y. (2000). Two dimensional microwave interferometer retrieval capabilities over land surface (SMOS mission). *Remote Sens. Environ.*, **73**:270-282.

Wigneron J.P., Laguerre L., and Kerr Y.H. (2001). A simple parameterization of the L-band microwave emission from rough agricultural soils. *IEEE Trans. Geosci. Remote Sensing*, **39**:1697-1707.

Wigneron J.P., Calvet J.C., Pellarin T., Van de Griend A.A., Berger M., and Ferrazzoli P. (2003). Retrieving near-surface soil moisture from microwave radiometric observations: current status and future plans. *Remote Sens. Environ.*, **85**:489-506.

Wigneron J.P., Calvet J.C., de Rosnay P., Kerr Y., Waldteufel P., Saleh K., Escorihuela M.J., and Kruszewski A. (2004a). Soil moisture retrieval from biangular L-band passive microwave observations. *IEEE Trans. Geosci. Remote Sensing Letters*, **1**:277-281.

Wigneron J.P., Parde M., Waldteufel P., Chanzy A., Kerr Y., Schmidl S., and Skou N. (2004b). Characterizing the dependence of vegetation model parameters on crop structure, view angle, and polarization at L-band. *IEEE Trans. Geosci. Remote Sensing*, **42**:416-425.

World Climate Research Programme.(1995). "CLIVAR a study of climate variability and predictability," WCRP-89, 1995.

Yakimov S.P. (1996). *Algorithms for the assessment of forest fire danger using remote sensing data. PhD thesis*, State Technical University, Krasnoyarsk, 155 pp. (in Rusian).

Yakubov V.P., Telpukhovskii E.D., Mironov V.L., and Kashkin V.B. (2002). Vector radio wave propagation through a forest canopy. *Journal of Radio Electronics*, http://jre. cplire.ru/jre/jan02/1/text.html (in Russian).

Yazerian G.G. (2000). *Monitoring of hydro-economic systems by means of microwave radiometry (by the example of Kuban rice fields). PhD thesis*, Institute of Radio-engineering and Electronics RAS, 110 pp. (in Russian).

Zarco-Tejada P.J., Ruweda C.A., and Ustin S.L. (2003). Water content estimation in vegetation with MODIS reflectance data and model inversion methods. *Remote Sensing Enviro.t,* **85(1)**:109-124.

Zhan X., Sohlberg R.A., Townshend J.R.G., DiMiceli C., Carroll M.L., Eastman J.C., Hansen M.C., and DeFries R.S. (2002). Detection of land cover changes using MODIS 250 m data. *Remote Sensing Environ.,* **83(1-2)**:336-350.

INDEX

Abdulla 23
absorption cross section 5, 93
account for vegetation effect
    single frequency
    measurements 207
  two frequency measurements
    214
  polarization measurements 227
  multi-configuration
    measurements 230
  in active microwave remote
    sensing 236
Allen 36, 78, 88, 119, 121
Ambarnikov 200
antenna 13
antenna temperature 14
application of the transfer theory
  161
Armand 54, 64, 250, 252, 256
Arsenin 232
Attema 36, 91, 238
attenuation 24
    by vegetation 119
    measurements 119

backscattering cross section 5
Barabanenkov 92, 94
Barton 180
Basharinov 36, 69, 88, 111, 148,
    149, 150, 151, 154, 178
Battle 242
Baturin 200, 202

Bazilevich 245, 254
b-coefficient 111
    frequency dependence 113,
        134
Birchak 30
    mixing model 30
bistatic scattering coefficient 53,
    54
    scattering cross section 5, 53
Bjorkstrom 244
blackbody 9
    emission 10
Bobrov 23
Bolten 240
Borodin 88, 168
bound water 22
    dielectric constant 35
Boyarskii 22, 23, 34, 35
brightness 9
  temperature 10
  of a bare soil 54
  of a vegetated soil 157, 159
Brown 91, 139
Brunfeld 36, 126, 142, 160, 182,
    222
Burke 69, 70, 112, 178, 211, 249
Bush 79

Calvet 23, 226, 234
Camps 18
Canadel 241
Castel 36

Chanzy 53, 70, 89, 112, 160, 190
Chauhan 39, 44, 70, 91, 191,
    240
Chen 252
Choudhury 61, 68, 70
Chuah 91, 96
Chudnovsky 248
Chukhlantsev 36, 44, 49, 51, 58,
    61, 69, 72, 75, 78, 80, 85, 87,
    90, 91, 95, 98, 99, 103, 105,
    107, 111, 112, 113, 115, 123,
    126, 129, 138, 151, 160, 183,
    204, 207, 209, 214, 217, 236,
    249
circular polarized wave 3
complex dielectric permittivity 5,
    23
conductivity 1, 34
conductivity of leaves 49
continuous approach 75, 78
Coppo 62, 70
Crichlow 121
Crosson 68
Crow 112, 160, 170, 196
Curry 91, 139
Curtis 23

Davenport 160
De Jeu 227
De Loor 33, 41
De Roo 43, 91, 122
Debye-type relaxation 34
Delaney 23, 25, 26
Dicke radiometer 17
dielectric permittivity 1
  of soils 23
  of vegetation elements 41
differential scattering cross
    section 5
Ding 91
discrete approach 75, 91
dispersion equation 92

Dobson 25, 29, 30, 32, 34, 39,
    44, 54, 71, 236
Dobson *et al.* model 32, 33, 35
Du 78, 88, 240
dual-dispersion model 41, 42
Durden 125

Eagman 70
effective permittivity 79
    of vegetation 88
elliptically polarized wave 3
Elagin 200
El-Rayes 41-44, 46, 48
emissivity 10
    of a layer with isotropic
        scattering 160
    of a layer with one-
        dimensional scattering
        159
    of a vegetation canopy 157
    of a vegetation layer 156
    of a bare soil 53, 71
    versus soil moisture 56, 71
energy flux density 9
England 68, 70
Eom 70, 91, 155, 162, 236, 240
extinction coefficient 94, 95, 110
    cross section 5, 93, 97
far-field zone 6
Ferrazzoli 36, 39, 69, 70, 91,
    110, 113, 115, 116, 154, 155,
    156, 161, 166, 169, 189, 240
Finkelberg 92
fluctuation-dissipation theorem
    11
Foldy-Twersky approach 92
forest fire risk 204
    transmissivity 116, 117
Franchois 43, 45
free water 22
frequency dependence of the
    extinction coefficient 134

Fresnel formulas 7, 55
Fung 41, 61, 70, 78, 91, 99, 155, 161, 236, 240

GIMS-technology 249
global carbon cycle 241
    simulation of microwave emission 173
Golovachev 64, 95, 126, 130, 160, 183, 204
Goita 198
Grankov 78, 88, 121, 122, 128, 138, 178, 200, 205
gravimetric soil moisture 22
Green function 6, 79
Guerriero 36, 39, 91, 110, 113, 115, 116, 154, 156, 166, 169

Hallikainen 23, 26, 27, 28, 56, 125, 198
Helmholtz integral equation 97
Herbstreit 121
Hertz's vector 6
Hilhorst 35
Hipp 23
Hoekman 125
Hoekstra 23, 25, 26
Hornbuckle 160, 193

Ilyin 23
incomplete coverage 165
integral representations 5
intensity of monochromatic wave 4
Ishimaru 5, 93

Jackson 23, 66, 70, 111, 112, 126, 129, 180, 192, 194, 196, 207, 212, 222
Jedlicka 41, 42-44, 121, 121
Judge 195

Karam 91, 97, 99, 100, 105, 112, 137
Kasischke 39, 167
Kerr 88, 115, 160, 195
Kira 39
Kirchhoff's law 11, 12
Kirdiashev 69, 91, 111, 115, 128, 129, 135, 151, 157, 160, 166, 177, 197, 200, 205, 207, 217, 222
Kleshchenko 23, 26, 27
Klochko 78, 88, 89
Kondratyev 127, 241, 246, 249, 252, 254, 256
Kong 64, 79, 80
Koskinen 125
Kosolapov 200
Kozoderov 200
Krapivin 243, 245, 249, 252, 254, 256
Krotikov 23, 30, 34
Kruopis 116, 117, 128, 139, 168, 198, 222
Kurvonen 198

Laymon 197
Lane 34
Lang 91, 94, 116, 236, 240
large plane particle model 98
Laursen 18
law of refraction 7
Le Toan 36, 167
Le Vine 18, 98, 99, 112, 137, 197
Leckie 36
Lieth 245
Leschanskiy 23, 25
Levin 11
Levin-Rytov's theory 12, 53, 156
Li 112, 121, 232
Liberman 200
Lin 70

linear polarized wave 3
Liou 68, 235
Liu 235
Lopes 105

Macelloni 91, 167, 169, 193,
    198, 200
Magagi 125
Marquardt 234
Mätzler 23, 36, 43, 61, 105, 138,
    157, 188
Mätzler's semi-empirical
    formula 43, 48
Maxwell's equations 1, 5, 6, 8
Maxwell's theory 1
McDonald 36, 39, 42, 44
Meesters 227
Melon 36
microwave emission from bare
    soils 53-73
  from forests 166-169, 197-204
  from vegetated fields 177-197
  from vegetation canopies
    147-175
microwave radiometer 12
Milshin 78, 88, 128, 198, 200,
    203, 205, 222
Mintzer 246
Mironov 23, 26, 28, 33, 35
Mo 61, 62, 112, 125, 129, 158,
    160, 207, 222, 236
modeling vegetation emissivity
    147
Moghaddam 125
monochromatic wave 3
Mougin 105, 122, 132, 138
multi-configuration
    measurements 230
Murata 125

Nashashibi 121
Nazarov 235

Nefedova 245
Nesterov 204
Newton 70
Njoku 64, 70, 112, 160, 197,
    232, 240
Normalized Difference
    Vegetation Index (*NDVI*)
    212, 213
Nyquist formula 14

O'Neill 36, 66, 70, 121, 181
Ogava 39
Olioso 246, 248
optical depth 111
  dependence on vegetation
    water content 129
  frequency dependence 134
  dependence on polarization 142
  dependence on vegetation type
    145
optical theorem 7
  thickness 111
Orlov 38, 39
Owe 59, 160, 195, 197, 227

Paloscia 59, 126, 129, 130, 135,
    137, 160, 185, 222
Pampaloni 59, 126, 129, 130,
135, 137, 160, 185, 197, 222
    and Paloscia's model 129,
    132
Pardé 117, 118, 139, 232, 234
Paris 121
Peake 41, 78, 88
Pellarin 58, 112, 160, 231
Peplinski 23, 26, 29, 33
phase function 94
    velocity 2
Phillips 250, 256
Pitts 125
plane wave 3
Plank 10

Poynting's vector 4
polarization indices 58-60
    plane 3
        properties of attenuation 142
        properties of emission 169
Polyakov 23, 27, 54, 57, 64
Press 234
Promes 67, 70
Pulliainen 125, 232

Quinones 125

radiometer sensitivity 16
radiometric measurements 18-20
Raju 70
Ranson 36
ray intensity 9
Rayleigh – Gans scattering 78,
    99, 166
Rayleigh scattering 78, 98
Rayleigh-Jeans radiation formula
    10
Redkin 78, 88, 89
reflectivity 8
  of a bare soil 53, 61
  and transmissivity of a
        scattering layer 149
  of a layer 8
  of the optically thick vegetation
        layer 159, 161-164
refractive index 24, 55
    model 30
retrieval algorithms 224, 230
Reutov 31, 64, 65, 69, 204, 248
Riabchikov 245
Richards 39
Rose 70
roughness effects 51
Ruf 18
Running 252
Rytov 1, 76, 80, 152
Ryzhov 75, 77, 79

Saatchi 36, 125
Sabburg 23, 29
Salas 44
salinity 25
Satoo 39
Savage 121, 122, 139
Savorsky 205
Saxton 34
scattered field 4
scattering amplitude 7, 93
    and absorption by pant
        elements 105
    cross section 5, 93, 97
Schiffer 97
Schmugge 22, 23, 25, 30, 33, 61,
    62, 68-71, 111, 112, 126,
    129, 178, 188
Selers 47, 246, 255
Senior 97, 105
Shadrina 44, 45
Shinohara 125
Shutko 26, 29, 31, 53, 54, 64, 65,
    69, 71, 112, 204, 207, 209,
    214, 217
Sidhu 91, 236
Simmonds 249
single scattering albedo 94, 113,
    156-158
Skou 18
small particle model 97
Smirnov 38, 39
soil brightness temperature 54
    dielectric properties 23-35
        emissivity 53
    hydrological constants 22
    moisture retrieval from
        radiometric
        measurements 209, 227,
        230
    physical properties 21-23
    profile effects 63
    reflectivity 53

root mean square height 54,
61
roughness correlation length
54, 61
roughness parameter 62
structure effect 66
Sosnovskiy 23
spatial variations of microwave
emission 171
Stakankin 169
statistical properties of
microwave emission 171,
204-206
Stogrin 41
Sume 105
Sun 39, 91
SVAT models 248

Tamasanis 78, 88
Tamir 121
Tamoikin 75, 77, 79
Tan 44
Tarko 245
tau-omega model 157
Tewary 121
Theis 70, 240
Thielheim 97
thin particle model 99
three component model 157
Tikhonov 22, 23, 232
Torgashin 38, 39
total-power receiver 16
transfer equation 9
theory 8
transmissivity 8, 117
of a layer 8
Tsang 70, 78, 80, 85, 91, 97,
162, 240
Tsargorodtsev 45

Ulaby 19, 25, 36, 39, 41-44, 46,
48, 54, 78, 88, 91, 92, 103,

105, 110, 118, 121, 121, 125,
126, 142, 160, 181, 182, 207,
222, 236, 238, 240

Valendik 205
Van de Griend 112, 126, 134,
135, 137, 143, 145, 160, 192,
222
Van de Hulst 98, 103
Varekamp 125
vegetation biomass retrieval
model analysis 217
practical realization 222
single frequency
measurements 222
spectral measurements 224
polarization measurements 227
multi-configuration
measurements 230
effect in active microwave
remote sensing 230
features 36-40
material's electrical
properties 41-51
screening effect 207
transfer coefficient 159
water content 129
water content and optical
depth 129
Vichev 128
Vilkova 245
Vinokurova 158, 160, 236, 238
volume density of electromagnetic
field energy 3
volumetric soil moisture 22, 56
Vorobeichik 126, 183, 219, 222

Wagner 125
Wang 22, 23, 25, 30, 33, 39, 59,
61, 67, 69, 70, 71, 92, 110,
112, 170, 180, 187, 197, 222,
234, 237

Wang and Schmugge model 30,
33, 35
wave equation 2, 3
vector 3
Way 44
Wegmüller 61, 70, 115, 138
Wen 96
Whale 121
Wigneron 36, 53, 61, 62, 88, 89,
112, 115, 126, 134, 135, 137,
144, 145, 157, 160, 170, 190,
198, 200, 224, 226, 230, 234,
238, 240
Wilson 121, 121, 141
Yakubov 121
Yakimov 172, 205
Yazerian 224

Zarco-Tejada 252
Zhan 252

Advances in Global Change Research

1. P. Martens and J. Rotmans (eds.): *Climate Change: An Integrated Perspective.* 1999
ISBN 0-7923-5996-8
2. A. Gillespie and W.C.G. Burns (eds.): *Climate Change in the South Pacific: Impacts and Responses in Australia, New Zealand, and Small Island States.* 2000
ISBN 0-7923-6077-X
3. J.L. Innes, M. Beniston and M.M. Verstraete (eds.): *Biomass Burning and Its Inter-Relationships with the Climate Systems.* 2000 ISBN 0-7923-6107-5
4. M.M. Verstraete, M. Menenti and J. Peltoniemi (eds.): *Observing Land from Space: Science, Customers and Technology.* 2000 ISBN 0-7923-6503-8
5. T. Skodvin: *Structure and Agent in the Scientific Diplomacy of Climate Change.* An Empirical Case Study of Science-Policy Interaction in the Intergovernmental Panel on Climate Change. 2000 ISBN 0-7923-6637-9
6. S. McLaren and D. Kniveton: *Linking Climate Change to Land Surface Change.* 2000
ISBN 0-7923-6638-7
7. M. Beniston and M.M. Verstraete (eds.): *Remote Sensing and Climate Modeling: Synergies and Limitations.* 2001 ISBN 0-7923-6801-0
8. E. Jochem, J. Sathaye and D. Bouille (eds.): *Society, Behaviour, and Climate Change Mitigation.* 2000 ISBN 0-7923-6802-9
9. G. Visconti, M. Beniston, E.D. Iannorelli and D. Barba (eds.): *Global Change and Protected Areas.* 2001 ISBN 0-7923-6818-1
10. M. Beniston (ed.): *Climatic Change: Implications for the Hydrological Cycle and for Water Management.* 2002 ISBN 1-4020-0444-3
11. N.H. Ravindranath and J.A. Sathaye: *Climatic Change and Developing Countries.* 2002 ISBN 1-4020-0104-5; Pb 1-4020-0771-X
12. E.O. Odada and D.O. Olaga: *The East African Great Lakes: Limnology, Palaeolimnology and Biodiversity.* 2002 ISBN 1-4020-0772-8
13. F.S. Marzano and G. Visconti: *Remote Sensing of Atmosphere and Ocean from Space: Models, Instruments and Techniques.* 2002 ISBN 1-4020-0943-7
14. F.-K. Holtmeier: *Mountain Timberlines.* Ecology, Patchiness, and Dynamics. 2003
ISBN 1-4020-1356-6
15. H.F. Diaz (ed.): *Climate Variability and Change in High Elevation Regions: Past, Present & Future.* 2003 ISBN 1-4020-1386-8
16. H.F. Diaz and B.J. Morehouse (eds.): *Climate and Water: Transboundary Challenges in the Americas.* 2003 ISBN 1-4020-1529-1
17. A.V. Parisi, J. Sabburg and M.G. Kimlin: *Scattered and Filtered Solar UV Measurements.* 2004 ISBN 1-4020-1819-3
18. C. Granier, P. Artaxo and C.E. Reeves (eds.): *Emissions of Atmospheric Trace Compounds.* 2004 ISBN 1-4020-2166-6
19. M. Beniston: *Climatic Change and its Impacts.* An Overview Focusing on Switzerland. 2004 ISBN 1-4020-2345-6
20. J.D. Unruh, M.S. Krol and N. Kliot (eds.): *Environmental Change and its Implications for Population Migration.* 2004 ISBN 1-4020-2868-7
21. H.F. Diaz and R.S. Bradley (eds.): *The Hadley Circulation: Present, Past and Future.* 2004 ISBN 1-4020-2943-8
22. A. Haurie and L. Viguier (eds.): *The Coupling of Climate and Economic Dynamics.* Essays on Integrated Assessment. 2005 ISBN 1-4020-3424-5